Tourism and Indigeneity in the Arctic

TOURISM AND CULTURAL CHANGE

Series Editors: Professor Mike Robinson, *Ironbridge International Institute for Cultural Heritage, University of Birmingham, UK* and Dr Alison Phipps, *University of Glasgow, Scotland, UK.*

Understanding tourism's relationships with culture(s) and vice versa, is of ever-increasing significance in a globalising world. TCC is a series of books that critically examine the complex and ever-changing relationship between tourism and culture(s). The series focuses on the ways that places, peoples, pasts, and ways of life are increasingly shaped/transformed/created/packaged for touristic purposes. The series examines the ways tourism utilises/makes and re-makes cultural capital in its various guises (visual and performing arts, crafts, festivals, built heritage, cuisine etc.) and the multifarious political, economic, social and ethical issues that are raised as a consequence. Theoretical explorations, research-informed analyses, and detailed historical reviews from a variety of disciplinary perspectives are invited to consider such relationships.

Full details of all the books in this series and of all our other publications can be found on http://www.channelviewpublications.com, or by writing to Channel View Publications, St Nicholas House, 31-34 High Street, Bristol BS1 2AW, UK.

TOURISM AND CULTURAL CHANGE: 51

Tourism and Indigeneity in the Arctic

Edited by

Arvid Viken and Dieter K. Müller

CHANNEL VIEW PUBLICATIONS
Bristol • Blue Ridge Summit

Library of Congress Cataloging in Publication Data
A catalog record for this book is available from the Library of Congress.
Names: Viken, Arvid, editor.
Title: Tourism and Indigeneity in the Arctic/Edited by Arvid Viken and Dieter K. Müller.
Description: Bristol, UK; Blue Ridge Summit, PA: Channel View Publications, [2017] | Series: Tourism and Cultural Change: 51 |
Includes bibliographical references and index.
Identifiers: LCCN 2016054817| ISBN 9781845416096 (hbk : alk. paper) | ISBN 9781845416119 (epub) | ISBN 9781845416126 (kindle)
Subjects: LCSH: Tourism–Arctic Regions–Sociological aspects. | Tourism–Economic aspects–Arctic Regions. | Arctic peoples.
Classification: LCC G155.A726 T68 2017 | DDC 338.4/79109113--dc23 LC record available at https://lccn.loc.gov/2016054817

British Library Cataloguing in Publication Data
A catalogue entry for this book is available from the British Library.

ISBN-13: 978-1-84541-609-6 (hbk)

Channel View Publications
UK: St Nicholas House, 31-34 High Street, Bristol BS1 2AW, UK.
USA: NBN, Blue Ridge Summit, PA, USA.

Website: www.channelviewpublications.com
Twitter: Channel_View
Facebook: https://www.facebook.com/channelviewpublications
Blog: www.channelviewpublications.wordpress.com

The policy of Multilingual Matters/Channel View Publications is to use papers that are natural, renewable and recyclable products, made from wood grown in sustainable forests. In the manufacturing process of our books, and to further support our policy, preference is given to printers that have FSC and PEFC Chain of Custody certification. The FSC and/or PEFC logos will appear on those books where full certification has been granted to the printer concerned.

Typeset by Deanta Global Publishing Services Limited.

Contents

Contributors

Suzanne de la Barre, PhD, is a faculty member with the Department of Recreation and Tourism Management, Vancouver Island University (Canada). She holds an adjunct position at Yukon College (Canada) and is a Research Associate with the Department of Geography and Economic History, Umeå University (Sweden). Her research uses place-based approaches to investigate change in peripheral regions, with a focus on northwestern and Arctic Canada. More specifically, her interests examine economic diversification, the cultural and creative sectors, tourism, community development and the role of social innovation in remote and rural areas.

Trude Fonneland is currently Professor in the Department of Culture Studies at Tromsø Museum, at UiT – The Arctic University of Norway. Her research interests revolve around contemporary religion in society, particularly Sami shamanism, tourism and popular culture. Her dissertation (2010) focused on the developments of modern Sami shamanism in Norway. She is also the author of several scholarly articles on the subject. Her latest publications include *Nordic Neoshamanisms* (2015: editor together with Siv Ellen Kraft and James Lewis).

Fredrik Hoppstadius is a PhD student in human geography at the Department of Geography, Media and Communication, Karlstad University, Sweden, where he is also affiliated with The Centre for Research on Region Building (CRS). Hoppstadius is working on a thesis on ecotourism and sustainable development in biosphere reserves. Before joining Karlstad University, he worked as a project assistant at Umeå University on issues of Sami tourism and northern development.

John S. Hull is an Associate Professor of Tourism Management at Thompson Rivers University in British Columbia, Canada, and a Visiting Professor at Buskerud and Vestfold University College, Norway and at the Harz University of Applied Sciences, Germany. He is also a member of the New Zealand Tourism Research Institute. His research addresses the sustainability of tourism planning and destination development in peripheral and rural regions. He has worked internationally with the United Nations Convention on Biological Diversity (UNCBD) on their tourism, technology and indigenous communities workshops in the Arctic and South Pacific, and with the Greenlandic Tourism and Business Council on strategies for cruise ship tourism. He has also consulted with First Nations in Newfoundland and Labrador, New Brunswick, Quebec, Alberta, British Columbia and the Northwest Territories in Canada.

E. Carina H. Keskitalo is Professor of Political Science at the Department of Geography and Economic History, Umeå University. Her research focus is on regional and environmental policy development under conditions of societal and environmental change (among other climate change adaptation and globalization issues). She is the scientific coordinator of the Mistra Arctic Sustainable Development program, Sweden's first social sciences and humanities program on the Arctic.

Yulian Konstantinov is a Social Anthropologist with a background in theoretical linguistics. He is known for his work in minority studies, post-socialist informal economies and, since 1994, with reindeer herding communities in the Russian Far North (Kola Peninsula). Currently, Konstantinov holds the position of Research Professor at the Department of Archaeology and Social Anthropology, University of Tromsø, Norway. Recent publications include: *Conversations with Power: Soviet and Postsoviet Developments in the Reindeer Husbandry Part of the Kola Peninsula* and *Socio-economic Developments in Murmansk Region: The Insiders' Point of View* (with Inna Ryzhkova, eds).

Britt Kramvig is Professor at the Department of Tourism and Northern Studies at UiT – The Arctic University of Norway. She takes a postcolonial position inspired by indigenous, feminist and science, technology, society (STS) debates and has done ethnographic work, written and made films on a range of different topics relating to indigeneity, tourism, gender, place as well as innovation. Britt insists upon working across disciplinary boundaries as people of the north perform these boundaries differently.

Patrick T. Maher, PhD, is an Associate Professor and Chair of the Department of Community Studies at Cape Breton University. His doctoral research examined the process of visitor experience in the Ross Sea Region of Antarctica and was funded by a Commonwealth scholarship. He played a critical role in founding the International Polar Tourism Research Network and has sat on its steering committee since its inception. He is also the lead for the University of the Arctic's Thematic Network on Northern Tourism, which recently received significant funding to develop a joint master's curriculum across the Circumpolar North. Patrick is a Fellow of the Royal Canadian Geographical Society and a 3M National Teaching Fellow (2014). He has published widely in scholarly journals and industry press, and is editor of the recent texts: *From Talk to Action: How Tourism is Changing the Polar Regions* (2013), *Polar Tourism: Human, Environmental and Governance Dimensions* (2011) and *Cruise Tourism in Polar Regions: Promoting Environmental and Social Sustainability?* (2010).

Dieter K. Müller is Professor of Social and Economic Geography and Deputy Vice-Chancellor, Umeå University, Sweden. His research specifically includes almost all aspects of second homes and second home-related mobility, Sami tourism, nature-based tourism, tourism labor markets, regional development and rural change particularly in northern peripheries and polar areas. Recently, Müller co-edited *Polar Tourism and Regional Development* (with A.A. Grenier) and *New Issues in Polar Tourism* (with R.H. Lemelin and L. Lundmark). Currently, he is chairperson of the International Geographical Union (IGU) Commission for the Geography of Tourism, Leisure and Global Change and a member of the steering committees of the International Polar Tourism Research Network (IPRTN), as well as a board member of the Arctic Research Centre at Umeå University (Arcum). Moreover, Müller serves on the editorial boards of *Tourism Geographies, Current Issues in Tourism, Scandinavian Journal of Hospitality and Tourism* and *The Finnish Journal of Tourism Research.*

Kjell Olsen is Professor in Cultural Studies at the University of Tromsø – The Arctic University of Norway. He has conducted fieldwork in Swaziland, Northern Norway, New Zealand and Guatemala. His research interests are focused on healing rituals, identity processes, the representations of indigeneity and tourism.

Albina Pashkevich is a Senior Lecturer in Human Geography at the School of Technology and Business Studies, Dalarna University, Sweden. Her research interest includes investigations concerning the processes of recent tourism development of Arctic Russian territories with the Arkhangelsk Oblast and Nenets Autonomous Okrug as case study areas. The focus of Albina's research lies in the study of institutional settings allowing the alternative development of local indigenous communities, including maritime and coastal tourism based on the cultural and natural heritage of these areas.

Jan Åge Riseth is a nature manager (MSc) and an institutional economist with a PhD thesis on *Modernization and Pasture Degradation. A Comparative Study of Two Sami Reindeer Pasture Regions in Norway 1960–1990* (2009). Riseth was Vice-Director for the Agency of Reindeer Management in the 1980s and later worked as a municipal planner, university college teacher/researcher and an institute researcher. Since 2002, he is a senior research scientist at the Northern Research Institute (NORUT) and also had an adjunct position at the Sami University College for 9 years. Since 2014, he is a judge at Finnmark Land Tribunal. Most of his research is interdisciplinary and includes reindeer management, nature protection, sustainability, modernization, impact analyses, climate change and traditional/indigenous knowledge. His most recent book is *Reindrift og ressursforvaltning* (2015).

Karina M. Smed, PhD, is Associate Professor at the Department of Culture and Global Studies at Aalborg University, Denmark. Her primary research interests are focused around tourism, consumption and identity, as the broad headings assembling a variety of publications and activities. Underlying these focal points is an interest in cultural processes as part of the tourist world and played out at tourist destinations. This has, within the last couple of years, sparked particular interest in Arctic destinations as very particular contexts for tourists' involvement and contribution to cultural processes due to the size and scope of tourism at these destinations. Most recently, she has conducted research in Ilulissat, Greenland, where particularly the Ilulissat Icefjord as a World Heritage Site has been addressed. She is a member of the Tourism Research Unit (TRU), the Culture & Consumption group and the Centre for Innovation and Research in Culture and Learning in the Arctic (CIRCLA) at Aalborg University.

Gaute Svensson, PhD, is a researcher within tourism at UiT The Arctic University of Norway. His background is in social anthropology, but also as a practitioner in the tourism industry. His interests primarily cover the social and cultural aspects of nature practices, among these outdoor recreation and tourism. He has written about reciprocity in nature based tourism, secrecy among hunters and anglers and encounter norms among anglers. He has also written about tourism destination organization in Northern Norway. He is currently working on a paper on secrecy in hunting and angling.

Seija Tuulentie is a Senior Researcher in the Natural Resources Institute Finland (Luke) and Adjunct Professor in Environmental Sociology at the University of Lapland. Her research fields include ethnic relations, nature-based tourism, rural development, second homes and conflicting land use issues. She has concentrated especially on the Arctic and northern questions in Finland and Scandinavia. Her dissertation dealt with the discussion around the legal claims of the Finnish Sami population from the national identity point of view. She has published about 30 refereed journal and book articles and 60 other scientific publications. She has been Acting Professor in Nature-Based Tourism for two years and in Sociology for one year at the University of Lapland.

Arvid Viken is Professor in Tourism at UiT – The Arctic University of Norway, where he is in charge of PhD education in tourism. Viken has been editor of a series books on tourism, and has written a series of chapters and articles in journals and books within the field. His research has primarily addressed questions related to northern issues as indigenous tourism, tourism on Svalbard and Arctic tourism in general. In recent years, he has also had a focus on festivals, place and borders, and he has edited a book called *Place Reinvention* (together with Torill Nyseth, 2009), and a book on border issues called *Grenseland* (together with Bjarge Schwenke Fors, 2014). Since 2010, Viken has focused on destination development and has managed several projects in this field. This has resulted in a book called *Destination Development. Turns and Tactics* (together with Brynhild Granås, 2015), and another called *Turisme: Destinasjonsutvikling* (2015).

Vladislava Vladimirova is a Senior Lecturer at The Centre for Russian and Eurasian Studies and at the Department of Cultural Anthropology and Ethnology at Uppsala University. Her publications include the monograph *Just Labor: Labor Ethic in a Post-Soviet Reindeer Herding Community* (2006), and different book chapters and articles, such as *Transnational Indigenous Organizations, Liberal Multiculturalism and Narratives about 'Indigenous Separatism' in the Russian North* (in Russian, 2015), *Traditions of Civil Society: The Role of the Council of Veterans of War and Labor in Northwest Russia* (in Russian, 2015).

Gro B. Ween is Associate Professor, Keeper of the Arctic Collection and Head of the Ethnographic Department at the Cultural History Museum, at the University of Oslo. Ween has done fieldwork in Aboriginal Australia, several places in Sapmi, in Nunavut and Yupik Alaska. She has written extensively on indigenous politics, natural resource management, natural and cultural heritage, human–animal relations, often within a postcolonial framework and based upon collaborative research practices. Ween is a member of several research groups and networks: the Centre for Indigenous Conservation and Development Alternatives (McGill), the ERC project Arctic Domus: Human–Animal relations in the North (Aberdeen), the NRC-founded Anthropos and the Material, and she is currently a member of the Norwegian Academy of Sciences and Letters CAS group Arctic Domestication: Human–Animal relations in the North.

Preface

This book is about Arctic indigenous tourism, in particular how tourism to indigenous communities is influenced by academic and political discourses and how these communities are influenced by tourism. Tourism in these areas is part of global tourism and cultural development, and indigenous tourism is shaped through internal ethnopolitical processes and in dialogue with majority populations and national states in the northern hemisphere.

Doing a book on indigenous tourism in Arctic areas is also born out of a certain insight, and one could call it frustration with available literature on the issue. This is because of the fact that, in particular, Nordic indigenous people fit poorly into the description of indigenous populations available in the literature. Many of them lead a modern life and engagement in tourism can be described as an outcome of agency, a lifestyle choice and just one alternative among many others. Hence, a fresh examination of the issue of indigenous tourism with a greater sensitivity toward geographical place and context appeared to be mandatory in order to overcome the superficial indigenous–non-indigenous divide. The Arctic seemed to be a reasonable area for such an approach, partly because it is our home region and an area we care for and partly because it offers diverging experiences of indigenous peoples in different kinds of states and political systems.

Hence, the book contains case studies from all around the circumpolar north: Canada, Greenland, Norway, Sweden, Finland and Russia. The book gives a glimpse of issues related to tourism in the area, and no encompassing overview is given. However, the issues discussed probably should have a general interest, focusing on issues related to the situations and challenges that many indigenous groups are experiencing in encounters with modern tourism. As there are very many indigenous groups in the Arctic, not all are focused on here. Due to the origin of the editors, Norway and Sweden, the majority of the chapters concern tourism in Sapmi. The other locations addressed are the Yukon in Canada, Illulisat of Disco Bay in Greenland, different places in the Sami area and in the north of Russia.

A major perspective is the ambivalence related to tourism as a modern force dealt with in ethnic groups strongly related to traditions and still including practitioners of nature harvesting and pastoralism. However, all these groups are also part of the modern world, affected by globalization and other transforming trends. This reality is confronted with a series of paradoxes. Basically, integration in a modern world ameliorates their living conditions, but weakens indigenous roots and traditional practices. Such modernizing processes are dealt with in most of the chapters of the book. We hope that the book will contribute to vital discussion about an important issue: where is indigeneity and indigenous tourism and how do they sustain in a world with strong ethnopolitical and commercial unification?

The book has gathered a series of scholars who have a lot of experience studying tourism to the Arctic and within indigenous groups. Almost all of the authors are located in the Arctic, and some have indigenous roots. The editors are grateful to them all for making this volume possible. We would also like to thank the publisher, Channel View Publications, for their interest in the topic, and their employees for their kind and professional handling of the process.

Arvid Viken and Dieter K. Müller
Tromsø and Umeå, June 2016

Part 1

Conceptualizing Arctic Indigeneity and Tourism

1 Indigenous Tourism in the Arctic

Dieter K. Müller and Arvid Viken

Indigeneity and Tourism

This book is about indigenous tourism in the Arctic. The term 'indigenous' is employed not to designate a natural category, but rather as an academically constructed term to describe peoples who are not organized into self-determined national states, yet represent culturally or ethnically distinguishable groups – normally minorities – within a state dominated by a majority people. Because of their cultural uniqueness, indigenous groups tend to be of interest to tourists and those organizing tourism. Uniqueness means difference. Tourism is basically founded on differences – different natures, cultures and geographies. Although uniqueness and difference are realities, indigenous groups are also celebrated in stories around the world, or in the academic realm, in discourses that circulate among the public and academics. Thus, there is not just one way to interpret a culture, but numerous interpretations that are based on different ontologies, epistemologies and discourses. One of these discourses is about indigenousness, indigeneity and ethnicity, and how these phenomena and concepts entangle and reflect other discourses. Indigenousness and indigeneity represent perspectives that are products of time and space. Throughout the previous century, the focus has been on civilization, primitivism, particularism, assimilation and so on related to indigenous peoples (Butler & Hinch, 2007). The previous decade's focus has shifted from the situation of indigenous groups to the indigeneity of all distinct cultural groups or peoples. These discussions are also important for the way that indigenous tourism is understood and practiced, as will be discussed in this and the next chapter.

Related to the discussion concerning who is indigenous and what is indigeneity, are questions both about the indigenous groups' self-determination, autonomy and separate institutions and regimes, as well as those about groups' integration into national state systems and market economy. Implicit are questions of power, participation and public acceptance. Indigenous groups have a particular position in the United

Nations system. In this context indigenous peoples are usually defined as follows:

> Indigenous communities, peoples and nations are those which, having a historical continuity with pre-invasion and pre-colonial societies that developed on their territories, consider themselves distinct from other sectors of the societies now prevailing on those territories, or parts of them. They form at present non-dominant sectors of society and are determined to preserve, develop and transmit to future generations their ancestral territories, and their ethnic identity, as the basis of their continued existence as peoples, in accordance with their own cultural patterns, social institutions and legal system. (Office of the United Nations High Commissioner for Human Rights (OHCHR), 2013: 6)

The dominant philosophy is that indigenous groups shall be treated both as special and as an integrated part of national states. Most Arctic indigenous groups, therefore, are part of the welfare state regarding education, health and social services, and in some countries, are given special rights to land and natural resources. But as minorities, they tend to be forgotten, neglected, counteracted or marginalized. For Nordic countries, northern tourism is not a matter of indigenous control. In Norway, for example, indigenous tourism is handled by a half-public destination marketing organization (DMO), and politically is principally a marginal activity (Viken, 2016). In Sweden, attempts to create an independent Sami DMO have been abandoned. As a result, indigenous tourism is not necessarily a strong and emancipating industrial or societal development factor. This issue will also be discussed in this chapter, as a frame for the book as such.

Indigenous tourism is part of a larger tourism picture. Tourism is a phenomenon that is a product of and a contributor to the global economy. It is a market economy with its ups and downs. In earlier times, tourism was strongly influenced by national politics and regulations, such as taxes on outbound travel, license requirements for businesses and regulations for travel. In many countries, air traffic control was state run and strictly regulated until the 1990s, and in some northern regions, regulations tied to subsidies still exist. Though tourism in the Arctic is by no means a new phenomenon, the penetration of northern space by infrastructure is in many cases rather recent and accessibility is still expensive (Lundgren, 2001). Public sector involvement is hence often significant, not least since tourism development is expected to be a remedy for economic and social problems (Hull, 2001; Jenkins *et al.*, 1998).

In some places, tourism has even been unwanted or banned in indigenous communities due to anticipated negative impacts. Nowadays, this seems strange as tourism is part of the economy and seen as a path for development. Such shifts can be related to the neoliberal turn in society, and a view of the economy as a free space for entrepreneurs, investors and every person. Therefore, it is argued that indigenous and ethnic groups and their interests have been commercialized. We shall elaborate on this perspective and its potential impacts.

This chapter is a presentation of the geographical area, and of the chapters in the book, whereas the next chapter gives a more thorough discussion of vital concepts and perspectives.

The Context of Indigenous Tourism in the Arctic

The Arctic is not a homogenous region when it comes to social, economic and political issues as the comprehensive overviews provided in the first and second volume of the Arctic Human Development Report (AHDR, 2004, 2015) powerfully demonstrate.

Arctic tourism cannot be understood without acknowledging the colonial practices and particular political economies related to the areas' history and presence as resource peripheries. In fact, it can be argued that it is exactly these relationships and their maintenance that construct the Arctic as a place worth visiting (Müller, 2013b).

Indigenous peoples are distributed all over the Arctic region; however, as Figure 1.1 indicates, they are seldom a majority. The opposite applies only in Greenland and in Nunavik and Nunavut. All these regions are, however, among the least populated areas of the Arctic. In the remaining parts of the Arctic, indigenous populations are outnumbered by non-indigenous populations, who, in some cases, have been present for hundreds of years. Furthermore, in the Nordic parts of the Arctic and western Siberia, besides Alaska, and in the most populated regions of the Arctic, indigenous populations seldom amount to more than 10% of the total population. Though European and partly Russian Arctic populations are in decline, this is not the case in North America where many indigenous groups are in fact increasing in numbers (AHDR, 2015). This, of course, creates distinct challenges in different parts of the Arctic where tourism is not only a means to create employment for the growing population in the American Arctic and in Russia, but it is also a means to attempt to hinder a further decline in population numbers. Additionally, tourism development implies not least an effort to offer employment for women in order to counteract skewed gender patterns.

The Arctic economy is traditionally based on three pillars (AHDR, 2004, 2015). First, large-scale resource extraction has been a major force behind interest in the north and has traditionally also been a major reason for infrastructure provision. Secondly, traditional activities have been another important part of the economy. And finally, transfers from southern governments have brought about public sector employment and various forms of support to northern residents. However, resource cycles repeatedly

Figure 1.1 Indigenous population in the Arctic regions 2010 (Source: Nordregio)

put pressure on Arctic labor markets as do economic crises and shrinking public spending (Clapp, 1998; Müller, 2013a). The demilitarization of the region after the cold war has further caused declining investment in the region. Hence, increasingly private entrepreneurial activities are anticipated in order to sustain and create new livelihoods in the region. In this context, it is important to realize that preconditions vary across the Arctic region. The Nordic parts of the Arctic are rather densely populated and have vibrant cities featuring institutions of higher education, a cultural life and a variety of public and private services (Keskitalo *et al.*, 2013). Moreover, they are accessible by airplane, by car and sometimes even by train. In contrast, remote regions of Siberia and North America largely lack these assets, which then presupposes different preconditions even for tourism development.

Though tourism in the Arctic is not a new phenomenon, interest has recently increased, as manifested by a growing number of scientific publications (Grenier & Müller, 2011; Hall & Saarinen, 2010; Lemelin *et al.*, 2013; Maher *et al.*, 2011; Müller *et al.*, 2013; Viken & Granås, 2014). In this literature, it is forcefully argued that recent tourism development in the Arctic is because of a growing awareness of Arctic issues. Climate change, vanishing sea ice, a renewed race for Arctic resources and new potential transport routes are some of the reasons why Arctic issues are covered in the media. Arctic tourism is hence sometimes depicted as 'last chance tourism' (Lemelin *et al.*, 2010). Still, some Arctic areas are relatively difficult to reach and the costs of getting there and being there are high (Hull, 2001; Lundgren, 2001). However, some Arctic destinations like the North Cape, the Icehotel in Kiruna, the Santa Park in Rovaniemi and the Gulf of Alaska have developed into mass tourism destinations regardless of these issues (Müller, 2015), while other regions are hardly visited and sometimes strive for increased development. As a consequence, the north is increasingly constructed as an international tourism space comprising not least a touristic wilderness (Pedersen & Viken, 1996; Saarinen, 2005).

Indigenous tourism has to be seen in the context of a changing Arctic vacillating between traditionalism and modernity. As argued above, resource extraction and military activity have certainly brought modernity to Arctic regions and though such industries can be seen as important preconditions for tourism, they compete for land use, labor and attention (Müller, 2011, 2013a). Hence, in many Arctic states, even members of indigenous populations have alternatives in terms of employment, education and lifestyles. They can stay or leave the area and their career is not contingent upon traditional activities (Leu & Müller, 2016). Thus, tourism is not the only option for making a livelihood and even indigenous peoples have the opportunity to avoid active involvement in tourism. Instead, tourism is a career choice for others; it is an economic decision, though it sometimes relates to expectations and the desire to disseminate knowledge about culture and history as well (Müller & Kuoljok Huuva, 2009; Tuulentie, 2006).

It is against this background that this book has been developed. While tourism development is feared by some as an intruder and a threat to indigenous culture, it is embraced by others as an opportunity to support indigenous livelihoods and to enable people to stay in their peripheral homelands. We think that this dichotomy is too simplistic. It reifies stereotypes of indigenous peoples as targets rather than as agents of development. Hence, not least in a welfare state context, indigenous people may choose the livelihoods they desire although this has to be done within a complex ethnopolitical context where ample expectations regarding the indigenous populations are voiced (Müller & Kuoljok Huuva, 2009; Müller & Pettersson, 2006; Tuulentie, 2006; Viken & Müller, 2006). Hence, this book aims at contributing to a more comprehensive debate on the nexus of tourism, societal change and indigenous people.

The Book's Profile, Authors and Chapters

Part of the criticism directed toward writings about indigenous peoples, particularly from people from the south, is that the authors of these works tend to be representative of the colonial sphere, even in the era of postcolonialism. They seek a real decolonization of academic writing, and for a more prominent place for alternative voices (Chambers & Buzinde, 2015). It is argued that it is the colonial perspectives, epistemologies and discourses through which indigenous groups tend to be observed and analyzed. Thus, indigenous tourism tends to be run, and academically analyzed, by representatives or ancestors of the earlier colonizers. With respect to the Arctic indigenous community, the notion of the North American Arctic still dominates public thought and scientific imagination and has led to the construction of a stereotypic image of the Arctic as remote, isolated and backward (Keskitalo *et al.*, 2013). Moreover, definitions of the Arctic are usually based on physical features like the tree line (Müller, 2013b). This causes the Arctic to be depicted as a natural region and mainly neglects the long-term presence and impact of people, particularly indigenous people, on Arctic landscapes. This also applies to tourism. As Viken (2013) claims, hegemonic structures imply that the narrative of polar tourism is constructed by scholars outside the Arctic, sometimes leading to peculiar accounts of problems and challenges. But also within the industry, the Arctic is mainly constructed as a wilderness, sometimes neglecting or obscuring the role and presence of indigenous peoples (de la Barre, 2013; Grimwood, 2015; Johnston, 2011).

Most of the contributors to this book live and work within the Arctic. However, only a few of the authors can claim to represent indigenous groups. Thus, some of the above criticism can also be applied here. And certainly, ontologies and epistemologies of the West are applied in the analyses. However, there is a difference from the critics' views concerning Africa and Asia; the indigenous groups analyzed are Western, and more or less modernized. It still, however, can be said that the contributions are made from outside an indigenous ontology and epistemology, if such entities exist. Moreover, though the ambition of the book had been to be circumpolar, the final result has a clear focus on the Nordic countries and the Sami. This mirrors the current scientific debate on indigenous tourism, which seems to be more prominent in the European part of the Arctic and, of course, also the professional networks of the editors who both work at universities in the European Arctic. Nevertheless, the book offers at least some material from outside Europe and aims at relevance for a circumpolar north.

The book consists of five parts. The first is called 'Conceptualizing Arctic Indigeneity and Tourism'. In this part, the Arctic and indigenous tourism are discussed in light of the discourses on indigeneity, neoliberalism and ethnification, colonialization and marginalization, commodification, othering and stereotyping. These are all well-known aspects of tourism, and will more or less be highlighted in the chapters that follow. Chapter 2 offers an introduction to central theoretical concepts relevant for the understanding of indigenous tourism in the Arctic. Arvid Viken and Dieter Müller illustrate the relationship of indigenousness and indigeneity and how these are employed in current scientific debate. The authors point not least to the fuzziness and randomness of indigenous self-categorization. As a consequence, they argue that being member of an indigenous group and leading a modern lifestyle go perfectly together. This is put into a geographical context as well in Chapter 3 where Carina Keskitalo discusses how the Arctic is perceived historically and currently – basically how it has been socially constructed and influenced by contemporary discourses in which indigeneity has had varying positions. The Arctic has been constructed in interactions between different ethnic groups and other major actors in dialogue with the entangling nation states. The focus of major stakeholders has traditionally been on natural resources and exploitation, but tourism and the tourists' views have increasingly received attention. Thus, the Arctic areas have been increasingly presented and represented as exotic and othered. In the eyes of the national public and media, the areas are seen as places for extraordinary people and activities. This also seems to be the case in social research, which focuses on the particularities of the regions, and less on common perspectives related to class, gender and everyday life. Keskitalo raises the question whether the aboriginalization which tourism strongly enhances, is a blessing or a curse. But, she also emphasizes that contemporary indigenous groups like the Sami are not 'passive victims', but partakers in local development and global trends, such as travel. More deeply, Keskitalo argues, there might be a conflict between a romanticizing tourism and other industries, be it reindeer herding or modern excavation.

In Chapter 4, Britt Kramvig writes about indigenous tourism in light of the theories of orientalism, neocolonialism and decolonization. The departure lies in the question of how the rights to self-determination are developing. Historically, Sápmi has been a victim of colonial repression. As an area where people adhered to a traditional livelihood until recently, Sápmi has been seen as a place to experience otherness. Kramvig discusses to what degree modern tourism represents a prolongation of this path, or is a means of decolonizing the area. She points to dilemmas and paradoxes via reference to cases where the cultural costs of tourism are significant. Should tourists be welcomed to local Sami events where the culture is exposed? It definitely can be exotic. Should the tourism industry – often in charge of non-Sami – be free to use Sami cultural remedies and signs in their promotion and production? She refers to cases where this has been seen as offensive, and has raised contests and public debates. However, as she also emphasizes, the neoliberal society is open and supports all sorts of groups, and otherness is a good selling point. However, neoliberalism is based upon colonialism, and so the celebration of otherness still has a colonial flavor. Sami tourism is a way of modernizing and perhaps decolonizing Sami affairs, albeit colored by its colonial past.

The second part is titled 'Arctic Contestations; Resourcification of Indigenous Landscapes'. As the title indicates, the focus is on processes through which Arctic indigenous resources are transformed into tourism products and experiences, or not. In Chapter 5, Dieter K. Müller and Fredrik Hoppstadius focus on Sami tourism development in northern Sweden. The chapter departs from recent global interest in northern resources like minerals, timber and wind power. These entail competing land uses not only for Sami reindeer herding, but also for Sami tourism. Against this background, Müller and Hoppstadius analyze how Sami tourism entrepreneurs assess these external challenges in comparison with other requirements related to the tourism business. The results indicate that Sami entrepreneurs are greatly concerned about the impact of the increasing exploitation of northern resources. However, it is not only industry that is creating these problems as even state policies on predators are perceived as a threat, thus indicating that Sami entrepreneurs see business and culture as two sides of the same coin. In contrast, business-related challenges are considered manageable.

In Chapter 6, Yulian Konstantinov shows how areas and rivers, traditionally resources for local people, among them the Sami of the Kola Peninsula, have been taken over by international tourist fishers and companies providing services to them. Red fish used to be an important 'currency' for Sami people. Since the Soviet period and thereafter, the rivers and their banks have been invaded by modernity – first the cold war, and then the leisure and pleasure industries. The resources – the red fish – have become a highly prized commodity. Tourists from abroad regard it as the last European wilderness. This has been a 'rewilding' supported by the Russian authorities and by environmental organizations. Obviously, the Sami and other local groups have been marginalized. Hence, indigenous people are the losers in a process wherein environmental concern and tourism development are pooled together for a seemingly successful development of the northern landscape.

In Chapter 7, Albina Pashkevich discusses the recent development of tourism in the Russian Arctic, with a particular focus on the Nenets *okrug* (region), an autonomous republic where the indigenous Nenets have a population of 7500. There has been a focus on new industries in this area of

Russia, of which tourism is an important part. In this chapter, Pashkevich shows how the authorities tend to prioritize tourism in their policies and plans, but most often without anything or much implemented. On the one hand, there seems to be an expectation from the industry of governmental facilitation, along with Russian traditions, and on the other hand, the government seems to operate in a neoliberal way, supporting those who are able to make business on their own. Thus, the development conditions for small firms are not good. Pashkevich also shows how a collaboration with Finnish entrepreneurs has failed. The conclusion is that tourism has not been a successful means of empowering the Nenets, and the classical fear of commoditization and touristification spoiling traditions and uniqueness exists.

In Chapter 8, Seija Tuulentie carries out an analysis of tourism planning as it relates to a small tourist resort called Kilpisjärvi, which is part of the Enontekio municipality on the border between Finland and Norway, and in the middle of Sápmi. Most of the tourists to the area are Norwegians, and snowmobiling is a very popular winter activity for them. The municipality has made a development plan for Kilpisjärvi, and a land use plan. The study essentially shows that the Sami are overlooked and marginalized in these processes, even though the Finnish authorities have assigned the Akwé: Kon guidelines for the protection of the nature and culture of indigenous peoples. Among the plans developed for the area, was the establishment of a national park. The tourism industry was against this, and the only stakeholder in favor was the Sami Parliament, whose voice was not heard. The Sami representatives are in favor of tourism, but the tourism industry seems not to be in favor of the Sami. Somehow, both parts want tourism, but the development still remains a matter of conflict.

In Chapter 9, Karina Smed discusses tourism development in Ilulisat (in the Disco Bay) in Greenland. This is primarily a nature destination, on the United Nations Educational, Scientific and Cultural Organization (UNESCO) list of preserved nature. She emphasizes how nature and culture are intermingled in the Greenlandic identity. The fact that Ilulisat is a major nature attraction, a UNESCO heritage status based on nature, somehow overshadows the cultural elements, both in representations of the place and in the tourism production going on. She also shows how an urge for visitability tends to under-communicate the cultural aspects of the place, due to legibility, standardization and cultural display. There is a massive emphasis on nature which is to live up to the tourists' expectations, and much more is arranged for nature than for cultural tourism. The tourists are led to natural tourism attractions, not to cultural elements.

The next three chapters are in Part 3 titled 'Touristification of the Arctic – Indigenous Wrapping'. This signifies that the outcome of the touristification processes observed is not entirely clear, or varies concerning indigenous content and motivation. In Chapter 10, John Hull, Suzanne de la Barre and Pat Maher make an analysis of aboriginal tourism as a creative industry. They analyze the web pages of two aboriginal organizations that promote tourism: Travel Yukon and a First Nation tourism website (YFNCTA). Their findings indicate that the First Nations are more or less in charge of much of the tourism development in Yukon, thus adding to the tendencies toward a stronger self-determination. The corresponding tourism also adds to a diminishing significance of the territory's colonial past. However, the account also shows a lack of signs of tourists being involved in the culture they visit, and they seem to play marginal roles in place-making, as the authors see it.

In Chapter 11, Vladislava Vladimirova discusses how festivals in northern Russia have been arenas for indigenous revitalization. She shows how several contemporary indigenous festivals seem to represent a continuity both from the Soviet and pre-Soviet eras. Political elites and government institutions tend to be in charge of the festivals, which are sponsored by international oil companies. Within the festivals, there exist hegemonic power structures, in which indigenous peoples like the Sami and Nenets are subordinated. However, at the same time, the festivals constitute arenas for cultural exposure and identity negotiations that are important for the indigenous peoples involved. As in the Soviet era, multiculturalism is celebrated at least in form, but, as then, the content tends to be controlled. Thus, the festivals tend to highlight traditions such as reindeer herding and folklore, to inscribe them and to legitimize indigenous cultures in the modern Russia.

In Chapter 12, Gro Ween and Jan Åge Riseth make a comparison between tourism in two Sami communities, one in the north of Norway, the other in the area called Trøndelag. The northern example relates to salmon angling in Tana, a border river between Finland and Norway. On the Norwegian side, the fisheries are strongly connected to local traditions, and tourist angling is limited. On the Finnish side, salmon angling is more industrialized and cheaper. Thus, there is competition for the fish, with much tension between the two sides of the river, no sustainable management and basically a lack of control over the development. This is in contrast to the southern case, in Namdalen, where tourism development is seen as more relaxed, better managed and involving local entrepreneurs who have an ethnopolitical agenda. Cultural preservation and mediation are taken seriously.

Part 4 of the book 'Tourism Negotiating Sami Traditions' focuses on how tourism is challenging Sami heritage and traditions. In Chapter 13, Kjell Olsen refers to a study of human behavior in encounters with a sacred stone, a *sieidi*. Today, the *sieidi* seems to be celebrated in a multitude of discourses and therefore has a variety of meanings. With tourism, the *sieidi* has moved from a peripheral position to the center of modernity – tourism. However, the *sieidi* can also be seen in light of discourses on nation building, neo-shamanism and the New Age movement, and can be inscribed in the discourse of authenticity. As part of the touristic discourse, the *sieidi* gives rise to reflections on life and sensual otherness. As the Sami had before, there are signs, from bones to beer boxes, telling of sacrifices. The *sieidi* adds a series of new dimensions to a landscape in which culture and nature are strongly intertwined, and in which tourism plays a part.

In Chapter 14, Trude Fonneland analyzes the element of spirituality among providers taking part in a development project for Sami tourism. The author shows how cultural and religious elements are aspects of the presentations and production that offer tourists a 'personal spiritual or religious experience'. The spiritual components are considered a strong element of many indigenous communities, and are protected by international (UN) declarations. There are other elements in the northern Sami areas that inspire spirituality, like the Northern Lights and the midnight sun. Many stories are told which reflect a spiritual frame of reference that differs from those of the tourists' homelands, thus providing a significant platform for destination development. However, as Fonneland shows, there is always a thin line between spiritual elements, marketing and commodification.

In Chapter 15, Gaute Svensson and Arvid Viken show how an old tradition in the Sami reindeer herding community of mixing friendship and working relationships is challenged by modernization. In former days (and sometimes even today), the reindeer herders needed help when crossing a fjord or a strait with their herd. Due to new ways of migration and new technologies, the needs differ and they are more closely associated with financial, bureaucratic or special practical skills related to modern technology. In the past, the payback typically was meat. Today, it may be an allowance to move around in the wilderness on snowmobiles or all-terrain vehicles (ATV), which formally is strictly limited to reindeer herders and their helpers. This is a challenge not only for the environmental management that tries to minimize such traffic, but also for reindeer herders who see tourism as a means to supplement their income.

Chapter 16 is an epilogue in which the editors, Arvid Viken and Dieter K. Müller, go back to the start and show how the contributions of the book are envisaging their starting points.

References

AHDR (2004) *Arctic Human Development Report*. Akureyri: Stefansson Arctic Institute.

AHDR (2015) *Arctic Human Development Report: Regional Processes and Global Linkages*. Copenhagen: Nordic Council of Ministers.

Butler, R. and Hinch, T. (2007) Introduction: Revisiting common ground. In R. Butler and T. Hinch (eds) *Tourism and Indigenous Peoples: Issues and Implications* (pp. 1–14). London: Butterworth-Heinemann.

Chambers, D. and Buzinde, C. (2015) Tourism and decolonization: Locating research and self. *Annals of Tourism Research* 51, 1–16.

Clapp, R.A. (1998) The resource cycle in forestry and fishing. *The Canadian Geographer* 42 (2), 129–144.
de la Barre, S. (2013) Wilderness and cultural tour guides: Place identity and sustainable tourism in remote areas. *Journal of Sustainable Tourism* 21 (6), 825–844.
Grenier, A.A. and Müller, D.K. (eds) (2011) *Polar Tourism: A Tool for Regional Development*. Montreal: Presses de l'Université du Québec.
Grimwood, B.S.R. (2015) Advancing tourism's moral morphology: Relational metaphors for just and sustainable Arctic tourism. *Tourist Studies* 15 (1), 3–26.
Hall, C.M. and Saarinen, J. (eds) (2010) *Tourism and Change in Polar Regions: Climate, Environments and Experiences*. London: Routledge.
Hull, J.S. (2001) Opening up the big land to the world – The role of the public sector in adventure tourism development in Labrador. In B. Sahlberg (ed.) *Going North: Peripheral Tourism in Canada and Sweden* (pp. 47–77). Östersund: Etour.
Jenkins, J.M., Hall, C.M. and Troughton, M. (1998) The restructuring of rural economies: Rural tourism and recreation as a government response. In R. Butler, C.M. Hall and J. Jenkins (eds) *Tourism and Recreation in Rural Areas* (pp. 43–68). Chichester: Wiley.
Johnston, M.E. (2011) Arctic tourism: Introduction. In P.T Maher, E.J. Stewart and M. Lück (eds) *Polar Tourism: Human, Environmental and Governance Dimensions* (pp. 17–32). New York: Cognizant Communication Corporation.
Keskitalo, E.C.H., Malmberg, G., Westin, K., Wiberg, U., Müller, D.K. and Pettersson, Ö. (2013) Contrasting Arctic and mainstream Swedish descriptions of northern Sweden: The view from established domestic research. *Arctic* 66 (3), 351–365.
Lemelin, H., Dawson, J., Stewart, E.J., Maher, P. and Lück, M. (2010) Last-chance tourism: The boom, doom, and gloom of visiting vanishing destinations. *Current Issues in Tourism* 13 (5), 477–493.
Lemelin, R.H., Maher, P. and Liggett, D. (eds) (2013) *From Talk to Action: How Tourism is Changing the Polar Regions*. Thunder Bay: Lakehead University Centre for Northern Studies.
Leu, T.C. and Müller, D.K. (2016) Maintaining inherited occupations in changing times: The role of tourism among reindeer herders in Northern Sweden. *Polar Geography* 39 (1), 40–57.
Lundgren, J.O.J. (2001) Canadian tourism going north – An overview with comparative Scandinavian perspectives. In B. Sahlberg (ed.) *Going North: Peripheral Tourism in Canada and Sweden* (pp. 13–46). Östersund: Etour.
Maher, P.T., Stewart, E.J. and Lück, M. (eds) (2011) *Polar Tourism: Human, Environmental and Governance Dimensions*. New York: Cognizant Communication Corporation.
Müller, D.K. (2011) Tourism development in Europe's 'last wilderness': An assessment of nature-based tourism in Swedish Lapland. In A.A. Grenier and D.K. Müller (eds) *Polar Tourism: A Tool for Regional Development* (pp. 129–153). Montreal: Presses de l'Université du Québec.
Müller, D.K. (2013a) Hibernating economic decline? Tourism and labor market change in Europe's northern periphery. In G. Visser and S. Ferreira (eds) *Tourism and Crisis* (pp. 113–128). London: Routledge.
Müller, D.K. (2013b) Tourism and the definition of the Arctic. In R.H. Lemelin, P. Maher and D. Liggett (eds) *From Talk to Action: How Tourism is Changing the Polar Regions* (pp. 9–20). Thunder Bay: Lakehead University Centre for Northern Studies.
Müller, D.K. (2015) Issues in Arctic tourism. In B. Evengård, J. Nymand Larsen and Ø. Paasche (eds) *The New Arctic* (pp. 147–158). Heidelberg: Springer.
Müller, D.K. and Pettersson, R. (2006) Sami heritage at the winter festival in Jokkmokk, Sweden. *Scandinavian Journal of Hospitality and Tourism* 6 (1), 54–69.
Müller, D.K. and Kuoljok Huuva, S. (2009) Limits to Sami tourism development: The case of Jokkmokk, Sweden. *Journal of Ecotourism* 8 (2), 115–127.

Müller, D.K., Lundmark, L. and Lemelin, R.H. (eds) (2013) *New Issues in Polar Tourism: Communities, Environments, Politics*. Dordrecht: Springer.
Office of the United Nations High Commissioner for Human Rights (2013) *The United Nations Declaration on the Rights of Indigenous Peoples: A Manual for National Human Rights Institutions*. Geneva.
Pedersen, K. and Viken, A. (1996) From Sami nomadism to global tourism. In M.F. Price (ed.) *People and Tourism in Fragile Environment* (pp. 69–88). Chichester: Wiley.
Saarinen, J. (2005) Tourism in the northern wildernesses: Wilderness discourses and the development of nature-based tourism in northern Finland. In C.M. Hall and S. Boyd (eds) *Nature-Based Tourism in Peripheral Areas: Development or Disaster?* (pp. 21–36). Clevedon: Channel View Publications.
Tuulentie, S. (2006) The dialectic of identities in the field of tourism: The discourses of the indigenous Sami in defining their own and the tourists' identities. *Scandinavian Journal of Hospitality and Tourism* 6 (1), 25–36.
Viken, A. (2013) What is Arctic tourism, and who should define it? Implication of an Anglo-American hegemony in academic writing. In D.K. Müller, L. Lundmark and R.H. Lemelin (eds) *Issues in Polar Tourism: Communities, Environments, Politics* (pp. 37–50). Amsterdam: Springer.
Viken, A. (2016). Reiseliv i samiske områder; nyliberalisme og marginalisering? In E. Angell and S. Eikeland and P. Selle (eds) *Nordområdene i endring. Urfolkspolitikk og utvikling*. (pp. 235–257). Oslo: Gyldendal Akademisk.
Viken, A. & Granås, B. (2014) Tourism destination development. Turns and Tactics. Farnham: Ashgate.
Viken, A. and Müller, D.K. (2006) Tourism and the Sami. *Scandinavian Journal of Hospitality and Tourism* 6 (1), 1–6.

2 Indigeneity and Indigenous Tourism

Arvid Viken and Dieter K. Müller

Introduction

Why do we focus on indigenous tourism? The answer is simple and relates to the core of tourism; as tourists we want to experience something that does not normally fill our everyday lives – we go to other places, perform other activities and seek out other cultures. Thus, tourists are, in general, celebrating otherness. However, the flip side of this is the mechanisms of othering, often implying some sort of distance-making or ranking of people or cultures. A few years ago, one of the authors of this chapter was convening a conference session on indigenous ecotourism with participants from all over the world. In his introduction, he emphasized the possible perils of othering in tourism (Viken, 2008). In the discussion afterward the participants, one after the other, stood up and claimed that they wanted to be othered, as otherness was their unique selling point as they saw it. They not only were aware of the attractiveness of their culture, but they were also familiar with the business rhetoric of uniqueness as a selling point. During the same session, a Swedish Sami tourism provider held a speech presenting his firm and told how he exposed his culture. He had a very modern and professional presentation, and he was clothed casually. After his presentation, a Maasai stood up and asked him: 'Are you really a Sami, you do not wear a costume? As a Maasai I never go anywhere without my traditional clothing and (with a smile he added) – a spear?' These examples go into vital questions touched upon in this book; how can and should indigenous peoples and cultures be exposed in a modern tourism context, and how do we deal with questions of authenticity and modernity in this context.

Maybe more interesting than what indigenous tourism is, is what it does. This question is also touched upon in a recent book on tourism and otherness: 'The various forms of Otherness consumed in tourism seem able (and are often purposely produced) to satisfy desires that are hidden or otherwise repressed in tourists' everyday lives', the editors (Picard & Di Giovine, 2014: 23) claim. Encountering indigenous cultures is seen as a way of connecting with history, traditions and cosmos, and to expose this otherness is a way 'to engage otherwise distanced and even alienated worlds' (Picard & Di Giovine, 2014: 22). The editors also unfold how the encounter with otherness is often a confrontation with the Self, too, both as an individual and as part of a community. This is a major point in the experience economy – that all practices, observations and impressions are part of the ongoing change of minds and hearts, and in fact, add to everyday life and social order (Picard & Di Giovine, 2014). Thus, there are both intellectual and emotional sides to meetings with others, and indigenous tourism therefore is more than enlightenment; it is about time and roots, contexts, alternative knowledge and grand narratives, myths and spirituality, and about alternative ways of performing modern lives. Also, it can make a difference for those performing this tourism; there are several examples of its role in processes of revitalization and the regaining of ethnic consciousness and pride, and a means of empowerment (Knight & Cotrell, 2016; Weaver, 2016).

This book is based upon different discourses that revolve around indigenousness, indigeneity and ethnicity. This is, as mentioned, a discourse in itself. But there are also other discourses which are more or less explicitly present in the texts, either as contexts or as narratives that give direction and content to indigenous tourism, such as discourses of traditions and their transformation, of spirituality and concerning cultural economy. There also exist important dialogues regarding modernization, globalization, commercialization and commoditization, along with those that touch upon indigenous tourism as a global phenomenon and tend to be influenced and changed as a consequence thereof. This chapter continues with a discussion of indigenousness and indigeneity and goes on to discuss tourism related to indigenous groups and indigeneity. Thereafter, indigenous tourism is viewed in light of a neoliberal economy, commercialization and neocolonial marginalization. Both editors' research is related to the Sami, the indigenous group of Scandinavia. This chapter and this book, therefore, focuses heavily on examples related to Sami tourism situations.

Indigenousness, Indigeneity and Tourism

Most often, tourism comprises an urge for experiences other than those of everyday life. People travel to experience different places and cultures, of which the lives and cultures of indigenous groups are a component. Indigenous groups are those having '... a historical continuity with preinvasion and precolonial societies that developed on their territories, [and are those who] consider themselves as distinct from other sectors of societies now prevailing in those territories'. This is the definition given by the United Nations (UN; Merlan, 2009). Indigenous groups do not traditionally regard themselves as such (Weaver, 2016). 'Indigenous peoples are not primarily self-defining populations, but categories that have been imposed by colonial orders, at least in the past centuries of the modern state' Friedman (2008: 43) claims. 'Indigenous' is therefore a modern term, and basically only a social, academic and often political construct (cf. Kuper, 2003). The idea of some groups being indigenous and others not has been criticized for being a way of essentializing tribal groups. Those particularly blamed for making such descriptions are missionaries, social anthropologists and travel writers (Islam, 1996), but travelers, tourists and the tourism industry are also part of this history. Thus, indigenous tourism and the writings thereof are not a neutral matter, but are activities filled with politics and ideology.

Non-indigenous cultures also provide tourists with otherness. Tourists want to experience local culture, be it indigenous or otherwise. There is no fundamental difference between an indigenous local culture and a non-indigenous local culture. Therefore, the academic discourse has moved more and more toward indigeneity, which '... impl[ies] first order connections (usually at small scale) between group and locality. It connotes belonging and originariness and deeply felt processes of attachment and identification, and thus it distinguishes "natives" from "others"', it is maintained (Merlan, 2009: 304). Thus, indigeneity does not only relate to peoples who constitute minorities within a majority society or a national state. The turn to indigeneity reflects a process of indigenization that in recent decades has more or less occurred in many parts of society (cf. Friedman, 2008). Thus, Greenlanders, Finns, Swedes, Norwegians and others with their own homeland can be seen as part of the indigeneity map.

As we see it, indigeneity refers to processes through which boundaries between ethnic groups are negotiated. Following Barth (1969), ethnic identity is a matter of self-ascription or choice. As most scholars see it today, we have a multitude of identities, and to various extents ethnicity is one of them. For indigenous groups, this identity is often more important than for others. Where the boundary is set is dependent on a series of contexts, and in particular on public debates, ethnopolitics and academic discourses. Thus, the perceptions of indigeneity vary significantly across the Arctic area. Whereas the policy of at least Norway and Sweden has been integration – that the Sami should be an integrated part of the welfare state – this is not necessarily the way that other indigenous groups are treated in other regimes. They may be handled through a politics of segregation, for instance to be protected from modernization. Or they may be given individual compensation for the fact that their land has been the subject of non-indigenous exploitation. The Nordic model has been different. Here, the policies – more or less due to processes of empowerment – have been to support the Sami society with the same institutions as exist for the majority population, concerning education, culture and arts, social care, healthcare and industrial politics. A result of the Nordic model is that the Sami communities are very similar to the Nordic communities. This is strengthened by the fact that most of the Sami live with and in-between the majority populations, not in designated Sami areas and never in reserves. Saminess is not based on territorial references. The difference, the ethnicity, is something highlighted in commemorative events, or when seen as appropriate in politics or commerce. Tourism is one such area, where the symbols of difference are exposed and the signs of similarity are moved to the backyard. However, for other Arctic indigenous groups, their ethnic label, and their otherness, may direct most parts of life, their life courses, careers and standard of living. This makes it difficult to write in general terms about indigeneity and indigenous tourism.

Tourism Related to Indigenous Groups and Indigeneity

'Indigenous tourism' and 'aboriginal tourism' are synonymous terms, and on occasion, 'ethnic tourism' is used with the same significance. Ethnic tourism, however, is a wider concept, which also includes tourism within minorities that are not indigenous. Indigenous tourism has been defined in different ways, but (re-)presentation of indigenous culture and way of life is normally part of it. Butler and Hinch (1996) also included the degree of indigenous control in their classification of different types of indigenous involvement. This is a normative stance, but it is difficult to disagree with it. As is the reality in many places, however, indigenous groups have no control over how tourism is undertaken, and are merely marginal partakers in the industry. What is most often displayed is indigenous people as the 'exotic Other', put together as a model culture (Bunten, 2010). As they tend to be portrayed, '[t]hese people wear somewhat traditional clothes, perform activities associated with their heritage ...' (Bunten, 2010: 290). For example, most exposures of the Sami present a model culture composed of people living in a tent (lavvu), herding reindeer and wearing costumes (colorful clothes), referring to life forms that are left. It is tourism scripted in history, heritage and traditions. The indigenous cultural tourism producers 'feel pressure to deliver a competitive product that appeals to perceived consumer desire for "the Other" ...' (Bunten, 2010: 288).

One way of classifying indigenous tourism could be to define it as tourism to an indigenous area or territory. This is the way that country-based tourism is defined – Norwegian tourism is about how tourists to and in Norway behave and are handled, as measured by the number of tourists in the country. Why should tourism to a region or a culturally delineated area not be termed and measured in a similar way? The argument against this will be that all visitors coming to a region with an indigenous population would then be indigenous tourists. These regions also have a majority population to whom this way of categorizing will seem odd, perhaps even upsetting, and this classification would include many people not even knowing that they were indigenous tourists. A more suitable definition would be as it is defined in Australia, where an 'indigenous tourism visitor is defined as one who participates in at least one indigenous tourism activity during their trip' (Ruhanen *et al.*, 2015: 74). This definition combines both production and consumption. All over northern Scandinavia, Northern Lights tourism tends to be combined with visits to indigenous camps. Many of the conference tourists to university towns such as Tromsø, Rovaniemi and Umeå would also be included in spite of them surely not being indigenous tourists in their own view. And what about individual car tourists stopping for half an hour to purchase souvenirs in a Sami souvenir kiosk? Should they also be seen as indigenous tourists? Altogether, there are many types of tourists for whom the indigenous component is small, but for whom it still exists. There is, of course, also tourism in which the indigenous component is much more prominent, such as tourism related to indigenous festivals and other events. In some of these festivals, indigenous people even constitute the majority of tourists (Viken, 2011). There are also special interest tours focusing on indigenous cultures; this was the way that Weiler and Hall (1992) once defined indigenous tourism, although it has a very small and unknown scale. Hence, Pettersson (2004) claimed that indigenous tourism is best seen as a continuum comprising specialist tourists and generalist tourists, where indigenous people and culture are the primary and secondary motivation, respectively.

There is an entire series of definitions that include additional points. Some of these focus on the tourists' ability to meet and experience indigenous people in their daily life (cf. Kunansekaran *et al.*, 2013). This requirement would be hard to fulfill in the Arctic areas where most indigenous people live modern lives very similar to those of the tourists. Thus, to be meaningful, indigenous tourism must relate to the cultural particularities of the ethnic group in question, be it contemporary or former traits. This is similar to the manner in which Smith (2001) defines it as related to four Hs of tourism; habitat (landscape), heritage, history and handicraft. Smith uses these labels not only for indigenous tourism, since they are vital to all localities or indigeneities that stand out as culturally based tourism destinations.

In their definition of indigenous tourism, Butler and Hinch (2007) see the indigenous product component and control as alternative assets. This opens up the possibility for defining much as indigenous tourism, if the control is indigenous. Such a definition makes the Canadian air company *First Air* an indigenous actor in the northbound tourism in the country, and likely open up the possibility for gambling tourism run by indigenous groups in the USA. This will also make much of the nature-based tourism of the north to indigenous tourism, as indigenous people are often involved in the businesses providing experiences. It also allows for many non-indigenous operators to be part of indigenous tourism. Thus, one may argues that there should be an 'and' between the requirement of indigenous experience components and control, as also discussed by Butler and Hinch (2007).

Many contemporary indigenous peoples live a modern life, much like other modern nationalities. Nonetheless, their tourism can still be called indigenous. Indigenousness is not based on external cultural traits, but on traditions, history and feelings. Since indigenous peoples constitute minorities, tourism in these areas, for instance in the northern Scandinavian regions, is not automatically designated as indigenous. The public nomenclature which designates the areas as Norwegian, Swedish or Finnish, is stronger. One therefore tends to end up saying that tourism in the area has an indigenous or ethnic component which varies in strength. Sometimes, the indigenous component is hidden, sometimes it is just an ingredient, sometimes it is the major focus and othertimes it is indigenous driven (cf. Nielsen & Wilson, 2012), which is the ideal model in some academic discussions.

The academic focus on indigenous tourism has mostly been on the negative impact of tourism. For a long time, tourism was seen as a huge monster invading the areas of indigenous peoples, introducing them to the evils of the modern world (cf. Castañeda, 2012; Yang & Wall, 2009). However, research has shown that this is not the correct way to perceive it. In most places, tourists are welcome and indigenous people see tourism as a path to modernity and economic development. But such development is always a two-edged sword. Tourism can mean progress, but most often also means the loss of traditions and cultural uniqueness. And, of course, there are examples of 'cultural pollution', 'bastardization', 'vulgarization' and 'phony-folk-cultures' (Yang & Wall, 2009). The background for such characteristics is often more or less romantic and the normative ideas of a former or prevailing authenticity. Ideally (to some) there should exist ancient cultures for modern consumers to gaze at, or even step into for a while, while travelling or on holiday. This is a cage model that is difficult to defend in a global world where we all, indigenous or not, are part of the same social fabric. However, all studies related to tourism's perversion of its surroundings have taught us to take care and to be aware of a whole series of developmental pitfalls.

Our point is that all definitions of indigenous groups, indigeneity and the related tourism are social constructs that tend to be biased and not neutral. They are all influenced by discourses and power. To set up strict requirements of a definition leads to a very small indigenous tourism, and by defining it too broadly, the term loses meaning and potential power. As our point of departure, we suggest that the term should rather be tied to the wider term 'indigeneity' than to 'indigenous groups', as this includes people who may be on the margin of the category, such as Greenlanders, some Sami groups and modernized or Sovietized Russian groups.

Neoliberalism and Commercial Ethnification through Tourism

All tourism destinations have territorial anchoring in a place or a larger region. Indigenous tourism is part of the neoliberal economy which was pushed by Margaret Thatcher and Ronald Reagan in the 1980s. Neoliberalism, however, dates back to the 1930s, and started when intellectuals perceived totalitarianism and Keynesianism as threats, and so emphasized the need for the state to act as guardian of the free markets (Birch & Mykhnenko, 2010). Hayek (1944) was one of the founders of this philosophical viewpoint. The neoliberal economy is touching and (re-)shaping not only indigenous peoples but also other minorities. In today's Western world there are many positive aspects related to ethnic or indigenous groups. Ethnicity is selling, particularly in the tourism sector. Comaroff and Comaroff (2009) even claim that ethnicity nowadays often emerges through commercialization and business processes.

According to Birch and Mykhnenko (2010), there are five core principles involved in a neoliberal economy: privatization of state assets, liberalization of trade, a focus on financial management, deregulation of labor and commerce, and marketization and commercialization of society as such. Neoliberalism is not just an economic philosophy, it is also a platform for the organization of social and cultural life where government's role is to 'steer rather than to row' (Foucault, 1978). Governments should not push a state forward, but rather give it direction. Many of the fields that used to be part of public policy have been depoliticized (Peck & Theodore, 2012). But authorities have an important role as facilitators, and are thus an active part in social and economic development contexts (Peck & Theodore, 2012). This has principally been the role of the state for the Sami community in Norway (Bjerkli & Selle, 2015). There are, of course, many additional critical points related to the neoliberal economy (Yeung, 2000). On the positive side, however, neoliberalism means that individuals and groups of people have a freer and strengthened status. As has been noted for some indigenous peoples, liberal policies have given relatively small groups the right to self-determination and to make demands on resources and land (Lawson, 2014).

Comaroff and Comaroff (2009) identify seven dimensions in the commercialization of ethnicity which in most cases means negotiation and/or challenges to cultures and traditions. Sami tourism development can be set into most of these facets. The first aspect mentioned is that indigenous groups claim exclusive rights to resources. These include the reserved right to reindeer herding in Norway and Sweden for Sami people (with few exceptions). The second factor that Comoroff and Comaroff (2009) mention is that trade or utility creates adherence to ethnic groups. Given the benefits of being Sami, is this a motive to appear as such? We recognize it as such from the discussions on whether a municipality should declare themselves Sami or not – a widely debated topic in Tromsø in 2013. This status could provide beneficial subsidies for Sami businesses. The third dimension that Comaroff and Comaroff mention is that much of what is created and stands out as important commercialization of indigenous communities, is often created or financed with capital from outside. In Kautokeino, the most successful attraction in recent decades has been a silver gallery established and run by a German–Danish couple and which is the local populations' major supplier of silver jewelry, an important part of their costumes. The fourth point of Comaroff and Comaroff is that many involved in ethno-based business are not particularly correct in their play with ethnic symbols and cultural expressions, but also that in such processes new expressions are created. There are also examples of cultural expressions having been adapted to tourism's aesthetics and other considerations such as finer knife blades and nicer *guksis* (cups), larger river boats and larger and more comfortable *lavvus* (tents), modern huts or *goahtis* (Lyngnes & Viken, 1998). Such adaptations are customarily motivated by profit.

The fifth element in Comaroff and Comaroff's analysis is the monopolization of local or traditional resources or production – preferably through legislation. As mentioned, reindeer herding in Norway and Sweden is a privilege of Sami people, regulated by law in both countries. In other Arctic areas, similar regulations regarding natural resources exist which affect modern industries such as mining, oil and gas exploitation. With regard to tourism, the principle has a weaker position, but as has been shown with fake Sami tourism in Finland, nowadays it does not happen without substantial opposition (Saarinen, 2001). The sixth aspect that Comaroff and Comaroff discuss is the right to land and water. Control over territory is the foremost means of increasing self-determination according to their research. This process, known as the devolution process, is the subject of huge debate in Norway and Sweden, and particularly in Canada. In the northernmost county of Norway, Finnmark Real Estate, a foundation of sorts manages what was formerly state land. This institution represents a kind of (pseudo-)privatization of approximately 95% of the land in Finnmark, the place where a large percentage of Norwegian reindeer herding takes place. The institution has a board with half of the members elected by and representing the county council, and the other half elected and representing the Sami Parliament. In contrast, however, Swedish and Finnish Sami still struggle for rights similar to those held by their

Norwegian neighbors. As a seventh point, Comaroff and Comaroff point out that commercial thinking has become an incorporated part of many ethnic groups. People have more or less been brainwashed into accepting the commercial use of cultural expressions that is taking place. Younger generations that have been surrounded by all the commercial applications react less negatively as this is the world they know. Generally speaking, many would say that a development characterized by commercial thinking is a step forward; one needs business and business operations for indigenous cultures to be sustained. The opposing view presents the negative effects, and relates to what is termed commodification, stereotyping and othering.

Commodification, Stereotyping and Othering

As previously mentioned, scholars have primarily focused on the negative aspects of tourism to and among indigenous groups and other ethnic minorities. Often, this has been amplified by a focus on indigenous peoples in developing countries, where challenges were related to significant differences in socio-economic conditions for the indigenous and colonizing population, respectively. More recently, positive traits have been identified and reported, and research stances tend to be more neutral. And whether the impact of tourism is positive or negative, most indigenous groups prefer to be part of the modern world. Tourism has shown itself as a path for development (MacCannell, 1992). Consequently, tourism development is not only a force from the outside, but it is also an internal drive for many indigenous or ethnic groups. Nevertheless, most groups want to preserve their identities as they relate to their indigeneity or ethnicity, and the research community, the majority societies and the minority groups themselves have a responsibility to support this. Tourism is a powerful modernizing phenomenon and should therefore be met with strategies and policies governing how one should cope with some of the impacts which may arise from it, namely commodification, stereotyping, othering and marginalization.

Due to the growing interest in indigenous cultures, most places have an increased awareness of the commercial opportunities in travel; more and more cultural expressions are transferred to goods and services for the tourist market. This is known as the process of commodification (Cohen, 1988; Frow, 1997). Commodification means that cultural expressions that used to have a particular meaning in a certain social and cultural context are transformed to something that is valorized and governed by market standards. The value of an expression is no longer an internal indigenous matter, but a matter decided upon by others. An obvious change that can be traced back to this, is the process of adaptation that cultural items and remedies undergo (Müller & Pettersson, 2006; Viken, 1997, 2008): some traditional artefacts such as knives, cups and furs become aestheticized; the tents and boats used to transport and accommodate tourists become bigger and more comfortable; the Sami song genre, joik, is turning into jazz, pop music and even hip hop, and is accompanied by drums and dance in ways not known before. Often, indigenous people also wear traditional costumes to be more exotic and to live up to the expectations of the tourists. All this adds to keeping the picture of a traditional culture alive. In the late 1990s, Kirshenblatt-Gimblett (1998) claimed that cultural markers tended to be unchanged in a tourism context, which MacCannell (1992) called 'frozen images'. It might have changed in many places, but there are still such expectations which, when fulfilled, add to tendencies such as stereotyping, othering and related phenomena. Nonetheless, as has been noted by some scholars, indigenous people are not automatically passive victims of such a process. Instead, they sometimes actively participate in this process in order to gain economic benefits (Tuulentie, 2006). What is permitted and what is forbidden in indigenous tourism is tested and negotiated in various ways among indigenous and non-indigenous stakeholders, and is furthermore dependent on current ethnopolitical discourses (Müller & Pettersson, 2006).

A stereotype is a rigid mental image of the characteristics of a culture or a group which is often also attached to individuals. Pickering (2001: 100) claims that stereotyping is to 'chain a person to a stunted abbreviation for a group or category in a way that denies or diminishes their individuality and agency'. In stereotypes there is no room for individuality. According to Pickering a stereotype is based on ideology and power. Very many of the stereotypes that exist make reference to unbalanced differences between social groups, in which the one is standard or normal, and from which the other is different, divergent or deviant. The term 'stereotyping' is closely related to othering; 'othering' is a modern term for stereotyping in Pickering's (2001) view. However, he also claims that othering has a wider perspective, '... bringing more clearly into frame, both for those involved in the process of othering as well as for the objects of this process' (Pickering, 2001: 69). Indigenous people are treated as something different, not as one of 'us', but as the 'Other'. The major societies and the national states have made themselves the neutral standard, and ethnic groups as 'ethnic' or 'indigenous' *minorities.* This is the process of othering (Fabian, 1983), a process to which tourism is obviously a strong contributor, as explorers, the missioners and social anthropologists before them (Fabian, 1983; Islam, 1996; Jordan, 1995; Mathisen, 2004). Through reifying or essentializing processes, the public has learned that indigenous people are different, although they today live similar lives to the people representing majority groups or nationalities.

Due to the processes of stereotyping, othering and even stigmatization related to the term and the perspective of indigenousness, there is an ongoing anthropological discussion of the fruitfulness and problems related to the term 'indigenous'. Kuper (2003), who raised the debate, sees the term as an ideological makeover of the old idea of 'primitive people' (Kuper, 2006). Others maintain that it also tends to be an exoticising term. Exotism is also seen as a malformation originating from anthropology (Kapferer, 2013). Kuper (2003) ties indigenous groups to hunter–gatherer communities, which he sees as the origin of all peoples. He claims that the status of indigenous is often based on ascriptions of how these people lived in earlier times, but that many of them now live modern lives. Still, they tend to be treated as different due to the attached indigenous – interpreted as primitive – label; such attributions implying a ranking of people and peoples. As a parallel to what Heller (1999) states for art genres, cultural differences exist, but to value some better than others is to make judgments based on prejudices, power hegemonies or ideology, and often even lack of knowledge. Similarly, therefore, it is meaningless and ideologically biased to rank cultures – they are only different and there is a diversity of cultures (Netto, 2015). In most territories, as in the Arctic, cultural diversity is significant. The authors of this book use the 'indigenous' label, but at the same time demonstrate the existing diversity and how modern trends and conceptions are part of this variety, or plurality as it is also called (Evjen & Beck, 2015).

Colonialism and Marginalization

Tourism has often had a colonial character. It is not only the West visiting the rest, it is also majorities consuming minorities. Indigeneity is somehow at the bottom of much tourism, as the folklore and everyday of the other is a bottom line. Therefore, it is not difficult to see and admit the colonial character of tourism. Simply defined, 'colonialism … is the conquest and control of other people's land and goods' (Loomba, 2015: 20), and indigeneity and indigenousness are phenomena shaped in this context:

> Indigenousness is an identity constructed, shaped and lived in the politicized context of contemporary colonialism. The communities, clans, nations and tribes we call Indigenous peoples are just that: Indigenous to the lands they inhabit, in contrast to and in contention with the colonial societies and states that have spread out from Europe and other centres of empire. (Alfred & Corntassel, 2005: 597)

Even if Arctic indigenous groups have not been colonized in the same way as African and Asian areas in the previous centuries, the same mechanisms have been in action; the national states have appropriated areas defined as nobody's land, based on a fixation of *terra nullius* that gave the then colonizers the 'right' to do so (Howitt, 2012). The politics which ensued were normally not to confer policy giving power and priority to indigenous groups. Although the appropriations of the national states have been apologized for in many countries, the lands tend to remain under the power regimes of the colonizing states. In the 1980s, Britton (1982: 347) maintained that tourism at that time tended to be enacted by those who once exercised colonial positions or their descendants, and that people from these countries were both in charge of the tourism operations and constituted the major tourist groups. Even if this may have changed since the 1980s, remnants of such structures still exist. Tourism development in the Sami areas of Norway, Sweden and Finland is basically in the hands of national and regional non-indigenous destination marketing organizations (DMOs).

In Norway where the Sami are said to be increasingly integrated into the welfare state and democratic system (Bjerkli & Selle, 2015), there are many examples of Sami interests and stakeholders being set aside in competition with modern technological or infrastructural projects such as the construction of roads, dams and electricity supply systems (Viken, 2016). Given the way that the politics have been and the way that the structure remains, it is evident that industrial policies are marginalizing the Sami in accordance with a pattern known from other indigenous groups (Altman, 2010). People who are not in the center of or adapting to the neoliberal policies and lifestyles, tend to be the losers.

A Canadian study indicates four factors that promote marginalization tendencies for indigenous groups (Marschke *et al.*, 2008). Firstly, such marginalization is grounded in the lack of knowledge about the groups in question. The second factor that the Canadian study encountered was not only negative attitudes toward indigenous people, but also an excessive romanticizing of former negative stereotypes. A third aspect of the Canadian study was the lack of consideration for the special needs of indigenous peoples. They are treated like anybody else; promoting equality through particular (and may be unequal) treatment is therefore not an accepted policy (Hernes & Hippe, 1992). As we shall see, this can also be applied to the public handling of Sami tourism. The final aspect mentioned in the Canadian study is that indigenous people feel that they are subject to specific management precisely because they are indigenous. To apply this principle to the Sami for example, one does not take into account Sami needs or requirements simply because they are Sami. That reindeer herders protest against a development that affects pasture land, is part of the political life. After consultation procedures are followed, projects or policies tend to be implemented, and reindeer drivers get some financial compensation but not sufficient to cover more troublesome future operations (Viken, 2016). Another way of formulating this is that reindeer herders tend to be purchased into tractability and silence. This is in line with what Cormaroff and Cormaroff (2009) have observed for ethnic groups worldwide; by renouncing their claim to land, indigenous groups in many places have gained access to modern goods and money in ways that give the majority society access to resources which have provided development and wealth.

Despite these power asymmetries, however, indigenous people have managed to change the game plan as well. By organizing themselves on an international level and by building global alliances which surpass the national level, indigenous groups are now able to put pressure on national governments, with the support, for example, of the United Nations Educational, Scientific and Cultural Organization (UNESCO) and various environmental groups (Barker & Soyez, 1994; Green, 2009).

Is there a True Indigeneity?

In this chapter, we have shown how indigenousness or indigeneity fits well with the tourism urge for otherness. But, at the same time, we have emphasized that the otherness of indigenous peoples and cultures is often less than their similarities with their neighbors. Tourist displays are often overloaded with signs of the particularities of the groups in question, be it representations of historical or contemporary realities. Therefore, the question of what is the truth about these cultures arises, particularly the question of authenticity. This is, of course, a tricky question. We strongly believe that societies and cultures are dynamic, relational and multiple, and that there is not just one, but a multitude of answers to the question, to be found in a series of disciplines.

The discourse of authenticity has alternatively focused on objects, tribes or cultures, and on the self. The object approach is the easiest one, it is often a question of fact – is it an object from a certain culture or produced in a particular period of time or not? But often what we are presented with are reconstructions or replica, and then the answer is not so easy. And what about new objects produced in the way that they have always been, are they authentic – and how about if modern materials have been used, and what about inventions made by indigenous people? The complexity of these questions has been discussed for decades, also related to tourism, for instance by Bruner (1994) and later by Theodossopoulos (2013) and many others. Concerning the second aspect, cultures, the question of authenticity becomes more of a moral issue; do inauthentic communities of cultures exist? We are tempted to reject this question, taking the risk or being blamed for political or 'anthropological correctness' (Theodossopoulos, 2013: 338). Thus, as we see it, a modern indigenous office employee is as authentic as one living off traditional (or modern) fishing or hunting. The question if a people is indigenous or not is mostly of academic or political relevance, and should not be important to the life of people. When it is, we are on the road from otherness to othering. Still, in the case of indigenous tourism, what most often is presented are life forms and artifacts that basically had significance in a former time period. But the old symbols may be important for the culture. What also remains is culture as ways of thinking, relating to people, nature, religious practices and customs. Thus, indigenous cultures exist, and they are authentic, but they are not necessarily tangible or visible. One of the interesting questions should be this: When tourism has become the livelihood, why should not the tourism provider's life be reckoned as an authentic way of life as that of a reindeer herder? The third aspect, authenticity of the self, is also a complicated matter, and touches upon the questions of who we are as indigenous or/and modern people, and how the contact with our inner selves are. This may touch upon issues of transparency, honesty and sincerity and about how we perform our daily life and how we act toward the 'Other' including tourists. Wang (1999) called this existential authenticity. However, as tourism providers, also indigenous people, in encounters with tourists, we perform through scripted roles (Viken, 2006), which in fact raises another question: Can we, as modern professional people, be authentic? This is a philosophical question far beyond the topic of this book.

Among the questions of indigeneity and authenticity is whether one can or should see it as a dichotomy or not. Are you either indigenous or non-indigenous? To us, this is rather simple to answer, as this is something one is more or less, and a matter of choice, whatever you are borne as. Some are born Inuit or Sami, and it is never a question, but for many their relations to these cultures are a great grandfather or a great grandmother or a great great grandparent. However, most peoples, cultures and kin are mixtures of ethnic and other backgrounds. In some places, membership of an ethnic group is a question of fulfilling a formally defined criterion. Concerning a Norwegian or Swedish Sami, it is also a question of feeling; do you feel that you are Sami? Some of those with a Sami grandparent have this feeling, some have not. You have a choice. Other places, you get the status of indigenous whatever your will is, due to a registered background. In a similar way, the question of authenticity can neither be answered by yes or no – there are degrees, and nobody has the authority to decide if an answer is right or wrong. Gradually, scholars are 'moving in the direction of acknowledging the existence of plural, multidimensional authenticities', it is maintained (Theodossopoulos, 2013). This can also serve as an answer to the question of whose authenticity? There is not only one, but many ways of defining indigenous, indigeneity and authenticity.

The questions raised in this chapter are essential for the development of indigenous tourism. We believe that this is a form of tourism that will endure, in the same way as those cultures this tourism represents will also exist in the future. Our point of view is that indigenous tourism is a modern, modernizing and globally integrated commercial activity. However, as such it also represents other values that are important to sustain; ways of thinking, practices and performances that contribute to a diversified global culture that also in the future, makes tourism an exciting activity.

References

Alfred, T. and Corntassel, J. (2005) Being indigenous: Resurgences against contemporary colonialism. *Politics of Identity* 9, 597–614.

Altman, J. (2010) What future for remote indigenous Australia? Economic hybridity and the neoliberal turn. In J. Altman and M. Hinkson (eds) *Culture Crisis: Anthropology and Politics in Aboriginal Australia* (pp. 259–280). Sydney: UNSW Press.

Barker, M.L. and Soyez, D. (1994) Think locally act globally? The transnationalization of Canadian resource-use conflicts. *Environment* 36 (5), 12–22.

Barth, F. (ed.) (1969) *Ethnic Groups and Boundaries: The Social Organization of Culture Differences.* Oslo: Universitetsforlaget.

Birch, K. and Mykhnenko, V. (eds) (2010) *The Rise and Fall of Neoliberalism: The Collapse of an Economic Order?* London: Zed Books.

Bjerkli, B. and Selle, P. (2015) *Samepolitikkens utvikling.* Oslo: Gyldendal Akademisk.

Britton, S.G. (1982) The political economy of tourism in the Third World. *Annals of Tourism Research* 9, 331–358.

Bruner, E.M. (1994) Abraham Lincoln as authentic reproduction: A critique of postmodernism. *American Anthropologist* 96 (2), 397–415.

Bunten, A.C. (2010) More like ourselves: Indigenous capitalism through tourism. *American Indian Quarterly* 34, 285–311.

Butler, R. and Hinch, T. (eds) (1996) *Tourism and Indigenous Peoples.* London: International Thomson Business Press.

Butler, R. and Hinch, T. (eds) (2007) *Tourism and Indigenous Peoples: Issues and Implications.* Amsterdam: Butterworth-Heinemann.

Castañeda, Q. (2012) The neoliberal imperative of tourism: Rights and legitimization in the UNWTO Global Code of Ethics for Tourism. *Practicing Anthropology* 34 (3), 47–51.

Cohen, E. (1988) Authenticity and commoditization in tourism. *Annals of Tourism Research* 15, 371–388.

Comaroff, J. and Comaroff, J. (2009) *Ethnicity Inc.* Chicago, IL: University of Chicago Press.

Evjen, B. and Beck, D.R.M. (2015) Growing indigenous influence on research, extended perspectives, and a new methodology. In K.W. Shanley and B. Evjen (eds) *Mapping Indigenous Presence: North Scandinavian and North American Perspectives* (pp. 27–56). Tucson, AZ: The University of Arizona Press.

Fabian, J. (1983) *Time and the Other: How Anthropology Makes Its Object.* New York: Columbia University Press.

Foucault, M. (1978) Governmentality. In G. Burchell, C. Gordon and P. Miller (eds) *The Foucault Effect: Studies in Governmentality* (pp. 87–104). Chicago, IL: University of Chicago Press.

Friedman, J. (2008) Indigeneity: Anthropological notes on a historical variable. In H. Minde (ed.) *Indigenous Peoples: Self-Determination, Knowledge, Indigeneity* (pp. 29–37). Delft: Eburon Academic Publishers.

Frow, J. (1997) *Time and Commodity Culture: Essays in Cultural Theory and Postmodernity.* Oxford: Clarendon Press.

Green, C. (2009) *Managing Laponia: A World Heritage Site as Arena for Sami Ethno-politics in Sweden*. Uppsala: Department of Cultural Anthropology.
Hayek, F. (1944) *The Road to Serfdom*. Chicago, IL: University of Chicago Press.
Heller, A. (1999) *A Theory of Modernity*. London: Blackwell.
Howitt, R. (2012) Sustainable indigenous futures in remote Indigenous areas: Relationships, processes and failed state approaches. *GeoJournal*, 77, 817–828.
Islam, S.M. (1996) *The Ethics of Travel: From Marco Polo to Kafka*. Manchester: Manchester University Press.
Jordan, G. (1995) Flight from modernity: Time, the Other and the discourse of primitivism. *Time & Society* 4 (3), 281–303.
Kapferer, B. (2013) How anthropologists think: Configurations of the exotic. *Journal of the Royal Anthropological Institute* 19, 813–836.
Kirshenblatt-Gimblett, B. (1998) *Destination Culture: Tourism, Museums, and Heritage*. Berkeley, CA: University of California Press.
Knight, D.W. and Cottrell, S.P. (2016) Evaluating tourism-linked empowerment in Cuzco, Peru. *Annals of Tourism Research* 56, 32–47.
Kunasekaran, P., Gill, S.S., Talib, A.T. and Redzuan, M. (2013) Culture as an indigenous tourism product of Mah Meri community in Malaysia. *Life Science Journal*, 10, 1600–1604.
Kuper, A. (2003) The return of the native. *Current Anthropology* 44, 389–402.
Kuper, A. (2006) The concept of indigeneity: Discussion. *Social Anthropology* 14, 21–22.
Lawson, S. (2014) The politics of indigenous identity: An introductory commentary. *Nationalism and Ethnic Politics* 20, 1–9.
Loomba, A. (2015) *Colonialism/Postcolonialism* (3rd edn). London: Routledge.
Lyngnes, S. and Viken, A. (1998) *Samisk kultur og turisme på Nordkalotten*. Bergen: Handelshøyskolen.
MacCannell, D. (1992) *Empty Meeting Grounds: The Tourist Papers*. London: Routledge.
Marschke, M., Szabiowski, D. and Vendergeest, P. (2008) Engaging indigeneity in development policy. *Development Policy Review* 26, 483–500.
Mathisen, S.R. (2004) Hegemonic representations of Sami culture: From narratives of noble savages to discourses on ecological Sami. In A-L. Siikala, B. Klein and S.R. Mathisen (eds) *Creating Diversities: Folklore, Religion and the Politics of Heritage* (pp. 17–30). Helsinki: Finnish Literature Society.
Merlan, F. (2009) Indigeneity Global and Local. *Current Anthropology*, 50, 303–333.
Müller, D.K. and Pettersson, R. (2006) Sami heritage at the winter festival in Jokkmokk, Sweden. *Scandinavian Journal of Hospitality and Tourism* 6 (1), 54–69.
Netto, M.N. (2015) From exoticism to diversity. The production of difference in a globalized and fragmented world. *Vibrant: Virtual Brazilian Anthropology* 12, 129–136.
Nielsen, N. and Wilson, E. (2012) From inviable to indigenous-driven: A critical typology of research in indigenous tourism. *Journal of Hospitality and Tourism Management* 19, 1–9.
Peck, J. and Theodore, N. (2012) Reanimating neoliberalism: Process geographies of neoliberalisation. *Social Anthropology* 20, 177–185.
Pettersson, R. (2004) *Sami Tourism in Northern Sweden: Supply, Demand and Interaction*. Östersund: Etour.
Picard, D. and Di Giovine, M. (2014) Introduction: Through other worlds. In D. Picard and M. Di Giovine (eds) *Tourism and the Power of Otherness: Seductions of Difference* (pp. 1–28). Bristol: Channel View Publications.
Pickering, M. (2001) *Stereotyping: The Politics of Representation*. New York: Palgrave.
Ruhanen, L., Whitfiord, M. and McLennan, C. (2015) Indigenous tourism in Australia: Time for a reality check. *Tourism Management* 48, 73–83.

Saarinen, J. (2001) *The Transformation of a Tourist Destination: Theory and Case Studies on the Production of Local Geographies in Tourism in Finnish Lapland*. Oulu: Department of Geography.

Smith, V. (2001) Tourism change and impacts. In V.L. Smith and M. Brent (eds) *Hosts and Guests Revisted: Tourism Issues of the 21st Century* (pp. 107–121). New York: Cognizant Communication Corporation.

Theodossopoulos, D. (2013) Laying claim to authenticity: Five anthropological dilemmas. *Anthropological Quarterly* 86, 337–360.

Tuulentie, S. (2006) The dialectic of identities in the field of tourism. The discourses of the indigenous Sami in defining their own and the tourists' identities. *Scandinavian Journal of Hospitality and Tourism*, 6, 25–36.

Viken, A. (1997) Turismens visualitet og estetisering av Sameland. *Sosiologi i dag* 27 (1), 49–71.

Viken, A. (2006) Sami tourism and Sami identity: An analysis of the tourism-identity nexus in a Sami community. *Scandinavian Journal of Hospitality and Tourism* 6 (1), 7–24.

Viken, A. (2008) Indigenous cultures challenged by ecotourism. In V. Surendar Reddy (ed.) *Ecotourism-Sustainable Development* (pp. 19–32). Hyderabad: The Icfai University Press.

Viken, A. (2011) Reinvention of ethnic identity: A local festival as a national institution on a global scene. In A.A. Grenier and D.K. Müller (eds) *Polar Tourism: A Tool for Regional Development* (pp. 179–206). Montreal: Presses de L'Université du Québec.

Viken, A. (2016) Reiseliv i samiske områder; neoliberalisme og marginalisering? In P. Selle, E. Angell and S. Eikeland (eds) *Brems eller drivkraft? Samepolitikken i den nye utviklinga i nord*. Oslo: Gyldendal Akademik (forthcoming).

Wang, N. (1999) Rethinking authenticity in tourism experience. *Annals of Tourism Research* 26 (2), 349–370.

Weaver, D. (2016) Beyond convention: Reimagining indigenous tourism. *Tourism Geographies* 18, 28–37.

Weiler, B. and Hall, C.M. (eds) (1992) *Special Interest Tourism*. London: Belhaven.

Yang, L. and Wall, G. (2009) Ethnic tourism: A framework and an application. *Tourism Management*, 30, 559–570.

Yeung, H.W. (2000) Organizing 'the firm' in industrial geography I: Networks, institutions and regional development. *Progress in Human Geography* 24, 301–315.

3 Images of the Northern and 'Arctic' in Tourism and Regional Literature

E. Carina H. Keskitalo

Introduction

The 'Arctic' has, in particular since the development of the Arctic Council as a cooperation organisation in 1996, come to be increasingly understood as an eight-state, circumpolar region consisting of Alaska, northern Canada, Greenland, Iceland, northern Russia and northern Norway, Sweden and Finland. The Arctic area is thus made to include not only the historically focused five-state Arctic rim area, but in total eight countries that had been implicated in national security issues during the cold war. This development of an Arctic region has increased the scope of what is regarded as 'Arctic' beyond most representations of the Arctic, and far beyond the historically important High Arctic, polar, as well as North American understandings of the Arctic (Keskitalo, 2004; cf. Shields, 1991).

However, including an increasing scope of areas into the 'Arctic' also results in extending the area to which Arctic representations are applied. This is a cause of concern as representations of northern areas in tourism as well as in 'Arctic' discourse have been criticised for focusing mainly on the pure environment as well as on indigenous groups primarily from the perspective of their links to the environment (e.g. Keskitalo, 2004). This has meant that a specific and unitary understanding of indigenous groups has been placed in focus, despite that both mixed and long-term non-indigenous settlers or blended groups are present; and it has also meant focusing on subsistence practices and on links to the environment rather than on the integration into capitalist system-based lifestyles, even in larger towns and cities (Keskitalo, 2004; Keskitalo *et al.*, 2013; Niemi, 1997; Olsen, 2006; cf. Shields, 1991). Contrasting with this picture, northernmost Europe – here, used to refer to northern Sweden, Finland and mainland Norway, apart from Svalbard – as areas which have been included in cooperation through the Arctic Council, diverge from the historical understandings of the Arctic; they also constitute at most sub-Arctic areas climatically. Studies have largely criticised that the image of the 'Arctic' was historically developed based on the experiences of settlers in the American New World, who drew upon frontier theory in order to position, and perhaps justify, their experiences: as civilisation opposed to non-settled pure lands, free for the taking (Keskitalo, 2004; Keskitalo *et al.*, 2013; Niemi, 1997; Olsen, 2006; Shields, 1991).

That such descriptions do not accurately portray northern areas – and may even exclude issues of economic development concerns that are relevant to the areas – has been recognised both in 'Arctic' studies in research as well as descriptions and organisation on the areas in policy. Variations between an Arctic description and northern Europe could be seen to be particularly large, and a number of points could be advanced to illustrate the divergence. As part of Old World Europe rather than a frontier-settled New World, northern Europe constitutes an area that has been settled much longer with a considerable blend of different groups (Keskitalo, 2004; Keskitalo *et al.*, 2013). Despite that Arctic-oriented literature focuses on indigenous Sami in northern Europe as almost a unitary interest, related to a large extent to reindeer husbandry interests, the Sami group is highly complex and includes multiple language varieties and highly diverging interests (indicated among other in Sami party politics and conflicts between reindeer husbandry, hunting and fishing factions, and numerous publications, e.g. Olsen, 2006, 2003; Tuulentie, 2006). Populations in the areas also include blends of Sami, Kven, Torne Valley Finn and national Swedish, Finnish and Norwegian identities, as well as multiple other identities (cf. Olsen, 2006). It is thus here a particular problem that much explicitly 'Arctic' work has excluded the history of the majority populations in the areas – which may in fact be extremely blended and constitute several different groupings – other than describing it in relation to the majority population at large in more southern parts of the countries, i.e. in national terms (cf. AHDR, 2004). Sami also cannot be conceived of as a reindeer herding identity only – while this is a culturally important occupation, presently it is practiced by a very small minority of Sami as well as others (cf. or e.g. AHDR, 2004). With regard to the focus on subsistence in Arctic literature, rather than constituting a purely subsistence-based traditional practice, the reindeer husbandry sector is also part of an industry where reindeer meat sales constitute the primary income (as sale of meat from mainly indigenous livelihoods is not, as in Canada, prohibited; cf. Mease, 2015; see also e.g. CBC, 2014). In Finland, reindeer husbandry is an equal right of all Finnish citizens, while in Sweden it is practiced mainly by Sami but with other identities involved as reindeer owners (Swedish *skötesrenar*): areas of exception also exist in Sweden where reindeer husbandry is practiced by other groups (Keskitalo, 2008). Livelihoods cannot thus be understood from a North American subsistence perspective, but must be understood within a market context and in relation to the specific nature and resource uses and practices for these as they have developed in northern European areas. The role of natural resource use must also be understood as evolving, as present employment in northern Europe is largely differentiated and only to an extent dependent on natural resource use.

There is thus reason to question the ways in which northern European areas and peoples have been represented in Arctic literature, and ask whether more accurate points of representation could be found to address this multiplicity of both areas and peoples. As a result, this chapter departs from an understanding of identities as constructed and intersectional, which it forwards as an important concept for understanding an 'Arctic' or northern area in change. The chapter exemplifies the role of these concepts through representations of indigenous as well as local groups as these have been criticised in tourism studies in northern Fennoscandia. Tourism is particularly relevant here as it constitutes perhaps the area where the use of simplified images may be most apparent; this is as the sector is dependent on the utilisation of touristic, relatively simplified images to sell a region or an area. This risk has been highlighted in tourism studies, noting for example that '[t]ourists are attracted by [what can be portrayed as] newness, strangeness and exotics ... and what can be called the otherness of places, regions or countries and the life of people living there' (Viken & Müller, 2006: 3). Or, as Olsen (2006: 37) notes in a study reviewing the representation of Norwegian Sami in local and regional tourism brochures, 'these representations give an impression of the Sami that perpetuates their image as radically different from Norwegians'. However, while descriptions in tourism can perhaps be seen as a most apparent manifestation of such descriptions, they are also and should also be understood as concepts that travel. If such concepts are used in representation e.g. in tourism, the risk is also that they are found in media or even come to structure understandings in policy and research. As a result, touristic descriptions should be understood not as an isolated phenomenon, but as a way of highlighting and exemplifying problematic representational patterns. As a result, used thus highlights criticism of the organising factors in descriptions of the north (Olsen, 2006: 37), which has been identified also in e.g. rural studies and in relation to descriptions of the Arctic in policy and research (Cruickshank, 2009; Keskitalo, 2004).

The chapter asks us to question what 'northern' is, and instead highlight the area as is, without preconceptions of 'Nordicity' or 'Arcticness' that may fundamentally detract and 'Orientalise' us from an understanding of the area (cf. Said, 1979). The following sections discuss the need to apply a broad social constructivist understanding that problematises, not essentialises, areas and identities; that intersectionality may provide a relevant understanding as it highlights multiple, and not circumscribed, singular identities; and that areas and groups have to be understood not as frozen in time but as changing, among other through globalisation processes. All three sections are interrelated and speak to the necessity of acknowledging fluid and intersectional rather than essentialistic descriptions, making examples relevant in their context as well as in this broader whole.

A Broadly Social Constructivist Perspective

One potential reason why northern regions have come to be understood in specific senses can be sought in processes of globalisation, whereby via increasingly international media and communications, simplified understandings of areas come to be used as shorthand for referring to complex and internally varying regions or areas. However, globalisation does not only bring about a need to relate to, or at least construct, concepts for more of the world, it also changes the world. While descriptions of the 'Arctic' were perhaps often formed for so called 'armchair travellers', those reading about exploration in the 1800s and early 1900s (see e.g. MacLaren, 1994), globalisation has since changed the world in that it has tied not only northern but all areas closer into capitalist societies, de-linked them from subsistence and direct links to the environment and accelerated urbanisation and resource development in northern areas as elsewhere. In a call for more complex and intersectional 'Arctic'-focused research, it is argued here that the level of complexity and also reflexivity in 'Arctic studies' would need to better relate to the general streams of social science that have criticised unitary descriptions and instead focused on increased reflexivity, such as in the cultural turn resultant from the 'crisis of representation' in the 1970s–1980s (e.g. Ray & Sayer, 1999). Such a perspective has also been seen as too little emphasised in tourism (e.g. Ateljevic *et al.*, 2005, 2013; Hall, 2004). In this, the argument with relevance to Arctic studies is that social constructivist, intersectionality based understandings and also an understanding of the impact of globalisation must be applied to northern or 'Arctic' areas in order to capture the complex areas they are today.

As a strongly developed orientation in the social sciences, social constructivism understands the world as not given, but constructed. It argues that while we are placed in a natural world, the way in which we understand it and ourselves, and in effect construct the meanings and use of our surroundings, is developed in interaction between numerous actors and the structure within which they are placed. In such an understanding, the statements of different groups thus must be challenged not as self-evident or as representing any objective truth but as made possible only within a very specific context in which we cannot take the way a group or area is constructed or presented as a given. This perspective thus highlights that no groups can be essentialised or understood as frozen in time (i.e. as having one given identity rather than developing), neither can any statement on behalf of specific individuals be assumed to fully represent a group at large. A further step in this direction is taken by postmodernism which would argue that no one truth about a given group or area can be found. A crucial common point, however, in this line of research is that it is important to ask whom a specific description gives preference to, and whom it leaves out. For example, Cheng *et al.* (2003: 97) note within the context of natural resource use that '[p]lace-based group identity is but one of several group identities one can assume in a natural resource controversy ... Groups intentionally manipulate the meanings of places hoping to influence the outcome of natural resource controversies'. Applying this perspective to location and not only to people, Cheng *et al.* further state:

> [e]very physical setting has multiple layers of meaning ... an open field is at once a potential residential subdivision, wheat field, or deer foraging area, depending on who is viewing the field – a developer, a farmer, or a hunter. What the field will eventually be used for, and symbolize, depends on the ability of each individual or affiliated group to manipulate and market its place meanings to policymakers. (Cheng *et al.*, 2003: 97–98)

This type of understanding of the roles that descriptions of peoples or areas play may be a crucial first step towards questioning descriptions that are either made about or by groups, locations or areas. Such a perspective in particular highlights that power perspectives – who is speaking and about what – are always present. Such power perspectives should be understood as not necessarily forcing or apparent but as potentially even naturalised or constituted by assumptions on appropriateness, or by ways of speaking and expressing in certain contexts. Tuulentie (2006) borrows the concept 'gaze' from, among others, Foucault and Urry in order to describe that:

> the production of pleasure is supported by professional experts who help to construct and develop our gaze as tourists ... However, not only experts but also locals are involved in directing the tourist gaze. Locals can socialize tourists to local traditions and manners in informal face-to-face interaction and thus practice social control. (Tuulentie, 2006: 31)

Such processes of a touristic or even an external gaze have numerous results: for instance, one specific identity could be placed in focus even by people who acknowledge multiple identities, if it is perceived that this is what the audience wants to hear, or if that is required to warrant inclusion into a specific context – either as an indigenous person, a researcher or a politician (despite that any one of these identities should not need to preclude the others). In this, a particularly important issue to comprehend is that existing practices are not always possible to display in the way they may be conceived by the practitioners themselves: rather, they may be 'structured in relation to the observer, for formal viewing' (Olsen, 2003). Any participant, or anyone who attempts to apply book knowledge – at the worst, drawn from other situations such as the North American to other situations, for instance in northern Europe – should thus be aware that such descriptions themselves preclude and simplify, as well as only apply (at most) to the context from which they are drawn. In addition, even in that context, the viewer or reader should be aware that the description is of something that is to some extent produced for consumption. Wang describes this well, exemplifying with a case from Japan:

> authenticity is a label attached to the visited cultures in terms of stereotyped images and expectations held by the members of tourist-sending society .... For example, what is the real Japaneseness is what has been marked; however, what is located in Japan without being marked is in a sense not the real Japaneseness and hence not worth seeing ... Authenticity is thus a projection of tourists own beliefs, expectations, preferences, stereotyped images, and consciousness onto toured objects, particularly onto toured Others. (Wang, 1999: 355)

The fact that actors may be expressing certain parts of their identity while excluding others, that certain parts of identity may be regarded as having to be excluded in relation to conceptions at different levels and by different actors, should thus be a basic understanding for any type of interaction with representations – in written or spoken representations, in newspapers or even when framing a research question (with the difference that while the tourist may seek a representation rather than question it, this is a luxury not afforded to researchers in northern areas – or any areas).

Tourism as a phenomenon may thus reproduce representations or need to relate to or problematise such reproductions, and both simplifying conceptions and criticism against these can be seen as relatively strongly expressed (the latter in tourism research). In general, tourism can be regarded as aimed at producing destination identity as a distinctive, as well as competitive place identity, increasingly so in an increasingly global marketplace (Dredge & Jenkins, 2003; Hubbard & Liley, 2000). The tourism sector is regularly made up of a number of different actors, ranging from national to sub-national/regional and local organisations who all try to market their specific level through a focus on distinctiveness compared to other units, creating 'a collage of images and place identities' (Dredge & Jenkins, 2003: 384). The extent to which these images are conceived of as acceptable locally, may to a large extent impact whether cooperation develops between stakeholders at different levels, and whether e.g. regional conceptions are seen as acceptable to local businesses

(Dredge & Jenkins, 2003). However, the ways in which tourism identities may be conceived at different levels and among different actors may be particularly divided with regard to the focus on the exotic in tourism. It has been noted that tourism often separates experiences into a potentially limited traditionality and a non-described modernity (e.g. Olsen, 2006). Given that such conceptualisations have often been applied in particular to groups that have been regarded as part of the exotic – part of nature and tradition rather than modernity, as discussed above – this issue may be seen as particularly relevant with regard to Arctic, northern or indigenous identities.

In one of the clearest examples discussing the branding of northern European regions, Olsen (2006: 37) criticises that the image of Sami as different from Norwegians is largely a result of the 'conceptual difference between tradition and a single all-embracing modernity'. Similarly:

> Features that are uncommon in the quotidian of most Sami people living in Norway make up the tourists' impressions of the Sami. This is an impression held by the pre-trip informants, but seems also to be confirmed in the encounters with the Sami tourist industry where seemingly traditional features are utilised as resources … What the tourists regarded as typical Sami were reindeers/reindeer herding/nomadism, indigeneity, a different culture and way of living, and traditions, traditional outfits, and the landscape and nature … In many ways these features are similar to what was emphasized … about 180 years earlier. (Olsen, 2006: 43; cf. Tuulentie, 2006; cf. Viken, 2006)

Seeing the need to thus describe and contextualise an understanding of Sami, and of a geographic area, Olson further states:

> This image [contrasting tradition and modernity] can appear puzzling as long as approximately 95% of the Sami population is said to live an everyday life that at least on the surface, does not differ dramatically from the modern Norwegian … The remaining 5% of the Sami mainly belong to the part of population that predominantly gains a living from reindeer herding. Even this traditionally nomadic adaptation has in most cases – if not always at the surface – become a modern way of livelihood in the sense that the practitioners utilize all suitable modern technology and devices and are fully integrated in the Norwegian welfare state. It is of interest to investigate how such radically different images between Sami and Norwegians can be upheld in spite of the similarities in the quotidian. (Olsen, 2006: 37–38)

This type of description, Olsen notes, may then gain more or less leverage depending on in what terms belonging or division into different groups is understood more generally. For example, Olsen (2006: 38) notes that 'the representations of the touristic image serve ethno-political purposes because other local organizing principles, among others the distinction between coastline and the interior, between separate Sami groups, classes and local communities increasingly has been replaced by ethnicity in the national political field'. As an example of how this situation has been created, he further illustrates how one single official Sami identity was developed from multiple groups as a process of aboriginalisation and to some extent development of Sami institutions (cf. Eidheim, 1997; Olsen, 2003, 2006).[1] Similar processes to this have been underlined as components of all nation-building as well as region-building processes (e.g. Neumann, 1999; Paasi, 1996), and are in such cases used to highlight and unify specific groups, necessarily by making them distinct from other groups. Such representations, once found in text or images such as tourism products (or even media or research), may then come to influence practice and create new divisions and needs for groupings.

As a result of these types of processes – as well as a potentially limited awareness of them in how we deal with representations – both tourism *per se* as a sector, and different constructions of the tourism identity by different organisations, may thus to a greater or lesser degree support misdescriptions that divide up tradition and modernity as well as indigenous and others, by utilising for example emblematic or expected signs to tradepost identity (Olsen, 2003; Tuulentie, 2006; Wang, 1999). The consequences of adhering to descriptions that choose and essentialise one identity over others may be that such identities no longer fit to reality experienced in what may be more naturalised understandings in the areas. For instance, Olsen (2003: 14) notes, 'Many of the local people live and have grown up in communities where the local culture is not, and has not been, inscribed as Sami but as local, as Finnmark culture, or as Norwegian. For them the "emblematic Sami" becomes a different category where they do not belong'.

An Understanding of Identities as Created Intersectionally

Extending upon a more general social constructivist or postmodernist understanding – understanding the layers of meaning in any policy development, natural resource conflict or other development as constructed and constrained by these types of representations – it would thus be important to also, in an 'Arctic' setting, understand each individual or group as constructed by multiple forces. In contrast with prominent foci in 'Arctic' conceptions, this could mean that neither traditional subsistence nor an unproblematised indigeneity, or in fact natural resource use, would be prefaced or assumed as given as the most important characteristics in a northern identity. Understanding the construction of intersectionality in such a way, Choo and Ferree (2010) note that no one is ever only oppressed or only empowered, but rather *both* oppressed and empowered by specific concepts. They note that, thus, 'lived experiences of oppression cannot be separated into those due to gender, on the one hand, and race, on the other, but rather are simultaneous and linked ... "[often, different groups] are situated within at least two subordinated groups that frequently pursue conflicting political agendas"' (Choo & Ferree, 2010: 132, with internal quote from Crenshaw, 1991: 1246). As a result, the perspectives of numerous groups should be understood, firstly, avoiding any assumption that these belong to only one group or can be described through only one perspective. Secondly, power should in this context be seen as relational, with 'the interactions among variables as multiplying oppressions at various points of intersection, and drawing attention to unmarked groups' (Choo & Ferree, 2010: 129). Thirdly and finally, intersectionality should also be regarded as 'shaping the entire social system' and thus not 'associating specific inequalities with unique institutions, instead looking for processes that are fully interactive, historically co-determining, and complex' (Choo & Ferree, 2010: 129).

Such a line of enquiry focused on intersectionality could provide opportunities for defining nuances as based in the specific local understandings that may exist, rather than based on overarching concepts, or concepts drawn from, or produced for, other areas. Echoing a broader move in cultural, gender and ethnicity studies, a focus on intersectionality would thereby target 'racialization rather than races, economic exploitation rather than classes, gendering and gender performance rather than genders—and recognize the distinctiveness of how power operates across particular institutional fields ... over time as well as between sites and institutions' (Choo & Ferree, 2010: 134). Taking such a broad focus on intersectionality and applying it to systems and not only to individuals would also necessitate discussing how different inequalities or power systems come to constitute path-dependent, historically and locally specific results (Choo & Ferree, 2010; Ray & Sayer, 1999).

This type of suggestion towards a broader intersectionality is, in fact, represented in much of the tourism research on northern areas, noting that conceptions that focus on only one line of identity result in increasing distances between the image that has to be portrayed in tourism and what are more complex realities. For instance, Tuulentie (2006: 26) emphasises for the Sami case, in line with an understanding of no group as either only victimised or empowered, that these descriptions are not created from outside only, but 'are also used by the Sami themselves in the field of tourism. More interesting would be not to regard the Sami as dominated "passive victims" and tourism as "subject-object" relationship but to realize that the subjective meanings and realities are constructed both by the tourists and the locals'. Thus, the fact that groups or individuals agree on, or participate in, specific constructions should not be assumed to make these genuine, but rather to acknowledge that concepts thus are not only victimising but also empowering – for instance, making it possible to create a tourism product. In addition, under the current level of globalisation, it should also be acknowledged that since a local can be a tourist elsewhere, and locals from other places are also tourists when they travel, 'the relationship between tourism and an ethnic group cannot be taken simply as one way effect but has to be considered more as reciprocal and renegotiated over and over again' (Tuulentie, 2006: 27).

As a result, the 'local' or assumed 'genuine' or 'authentic' representation may more as a rule than an exception be influenced by global trends and experiences from elsewhere. The 'local' nature of 'local residents' thus has to be questioned, and regarded not as authentic representations of the specific area but rather as multifaceted, contextually developed identities. As such, these identities may differ across almost all characteristics that can be highlighted under intersectionality focused analyses, with the aim of studying the importance of how these are constructed in the specific context. Thus, for instance, a female forest owner or part of a self-identified Sami household may have strongly different understandings of the role of forestry, reindeer husbandry or even the locality, compared to the male forest owner or reindeer herder in the same family.

Taking on an understanding such as this would also mean to further question the often gendered role of 'northern work' in literature. While many traditional descriptions of the 'Arctic' focus largely on male occupations in hunting, fishing, reindeer husbandry, mining or forestry, women's labour fundamentally makes possible and empowers these. In particular, it has been noted that reindeer husbandry is often supported both by women's work in other sectors such as the service or public sector and by men's seasonal work in other sectors such as forestry. Parallels can be drawn to studies in other areas where such relations have been identified. For instance, in agriculture, 'work has been synonymous with the labour undertaken by men, meaning that women's labour, both in the household and on the farm, was considered of less importance' (Andersson, 2014: 67). As a result, we need also to be sensitive to whose gendered understandings of the 'Arctic' we are perpetuating when taking up or utilising specific descriptions. Do general descriptions of forestry, fishing and mining really capture the activities in the areas? In undertaking studies that highlight any one perspective, do we then increase this focus beyond its importance in daily life? For instance, while northern areas are often discussed in terms of primary production, the service sector is in some areas larger, without this always reflected in descriptions of regions (see e.g. AHDR, 2004).

A focus on intersectionality would thus result in understanding northern areas and subjects not as given but as constructed in power dimensions and descriptions including among others centre, periphery and processes of racialisation, gendering and economic exploitation. In particular in this regard, it may seem rather surprising that despite the fact that most 'Arctic' areas have historically constituted resource-rich areas with a considerable focus on resource extraction, very few studies that use the 'Arctic' labelling for their cases have focused explicitly on the role of class (for one illustration of the orientation of Arctic-focused work, see Harrison & Hodgson, 1987). This is despite the large role that concepts of class have historically had in particular in the Nordic areas, in order to analyse the role of economic exploitation and economic systems in the co-construction of opportunity.

An Understanding of Globalisation as a Context of Change

Finally, including understandings of the context of change is a prerequisite for developing an understanding of any area; literature or preconceptions from the past cannot necessarily be used to describe the present. Today, globalisation is fundamentally changing the context of our lives, wherever we live in the world. In the 'Arctic', globalisation is fundamentally the basis for 'Arctic' organisation occurring at all, as increasing scales of organisation and the possibilities of use of modern communication technology to span these made it possible to develop an Arctic Council spanning eight states and some 10% of the world's area. Globalisation is also changing the context of resource use, from local production and consumption towards increasingly large-scale production and far-reaching production and consumption pathways, where decisions on local jobs may be taken far away. Migration and mobility both inside, out of and into the 'Arctic' or northern areas is also increasing with, for example, Thai berry pickers seasonally employed in Swedish and Finnish forests, and 'Arctic' residents more rarely being purely locally based (e.g. Keskitalo & Southcott, 2014). For example, in the Swedish context, forest owners are, to an increasing extent, female and urban, often having inherited their forest land while living and working at other locations and in other sectors than forestry (see e.g. Nordlund & Westin, 2011).

Understanding any place as 'traditional', we thus have to ask what types of descriptions are excluded, and whether this understanding really captures the area and the dynamics within it. Keskitalo *et al.* (2013) highlights the fact that the descriptions in the first Arctic Human Development Report (AHDR I) to a large extent misdescribes northern Sweden by ignoring substantial existing sub-regional literature, which has both other foci than traditional 'Arctic' literature and also does not brand these areas as 'Arctic'. Much of this domestic literature highlights the challenges of highly industrialised and even post-industrial areas of northern Europe which are strongly impacted by global trends of urbanisation and an aging population.

As a result, it becomes a problem if globalising tendencies that impact not only on resource industries (such as forestry, mining, oil and gas) but also on the service and public sectors including healthcare and education are not taken into account. In fact, as livelihoods in these areas are largely dependent on the provision of health, education, energy and technological employment infrastructure – and almost any other infrastructure related to modern life – one major issue may relate not only to primary production where jobs have largely been substituted by technology, but also to the larger-scale conditions that determine whether local or regional growth (or, indeed, maintenance of existing municipal structures) is possible under urbanisation and with an aging population.

Thus, it may be relevant to examine change by moving beyond the local, and also study structural changes and production and consumption chains – as well as political decision-making – at regional, national and international levels. One specific question relevant to at least the northern European areas of the 'Arctic' might thus be the following. In a situation of aging populations in increasingly urbanised welfare states under a globalised market and increasing flexibilisation of labour (with perhaps fewer permanent jobs in the future), how can the countryside be sustained?

In this respect, it may thus be crucial that depictions of these areas do not fall into the trap of describing them purely as wildernesses, for which such infrastructural issues would, almost by definition, not be relevant. Such descriptions have been prominent in tourism, but perhaps also in Arctic studies and policies and more broadly in rural studies (Keskitalo, 2004; for a contestation of common rural studies concepts applied to Norway see e.g. Cruickshank, 2009). In the example of tourism related to the case of reindeer husbandry, it has been noted that:

> there is a conflict between tourists' romantic views of nature and their nostalgic view of indigenous peoples versus economic and viable reindeer herding. Protected areas are sites in a global system of place imagination based on the division between civilized social spaces (tourist home area) and natural exotic spaces (tourist pleasure area). This travel discourse is particularly evident in nature-based tourism and ecotourism, which rely on ideas that places and local people are pristine and untouched by Westernization and mass-tourism, with the possible consequence of forced conservation of people and cultures as wild and premodern. (Wall-Reinius, 2012: 629–630)

It thus needs to be understood that similarly to other concepts highlighted here, 'in most cases "wilderness" involves one or several dualisms, such as ideas of nature as the creation of humans versus humans as the epiphenomena of nature itself or moral dichotomies of bad versus good nature' (Øian, 2013: 179; see also Arts *et al.*, 2012). Wilderness can thus – again, similarly to many of the other concepts discussed above – be regarded as a construction created for consumption from the outside, rather than a construction primarily reflecting local understandings (Sæþórsdóttir *et al.*, 2011). The issue of whether concepts are internal or external to an area may also be relevant in this consideration; the view of wilderness as pure nature apart from the social seems to be particular to an Anglo-Saxon construction (e.g. Vepsäläinen & Pitkänen, 2010). However, as Vepsäläinen and Pitkänen (2010: 196) note, '[i]n Finnish culture and traditions wilderness has not been an evil or bad thing, an object to win, tame or change into something else ... Instead, it has been appreciated for its importance for survival and experienced as an inevitable part of everyday life'.

Conceptualisations that focus on the countryside, or wilderness, or any other concept that separates the natural from the cultural, urban or modern, can thus, as seen for many of the concepts above, exert political consequences both within specific groups and for how groups and areas are treated. Under the current extent of globalisation (and related concepts such as the Anthropocene), it could be argued that few if any areas apart from human influence exist today, and that such distinctions are thus outdated and incorrect. It has been noted in particular in rural studies literature, that the use of such concepts to understand 'rural' areas as differentiated from the urban and from centres of development, may itself lead to infrastructure and economic development decisions focusing away from such areas, even if these areas may in fact include urbanites or be highly relevant to such infrastructure and economic development decisions and localisation (e.g. Sherval, 2009). Therefore, the dynamic and changing nature of areas must thus be taken into account in any descriptions of the 'Arctic': comprehending of 'the landscape as both socially constructed and physically concrete ... a complex interaction between nature and culture, to recognize the continuum of nature and culture in the landscape, rather than reconstructing the dichotomy of nature versus culture' (Wall-Reinius, 2012: 629–630). However, despite this 'the vexed nature of production of "the local destination" at different scales has remain[ed] relatively neglected ... few studies have been concerned with the implications of different definitions of "destination" upon policymaking' (Dredge & Jenkins, 2003: 385).

Discussion and Conclusion

This chapter has highlighted that representations of areas and peoples, whether it is in 'Arctic' research or within tourism, are never only a representation or politically innocent. Rather, as many of the authors cited above note, representations have effects with regard to what you need to represent yourself for or in relation to, and in effect, what you can be. While such conceptions may be most apparent in tourism, they also apply more broadly. Olsen (2003: 7 and 3, respectively), for instance, notes that 'for the

Sami it is an increasing problem that everyday life does not fit the idea about indigenous people, and this is a problem not only found in tourism, but also in other realms' and that the 'touristic way of exhibiting culture is also found in other fields than tourism, and is in danger of reinforcing clear-cut ethnic boundaries in an area that should rather be understood by concepts such as hybridity' (or here, intersectionality). It is also acknowledged that descriptions have explicit political effects: an area regarded as 'wilderness' will hardly be seen as requiring urban infrastructure detailed planning, and a government or policymaker that – unthinkingly or not – subscribes to these images so presented will be led along specific paths. Notably, this has resulted in 'Arctic' policy area description highlighting different issues than do policies developed for the same areas in other fields (cf. Larsson *et al.*, 2015).

If concepts such as these thus provide an incorrect understanding of areas – necessitating such conceptualisation to be contested in numerous publications among other in tourism research – there is also a reason to ask why these conceptualisations have come to be. A number of explanations can be found historically. One of these relates to the political nature of descriptions, formed for and by specific authors and audiences and dominated by a focus on the North American context (see e.g. Keskitalo, 2004; MacLaren, 1994). It has also been noted that many organisational frameworks historically utilised in the Arctic area studies field have focused on the individual, as the individual constitutes an important locus e.g. in anthropology (see Béteille [1998], who also notes the focus in this discipline on indigenous people). Institution-level focused disciplines such as political science, to some part geography, and legal studies have historically not been as present (see e.g. AHDR, 2004). This may have led to a lesser focus on specific topics: for example the role of the municipal- or county-level systems as planning bodies with regard to natural resource rights, or even a focus on the distribution of rights at constitutional level that may have led to certain results at local level. Studies or descriptions of the Arctic as well as indigineity have thus perhaps not been aimed at fully covering an area, but rather at representing it from specific interests, angles and disciplines. Applying social constructivist, intersectional and globalisation-related perspectives may then make Arctic or northern studies both more relevant to northern areas and to developments in social science at large, as well as serve to place the spotlight on the risks of simplifying descriptions, relevant not only to tourism practice or the Arctic but also to other practices, disciplines and areas of the world (see e.g. Carter & Hollinsworth, 2009). In focusing on the importance of integrating 'Arctic' and tourism studies with the critical turn in social sciences at large, we may start to escape the parochialism of Arctic area studies as well as better understand the ways in which characterisations in tourism – as well as in policy – are developed.

Acknowledgement

The author is grateful for funding from FORMAS through the FORMAS SAMI call and from MISTRA through the MISTRA Arctic Sustainable Development Programme.

Note

(1) For instance, 'Such an image of the Sami as a folk with clear-cut boundaries to a Norwegian population that is often found in a national discourse, can easily be contested by Norwegians with a similar way of living and a descent that can be labelled Sami as well as Norwegian, Kven, or Russian' (Olsen, 2003: 5).

References

AHDR [Arctic Human Development Report] (2004) *Arctic Human Development Report*. Stefansson Arctic Institute: Iceland.

Andersson, E. (2014) *Doing Gender (in) Equality in Swedish Family Farming*. Umeå: SLU.

Arts, K., Fischer, A. and Van der Wal, R. (2012) The promise of wilderness between paradise and hell: A cultural-historical exploration of a Dutch national park. *Landscape Research* 37 (3), 239–256.

Ateljevic, I., Harris, C., Wilson, E. and Collins, F.L. (2005) Getting 'entangled': Reflexivity and the 'critical turn' in tourism studies. *Tourism Recreation Research* 30 (2), 9–21.

Ateljevic, I., Morgan, N. and Pritchard, A. (eds) (2013) *The Critical Turn in Tourism Studies: Creating an Academy of Hope*. London: Routledge.

Béteille, A. (1998) The idea of indigenous people. *Current Anthropology* 39 (2), 187–192.

Buultjens, J., Brereton, D., Memmott, P., Reser, J., Thomson, L. and O'Rourke, T. (2010) The mining sector and indigenous tourism development in Weipa, Queensland. *Tourism Management* 31, 597–606.

Carter, J. and Hollinsworth, D. (2009) Segregation and protectionism: Institutionalised views of Aboriginal rurality. *Journal of Rural Studies* 25, 414–424.

CBC (2014) Kitikmeot Inuit warned to stop selling caribou meat online. CBC News. See http://www.cbc.ca/news/canada/north/kitikmeot-inuit-warned-to-stop-selling-caribou-meat-online-1.2802620 (accessed 17 October 2014).

Cheng, A.S., Kruger, L.E. and Daniels, S.E. (2003) 'Place' as an integrating concept in natural resource politics: Propositions for a social science research agenda. *Society and Natural Resources* 16, 87–104.

Choo, H.Y. and Ferree, M.M. (2010) Practicing intersectionality in sociological research: A critical analysis of inclusions, interactions in the study of inequalities. *Sociological Theory* 28 (2), 129–149.

Cruickshank, J.A. (2009) A play for rurality – Modernization versus local autonomy. *Journal of Rural Studies* 25, 98–107.

Dredge, D. and Jenkins, J. (2003) Destination place identity and regional tourism policy. *Tourism Geographies* 5 (4), 383–407.

Eidheim, H. (1997) Ethno-political development among the Sami after World War II: The invention of selfhood. In H. Gaski (ed.) *Sami Culture in a New Era* (pp. 29–61). Karasjok: Davvi Girji OS.

Hall, C.M. (2004) Reflexivity and tourism research: Situating myself and/with others. In L. Goodson and J. Phillimore (eds) *Qualitative Research in Tourism: Ontologies, Epistemologies and Methodologies* (pp. 137–155). London: Routledge.

Harrison, R. and Hodgson, G. (1987) Forty years of Arctic: The journal of the Arctic Institute of North America. *Arctic* 40 (4), 321–345.
Hubbard, P. and Lilley, K. (2000) Selling the past: Heritage-tourism and place identity in Stratford-upon-Avon. *Geography* 85 (3), 221–232.
Keskitalo, E.C.H. (2004) *Negotiating the Arctic: The Construction of an International Region*. Routledge: London.
Keskitalo, E.C.H. (2008) *Climate Change and Globalization in the Arctic: An Integrated Approach to Vulnerability Assessment*. London: Earthscan Publications.
Keskitalo, E.C.H. and Southcott, C. (2014) Globalisation. In J. Nymand Larsen and G. Fondahl (eds) *Arctic Human Development Report* (pp. 397–421). Copenhagen: Nordic Council of Ministers.
Keskitalo, E.C.H., Malmberg, G., Westin, K., Wiberg, U., Müller, D. and Pettersson, Ö. (2013) Contrasting Arctic and mainstream Swedish descriptions of Northern Sweden: The view from established domestic research. *Arctic* 66 (3), 351–356.
Larsson, L., Wiberg, U. and Keskitalo, E.C.H. (2015) Contrasting territorial policy perspectives for Northern Sweden. *Journal of Barents Studies* 2 (1), 11–33.
MacLaren, I.S. (1994) From exploration to publication: The evolution of a 19th-century Arctic narrative. *Arctic* 47, 43–53.
Mease, A. (2015) Wildlife Management in Canada. Northern Research Portal. See http://scaa.usask.ca/gallery/northern/content?pg=ex12-2 (accessed 9 February 2015).
Neumann, I.B. (1999) *Uses of the Other: 'The East' in European Identity Formation*. Manchester: Manchester University Press.
Niemi, E. (1997) Sami history and the frontier myth: A perspective on Northern Sami spatial and rights history. In H. Gaski (ed.) *Sami Culture in a New Era. The Norwegian Sami Experience* (pp. 62–80). Karasjok: Davvi Girji OS.
Nordlund, A. and Westin, K. (2011) Forest values and forest management: Attitudes among private forest owners in Sweden. *Forests* 2011 (2), 30–50.
Øian, H. (2013) Wilderness tourism and the moralities of commitment: Hunting and angling as modes of engaging with the natures and animals of rural landscapes in Norway. *Journal of Rural Studies* 32, 177–185.
Olsen, K. (2003) The touristic construction of the 'emblematic' Sami. *Acta Borealia* 20 (1), 3–20.
Olsen, K. (2006) Making differences in a changing world: The Norwegian Sami in the tourist industry. *Scandinavian Journal of Hospitality and Tourism* 6 (1), 37–53.
Paasi, A. (1996) *Territories, Boundaries and Consciousness: The Changing Geographies of the Finnish-Russian Border*. Chichester: Wiley & Sons.
Ray, L. and Sayer, A. (eds) (1999) *Culture and Economy after the Cultural Turn*. Thousand Oaks, CA: Sage Publications.
Sæþórsdóttir, A.D., Hall, C.M. and Saarinen, J. (2011) Making wilderness: Tourism and the history of the wilderness idea in Iceland. *Polar Geography* 34 (4), 249–273.
Said, E.W. (1979) *Orientalism*. New York: Vintage Books.
Sherval, M. (2009) Native Alaskan engagement with social constructions of rurality. *Journal of Rural Studies* 25, 425–434.
Shields, R. (1991) *Places on the Margin: Alternative Geographies of Modernity*. London: Routledge.
Tuulentie, S. (2006) The dialectic of identities in the field of tourism: The discourses of the indigenous Sami in defining their own and the tourists' identities. *Scandinavian Journal of Hospitality and Tourism* 6 (1), 25–36.
Vepsäläinen, M. and Pitkänen, K. (2010) Second home countryside: Representations of the rural in Finnish popular discourses. *Journal of Rural Studies* 26, 194–204.
Viken, A. and Müller, D.K. (2006) Introduction: Tourism and the Sami. *Scandinavian Journal of Hospitality and Tourism* 6 (1), 1–6.

Viken, A. (2006) Tourism and Sami identity – An analysis of the tourism–identity nexus in a Sami community. *Scandinavian Journal of Hospitality and Tourism* 6 (1), 7–24.
Wall-Reinius, S. (2012) Wilderness and culture: Tourist views and experiences in the Laponian World Heritage area. *Society & Natural Resources* 25 (7), 621–632.
Wang, N. (1999) Rethinking authenticity in tourism experience. *Annals of Tourism Research* 26 (2), 349–370.

4 Orientalism or Cultural Encounters? Tourism Assemblages in Cultures, Capital and Identities

Britt Kramvig

Introduction

What challenges lie ahead for indigenous tourism projects? What should be considered when using culture as a basis for economic development? Which fundamental paradoxes and challenges will be encountered when aiming to grow Sami tourism and the creative and cultural industries? It is believed that Sami culture has undeveloped potential as a product and an attraction in new major tourism initiatives in the northern areas of the Nordic countries. The autonomy and participatory rights of the Sami people are ensured through international conventions, and national governmental recognition of the right to participate makes it difficult to develop the products of Sami tourism without doing so in collaboration with Sami actors and institutions.

This effort has been an important element of the ongoing decolonisation in Sápmi. 'Decolonisation' is a concept attracting a multiplicity of practices relating to different actors, communities and institutions, as well as cultural projects concerned with the Sami people's rights and efforts to define themselves and shape their own future. In many areas of Sápmi, people retain a nomadic lifestyle and the knowledge and traditions gained from herding, hunting and gathering. Traditional knowledge practices and livelihoods are now carried out differently due to new political conditions and regulations. There is pressure from the growth in extraction industries; the expansion of cities and villages; the construction of roads, windmills and electricity lines; as well as increasing travel into Sápmi. In this complex situation, many look to tourism as an opportunity, not only as a business activity *per se*, but also as a way to practice the knowledge of nature, and as a gentler way to utilise the land and engage in traditional cultural practices in a new way. This entails some form of exotification and commercialisation of culture and identity, which people perceive and with which they struggle. This also then means the emergence of new tensions, both internally and externally.

On the one hand, tourism at its best gives access to articulations of self-respect and pride in a cultural background, the possibility of preserving knowledge and making a living from the handicraft (*duodji*), music, theatre, film-making and design from Sami traditions and, at the same time, cope with contemporary issues and ways of knowing. On the other hand, this effort means a growth in new forms of ownership, where important elements of local identities and cultures become resources used in products, objects, performances and events, transformed into products for a market. Some of these products are meant for external markets, others are produced for local Sami audiences or consumers. In this chapter, I will argue that these products often come with disturbance and passion, which creates new areas of conflicts and debates, and new encounters. However, this is not a debate which only happens within the Sami communities, and these communities are hardly ever solely Sami. When new actors seek to establish themselves in this area or seek to adopt important Sami symbols for commercial purposes, it creates conflicts at the minority–majority interface, along with new inter- and intra-ethnic ties.

In this chapter, I use the term 'Ultima Thule' to highlight the travelling that has taken place for centuries in the territory of the Sami indigenous people. Tourists enter into complex relationships with the communities they travel through and fuel both colonial encounters but possibly also contemporary decolonisation. In addition, I argue that Sami tourism creates new and ecologically based methods of production and products, new conversion brands and new arenas for claiming autonomy and recognition of traditional knowledge that provides an effective articulation of ownership and belongingness. However, to sell 'otherness' relies on colonial imagery, from where the dilemmas emerge.

Ultima Thule: The Journey Towards the End of the World

We know the term 'Ultima Thule' from the very first travel descriptions of the very first tourists who travelled north and then also to Sápmi. These travellers were explorers on expeditions, who journeyed towards what for them was the end of the world. The first written records from Sápmi were made by these early, mostly European, explorers. Travelling, discovering, documenting and claiming new lands and resources were all part of the European need for expansion, a need driven by capitalism, technology and religion. Said (1985) has argued that the records made by the Europeans were not based on the empirical realities of the land these travellers entered. As Said (1985) points out, we should instead read these records as expressions of a collection of European aspirations and desires, repressions, initiatives and projections. The 'others' constituted mirror images for detecting something in one's own self, the entire population or civilisation. The Samis were, like other indigenous people of the world, described as wild, unruly and uncivilised, and their religious beliefs were seen as an expression of barbarism that had to be overcome. The first expressions of 'civilisation's' entry into Sápmi often came with the construction of churches, followed by the establishment of trading posts. Then came the establishment of military fortifications, educational institutions, trade routes, etc.

The fascination with indigenous peoples, with the 'noble savage', has been described as deeply rooted in the Western modernity project. The authentic 'we' that the West lost access to as a result of modernity's disintegrating effects could be recovered as a nostalgic vision of the past that could be obtained from contact with others.

The 'otherness' duality can be seen as a prerequisite for the colonial projects, and disciplining the Sami people in terms of law, religion, education and language. They were part of 'the good will', where the otherness should be civilised and the Sami people were thus given access to the benefits of the modern welfare state, which was starting to emerge at the beginning of the 20th century. The first colonial era started as early as the 1500s. In the 1800s, the major powers of Europe expanded even more. Recent research has shown that Norwegians also participated in this expansion, even though Norway has never seen itself as a colonial power (Kjerland & Rio, 2009). Likewise, the term 'colonisation' has not been used as a working category to describe the Norwegian intervention in, and annexation of, the land of the Sami people, of which they still do not claim private ownership. Land was used by different *siida*, flexible communities of kin and others, or Sami villages, that migrated on east–west lines before the national borders were set on north–south lines, crossing and closing down the traditional migration routes, and access to pasture, land, resources, 'secret' places (*sieidies*), kin and communities.

Decolonisation in Sápmi

The first joint Sami congress was held in Trondheim on 6 February 1917.[1] This was the start of collective articulation of the challenges faced by the Sami people and their demands for recognition, political participation and rights. The movement for Sami rights was not seriously debated on the international agenda until the *Stilla* civil disobediences during the period 1979–1981 and the hunger strike outside the Norwegian Parliament. These events made the Norwegian state's colonial practice towards the Sami visible, nationally and internationally. In the wake of these events, in the 1980s and the 1990s, the effects of the colonial processes were brought into the national as well as the international consciousness and the Sami political movement became connected to the international indigenous movement. The new generation of Sami became aware of and disputed the colonial politics, making Sami identities increasingly relevant in the public political debate on indigenous autonomy, respect for differences and land claims. The Sami political revival that began in the 1960s (Eidheim, 1971) has, in the decades that followed, led to an articulation of what constitutes significant Sami values and to efforts to identify how these values can be made significant in everyday life as well as on the political scene of the Sami people (Hirvonen, 2008; Kramvig & Flemmen, 2010). A new Sami nation emerged, along with pride in indigenous values, cultural expressions and new symbols (Stordahl, 1994). In this process, reindeer husbandry, significantly different cultural expressions, such as *joik* and *doudji*, and the language became significant arenas for the articulation of Sami community and identity.

These processes have been painful for many people. At times, they have created struggles and conflicts about how to refer to the past and present and to places. Families, as well as places, have been subjected to several reinterpretations, and 'forgotten' Sami pasts have been retrieved and again made the subject of often turbulent debates. This makes Sami objects fragile because they contain highly condensed and, for many, valuable, repressed and painful knowledge, experiences and stories. The challenge in today's Sápmi is, among other things, related to the effort to reclaim and revitalise Sami traditional culture and knowledge, as well as to explore how to live in the contemporary world. These efforts are central to the international political arena, such as expressed in the United Nations Permanent Forum on Indigenous Issues and the International Labour Organisation (ILO) Convention No. 169 of Indigenous and Tribal People. At a national level, this is done through the construction of Sami institutions and the development of autonomous administrative units in areas such as environment and resource management, culture, education and health. In addition, the same efforts are made in people's everyday lives, in the challenge of recapturing the Sami's traditional knowledge, and in the challenges related to self-articulation in a turbulent and complex landscape. In recent theoretical thinking, the concept of decolonisation has been introduced to highlight the ways in which indigenous people meet and seek to renegotiate previous colonial structures and all of its paradoxes. Kramvig and Flemmen (2008) and Ween and Lien (2012) highlight how decolonisation affects everyday relations within local communities. Ween and Lien (2012: 93) are concerned with how the processes of decolonisation draw attention to, and play off, essentialised ethnic identities, serving to deflect attention from concurrent bureaucratic processes, such as environmental regulation, often with adverse, recolonising tendencies.

Kramvig and Flemmen (2008) outline that recapturing and respectfully managing the Sami's traditional knowledge is something that individuals must take responsibility for and to which they must find solutions. We need to approach an object as something that people act towards and with (Star, 2010: 603). Its materiality derives from action, not from prefabricated stuff or 'thing-ness'. Kramvig and Flemmen (2016) argue that these actions are multiple and (some) Sami objects therefore come with multiple and conflicting stories. These objects differ in the sense that they do not travel easily from one place to another. Decolonisation should be considered as an ongoing process, locally situated and with diverse practices. This is in line with Ahmed's (2000) formulation of the post-colonial concept; for Ahmed, post-colonialism is about rethinking how colonialism operated at different times in ways that permeated all aspects of social life. It is hence about the complexity of the relationship between the past and the present. Encounters are meetings that are not simply in the present: each encounter reopens past encounters. We need to ask how contemporary models of proximity reopen prior histories of encounters.

Sami Creative Industries and Politics

The Sami tourism and cultural industries, which make use of materiality and objects associated with the Sami revitalisation process, are interwoven with the political arena. Sami artists and cultural actors, regardless of whether they are musicians, writers, filmmakers, festival organisers or entrepreneurs in the growing fields of tourism and culture, articulate and explore new Sami forms of expression. These artistic and cultural actors establish themselves as entrepreneurs offering creative products but, in addition, play a prominent role as guides for Sami adolescents and as tutors for the future of Sami communities. At the same time, Sami cultural expressions and art have been performative elements in politics ever since the start of the political movement. The cultural expressions contribute to, and become arguments in, the articulation of respect for cultural differences. This has the potential to form the basis for the articulation of indigenous autonomy and political rights. In addition, tourism could be considered a hybrid practice, involving heterogenous elements, such as hotels, restaurants, travel companies, tour guides, buses, boats, headphones, films, cameras, people, fabrics, music and images, all enabling a redirection to other networks and flowing between inside and outside of that which is defined as specific to tourism (Certeau, 1988). Interventions are found in the expectations imposed by the market, by local and global industrial cooperatives, and by financial investors who set goals for innovation, growth and company operations. This leads to the necessity of simultaneously addressing a multitude of considerations. Quite possibly, the range of tensions for Sami participants is greater than the range faced by other cultural actors, although the challenges of making any culture into an industry are similar (Olsen & Kramvig, 2009).

Theoretical Considerations of Indigenous Tourism

Both tourism and the development of cultural industries have a basis in commercialisation and consumption. They entail highlighting and facilitating cultural expressions, places and traditional knowledge extracted from everyday practices and made available to the market (Viken, 2006). This implies a movement from everyday practices into a space where traditional knowledge can be presented in ways that create demand and a willingness to pay. These are the cultural arrangements necessary for making tourism an independent phenomenon, as well as creating perceived significance out of certain social phenomena (Olsen, 2002: 160). While from a global perspective we see that ethnic consciousness is strengthened as part of a more existential orientation, to an increasing extent, we also see that what constitutes the basis for ethnicity also constitutes the basis for commercialisation. Cultural identity comes to simultaneously reflect both affect and interests, or emotion and commodities. In this field, culture, identity and neoliberal interests have been compiled in new ways (Comaroff & Comaroff, 2009).

Research into indigenous tourism has examined whether tourism represents an opportunity for economic independence and the cultural revitalisation of traditional knowledge or whether it represents a threat to indigenous communities through cultural degradation and hegemonic expropriation of vulnerable cultural expressions (Butler & Hinch, 1996; Graburn, 1976; Halter, 2000; MacCannell, 1999). Recent studies reflecting on active involvement argue that it enhances economic sustainability and real contact (Butler & Hinch 2007). In this chapter, I set out to investigate whether or not postcolonial theory has the potential to redeem an understanding of the more fundamental paradoxes, where the actors within the Sami tourism and cultural industry operate. Said's (1985) book *Orientalism* is a classic work within postcolonial studies. Said describes how European writers have acted as cultural victors by authoring the narratives that, for centuries, managed to construct 'the Orient' as the mirror image of 'the West'. During the colonial expansions of the 1800s, the production of knowledge about the Orient was monopolised by the West. These contributions should, according to Said, not be seen as knowledge of the East, but rather as a part of the West's own self-understanding. In addition, Said reflects on how these performances have imparted deep and lasting imprints in the Orient's self-understanding. Spivak (1999) is inspired by but, at the same time, is writing in contrast with Said's project. If the Orient's self-understanding is also imbued with Western perspectives, the objective cannot be a quest for purity or, as in our case, a 'real' sense of

'Saminess', when otherness is inscribed. On the contrary, the starting point must be based on respect for the contradictions embodied in and articulated by the 'colonised'. Ambiguities and contradictions must be brought into the analysis and respected.

There is a need for commercial actors to act on Sami objects on behalf of this knowledge and complexity. In addition, this assumes that the 'others', i.e. those who purchase Sami cultural products, are familiar with and have respect for the complexity of contradictions from which Sami cultural expressions are created. This complexity sets the conditions for a variety of arenas of tension, to which I will return later in the chapter.

National Strategies and International Agreements on Indigenous People and Indigenous Tourism

Tourism based on Sami nature and culture is viewed by both the Norwegian Sami Parliament and by the Norwegian national authorities as a promising starting point for entrepreneurship as well as for developing new, ecologically and culturally sustainable industries for Sami communities. This ambition is reflected in the government's planning document, 'New Building Blocks in the North – The Next Step in the Government's Northern Strategy' (2009),[2] which addresses strategies for the northern areas. The document states that if indigenous communities manage to cope with the global processes of change without having to give up their own culture and industries, favourable conditions for business development must be facilitated in smaller communities.

Culturally based industries within tourism, trade, small industry, Arctic food, design, etc., are described as industries that can ensure positive development. The tourism industry highlights that culture and adventure are included as increasingly important motivational factors for travellers in a market where tourists want a greater degree of tailoring with shorter, more experience-based tourism products. Sami cultural expressions are given legal protection through the Norwegian ratification of international conventions. The ILO Convention No. 169 provides indigenous people with a special status, where their culture, lifestyle and traditions must be protected (Minde, 2008). The convention highlights that indigenous people have 'the right to continue to exist with their own identities and the right to determine their own way and pace of development'. This legal protection applies to all areas of society, including entrepreneurship programmes, tourism and the development of other cultural industries. Åhrén (2010) discusses to what extent indigenous peoples have the right to own or determine their collective creativity. His doctoral thesis concludes that indigenous people have a right to be free from the utilisation of their cultural elements, which seriously harms their cultural identity. The challenge is then to judge whether, and in what situation, a use causes harm (Åhrén, 2010: 277). These various documents do, however, imply some dilemmas. On the one hand, cultural identity is given a specific status and protection. On the other hand, cultural identity is viewed as a basis for new business developments.

The attempts to resolve this within a Norwegian context look to ensure that the development of Sami tourism is based on international law considerations, and that the Sami administrative agencies, organisations and industrial actors are involved. This reflects an attempt to anchor Sami tourism products, organisational structures and programmes in collaborative efforts with the Sami Parliament and other Sami actors with knowledge of Sami traditions. Butler and Menzies (2007) argue that the goal of having commercial activities in indigenous communities built on sustainability and the indigenous people's choice requires tying traditional ecological knowledge (TEK) to development plans. For TEK to be maintained, one must be aware of and ensure that the relations between groups of indigenous people and their land are safeguarded. In addition, the knowledge that people have built up over time about the land and its resources must be used to create an ecological and culturally sustainable form of development. Butler and Menzies (2007) go on to say that the inclusion of TEK in planning and developing tourism can decolonise and support indigenous groups' autonomy and cultural revitalisation and, at the same time, encourage economic development. However, what planning tools would this require, and will it solve the fundamental paradoxes inherent in this field? I will address this essential question later. First, I will introduce a selection of cases that, in a variety of ways, illustrate the multitude of challenges and paradoxes that unfold at a time when Sami cultural expressions are commercialised in a number of different ways.

Sami Cultures and Natures: Encounters with Commercialisation

Over a period of two years, I worked with seven Sami tourism companies to develop educational programmes and new products, along with establishing a joint, web-based information and booking portal.[3] All the companies were small-scale family businesses. They were responsible for a variety of functions, such as marketing, product development, sales, financial management, budgeting and accounting, networking, organisational work and implementing all the elements of traditional knowledge expected of a host. Several of these companies were built up around old mountain lodges. Many companies worked to identify how to reorient former operations into new markets and market needs. A keyword for them was 'innovation', or perhaps 'cultural orientation' is a more precise term to describe the challenges they faced. In short, this was a way to create knowledge of new commercial adaptations and, at the same time, maintain traditional practices and the community's value-based orientation. For the actors specifically, this meant spending time reflecting on whether or not they should let the tourists know where eagles, falcons and other rare birds of prey build their nests. Should they mark or let the tourists know about Sami sacrificial sites? What would be the consequences of this? Should they reveal the good fishing grounds or cloudberry fields, and give travellers information and therefore access to other resources that everybody agreed belonged to the community? Would this increase the tourists' sense of adventure and thus the product's attractiveness? There was no consensus on this matter, and the various operators had different strategies for meeting these challenges.

This meant that Sami tourism operators had to have a double awareness and a double set of practices. They had to be aware of the preferences of tourism operators, public tourism agencies and visitors and, at the same time, manage these preferences in line with the basic local value-related orientations and practices in the administration of the land. This tension was highlighted through stories of how other contractors in the field had stepped over the threshold by filling a boat with fish, showing a *sieidi*[4] or a *sáiva* lake,[5] or telling stories about how a *lavvu* or a *goahti* is traditionally organised and have sacred and sacrificial sites.[6] These stories formed the basis for promoting their own views on how much of what was valuable, fragile and sacred, could be sold, by whom it could be sold and who would have overall responsibility for clarifying, and possibly sanctioning, violations of what were local, tacitly ethnic and administrative practices.

In a project where we examined gender equality in the everyday life of Sami-Norwegians (Kramvig & Flemmen, 2010), we talked with a mother and a daughter from a reindeer herding family, who hosted tourist groups as an additional business during periods when the reindeer herding economy was in a downturn. They set up a *lavvu* where they hosted the tourist groups and served traditional Sami food. As the woman said:

> I remember that we got guests at the farm, and he [her husband] was going to participate. However, [this was] just before they were leaving for the mountains, then he had to go. It was right in the middle of the spring migration, and the flock stood and hopped on the mound and were ready to begin. Then my daughter and I had to take responsibility for feeding the guests that came here. So they asked us if we could tell [stories]. And we could then talk without being affected. Then I learned that Aina [the daughter] had thought about that part, that she had thought of what it would be like to be a woman involved in reindeer herding, and then I cried.

In this situation, the presence of the tourists provided an opportunity to articulate important, but fragile knowledge and opinions, because the situation was open. The latitude for experimentation flowed from the openness that prevailed from their distant and fleeting relationships with the tourists. Aina could use the openness of this situation, saying aloud that she had thought about choosing a life as a reindeer herder. Thus, she could articulate the argument among a group limited to herself, her mother and a handful of others. The tourists' presence gave Aina an opportunity to present the position of a reindeer herding woman. By doing this, she could also examine more closely the meaning of that position. Aina's mother witnessed this performance, a performance that brought forth a loss of connection to reindeer herding that she herself was feeling. Earlier in the conversation, the mother had talked about her experience of loss and the shame of not having the knowledge that her own mother had. Thus, she lacked the knowledge that a woman in a reindeer herding family should have. The interaction with the tourists and appearing as a reindeer herding woman in the eyes of others helped her to gain insights into what it meant to be a reindeer herding woman. Formulating this loss also made the shame a collective experience and thus part of a collective responsibility. This is why this situation can be described as a postcolonial moment because it can reconcile past injustices (Verran, 2002). In this specific context, the moment takes place in the arena of tourism, where the tourists are confronted as 'the others' of the dialogue, still participating as witness in the ongoing decolonisation. However, this is a double-edged sword. The fact that the Sami women continue their traditional knowledge and receive recognition for this knowledge is necessary for their knowledge to survive. At the same time, this is a survival that occurs through the presentation of culture. It is the most intimate and sensitive aspects of cultural identity that are expressed in these relations, and are made available to those who buy a seat in the *lavvu* and access to the stories. When traditional knowledge, whether it is related to nature or culture, is used in new relations and networks, culture, identity, marketing and reputations are brought together in new and very complex ways. Cultural differences are in circulation, which at the same time and in ambiguous ways give access to similarities.

Comaroff and Comaroff (2009) point out that ethno-commercialisation opposes mainstream economic rationality. This is partly because the 'differentness' can apparently be reproduced and sold without losing value. The reason for this is that the value of the commodity, which is the culture, identity and differentness, does not decrease in price due to mass production. The authors argue that, on the contrary, mass circulation confirms ethnicity, both in general and in particular, as in the case of the mother in the *lavvu*. This contributes to ethnic incorporation, and makes the status of the ethnic body a source and a means of identity. These are examples of what Comaroff and Comaroff (2009) regard as 'commercial ethnicity', which they believe is rapidly growing in many parts of the world. Without doubt, this opens up new and (perhaps) more ecologically based methods of production and products, new conversion brands, new arenas for claiming autonomy and recognition of traditional knowledge, that provide effective articulation of ownership and belongingness, while often happening in the absence of other options. To sell otherness 'relies heavily on colonial imagery that is inherently (if implicitly) racist' (Mathers & Landau, 2007: 253). The dilemmas that emerge, and the costs over time of strategies that implicate the commercialisation of traditional knowledge and ethnic identity, may turn out to be complex challenges.

The Price of Exotification

A confirmation is taking place in Guovdageaidnu and there are tourists visiting the village. This is a day of celebration, and all Sami wear the Sami national costume, called *gakti*. I attend the service. The church gallery buzzes with the sounds of voices and flashes from cameras rain down over the people assembled below, with their bowed heads and their hymn books. The tourists do not stop to participate in the worship, but instead they take pictures and walk in and out of the gallery. I am a tourist like them, yet I feel a creeping sense of irritation at the constant disturbances of what, for many, is an important religious and ritual event. It gets even worse when the service is over and the congregation is leaving the church. I feel a bit like a movie star. Below the church steps, there are several layers of tourists ready with their cameras. They want to capture the moment, those moments that are currently found on many postcards. Feeling ashamed, I slide away in my black coat. The hoard with their flash lamps is not accosted or expelled. In Guovdageaidnu, there is simply no tradition for expelling people. A bit dazed, I walk down the hill from the church and ask myself whether or not intervention in a public, but important, religious ritual is something one must live with if one wants to attract tourists to Sápmi. How should restrictions or limitations on the consumption of cultural events be established, and by whom?

Indigenous tourism, or ethnic tourism, always involves a certain separation or fixation followed by the necessity of 'exotifying' the other. For culture to sell, market actors must view the products as attractive, and the marketing practices must raise the interest or the desire that drives tourists to Sápmi. Making a difference in a highly competitive international tourism market requires clarity in communication, and it is usually done by adopting the most exotic facets of Sami culture. The tourist brochures from Sápmi contain many Sami jackets and lots of reindeer. The tourist organisations have, among other things, been at odds with Sami institutions because they wanted to use the terms 'Lap' or 'Lapland', but locally the term 'lap' is seen as laden with the prejudices and shame of the colonial past. These terms, compared to Sami and Sápmi, are more widely recognised in Europe. Commercialisation, which satisfies demands and remains engaged with market needs, will inevitably balance out between exotification and banalisation. This is one of the fundamental dilemmas of indigenous tourism.

Annually, around 500,000 tourists visit Rovaniemi in Finland. A large number of products and experienced manufacturers have emerged in the area. Rovaniemi markets itself under the collective term 'Lapland', a town in the wilderness. The adventure products cover a wide range of activities, from boat trips to tours driven by huskies, skiing, the ice hotel, to the Santa Claus village, in addition to a variety of activities that lean on Sami cultural expressions. Many controversial products, which have been criticised for their limited respect for Sami cultural expression and Sami autonomy, have emerged in Rovaniemi (Saarinen, 2001). As recently as 2008, this resulted in protests and news reports about the tourism industry's abuse of Sami culture. Sami youth organisations protested in Rovaniemi, under slogans such as 'Burn the Fake' and 'Respect our Culture'. They pointed out that Sami cultural expressions were abused, and that many Sami tourism products were neither made by the Sami nor had any reference to or respect for Sami *gakti* and their local patterns[7] (Sivertsen, 2009). The news report that acted as the tipping point for Sami youth organisations was a feature on NRK Sami Radio[8] that described a Sami baptism adapted for tourists as a kind of initiation process where the tourists had to kneel down and a knife was pressed against their neck. In addition, Finns conducted this 'ritual' in fake, Sami-styled jackets. The Sami youth called for a change, pointing out that to showcase authentic Sami life is the best experience for tourists. They demanded respect and pointed to the tourist industries tendency to use Sami objects in a manner that discriminated against the Sami people, and argued that they would no longer accept the use of non-native Samis in the tourist industry. In addition, the youth organisation pointed out that this added to the creation of stereotypical images of Sami culture, which was no longer acceptable (Sivertsen, 2009).

Since Finland has not ratified ILO Convention No. 169, the legal rights of the Sami are less tightly integrated with existing legislation when compared to legal provisions in Norway. From a historical perspective, this has allowed greater space for exotification within the Finnish tourism industry. It was the particular way in which Sami identity was fixated upon and presented that served as the focal point for the protests by the Sami youth organisations. Sivertsen (2009) describes how (urban) Sami youth are searching for a means to express identity and culture in ways that are flexible and relevant to their ongoing challenges. In addition, cultural flexibility has been described as one of the overarching values that guides everyday Sami practices (Kramvig, 2005). This flexibility tends to be lost in the creation of commercial expressions and products. This becomes a part of the public arena, or the incorporation of Sami identity in relation to which young people must negotiate their own cultural expressions. This discomfort was explicitly expressed through the demonstration. The complex assembly for which postcolonial theorists such as Spivak (1999) argue, one that lays the foundation for shaping future policy, is lost in negotiations when indigenous people continue to be portrayed as exotic characters who belong to a lost past. At the same time, there are reasons to question whether the Sami right to self-determination over what can be sold and who can sell it, actually solves the fundamental dilemmas implied by Comaroff and Comaroff (2009), such as capital interests entering and articulating themselves and their affairs in a Sami world, in everyday Sami life and in the self-articulation of the Sami. There is a constant risk that all forms of commercialisation, including those in which the Sami people are themselves involved, must lean on the exotification of belongingness, culture, identity and boundaries.

Commercialisation and Violations

There have been several public encounters in the last few years regarding the commercial (and violating) use of the Sami national costume, the *gakti* (Åhrén, 2010; Kramvig & Flemmen, 2016). One of the most recent cases is the dispute around Bergen-based toy and costume company, Ruben Treasury, which imports and sells 'Sami costume, including hat, belt, gloves and boot covers for ladies' (see Figure 4.1).[9] Social media has become an area for debate around such objects. The artist and academic Maria Kvernmo went public with her criticism, establishing a Facebook group 'for those that do not feel honoured by Rubens toyshop Sami costume'. She said that she had contacted the owner of the shop to say that the Sami felt offended by the 'fake costume' and that it was disgracing the Sami people. The owner's reaction to the criticism was that he did not mean to offend the Sami people, on the contrary, it was meant to honour the Sami.

> I'm in the costume business, a business for people who want to celebrate different people, times, heroes and Hollywood stars and to create parties around specific themes. These are costumes, not copies. They are not meant to be dresses, but used as entertainment for a night party .... I do hope that you can consider these costumes as an honour and not disdainful of your culture, as with the Scottish, German, Irish, Arabic, French, etc ... The costume is made for fun and play and to create good vibes at party events.[10]

Figure 4.1 Sami costumes supplied by the Ruben Treasury

In Facebook discussions, he expressed that 'he had never heard of anyone that used this costume to offend anybody'. The president of the Sami Parliament commented that she considered this an unworthy approach towards Sami culture, and that she understood why many spoke out against this. She argued that the indigenous and minorities' ceremonial dress is often appropriated as carnival costumes, and that ceremonial clothing represents a genuine culture, not entertainment.[11]

This sparked encounters that went in different directions, both a debate among the Sami on how (again) to approach the ongoing tendency of appropriation of the Sami *gakti* and other objects of importance with references to the ongoing indigenous debate on this matter. There was also a debate about the tendency towards the utterance of racist comments, as well as questioning the Sami as indigenous. Other voices argued that the Sami should not voice this criticism as it reinforced public Norwegian stereotypes of the Sami as 'whining' and 'without a sense of humour'.

My argument is that it is demanding for individual people to take on the responsibility of raising this again and again and that political actions, regulations or guidelines need to be debated in relation to tourism. There is a need for commercial actors to act on Sami objects knowing the postcolonial condition and the multiplicity of expressions, but also the difficulties of coming to terms with the past, and how the colonial past is still enacted differently in social situations. In addition, this assumes that 'the others', i.e. those who purchase Sami cultural products, are familiar with and have respect for the complexity of contradictions from which Sami cultural expression is created. This complexity sets the conditions for a variety of arenas of tension that need to be known and respected.

Summary

Indigenous tourism, or adding cultures and identities as input factors in economic activities, creates many new interconnections and new paradoxes, which I have sought to open up in this chapter. The link between identities, culture and industry creates new value conversion chains that are woven together with demands for autonomy and self-determination, or that express a sense of belongingness and the revitalisation of (partially) lost traditions and knowledge. This often happens in the absence of alternatives in marginalised Sami villages, especially in connection with reindeer herding, which is an industry with one of the lowest average incomes. The economics of culture and identity can be seen as the economics of difference and desire, where affect and interest, the emotion and merchandise, are linked together in new ways. According to Comaroff and Comaroff (2009), this situation offers the possibility of highly uncertain political, economic and ethical impacts. This uncertainty is a prerequisite for neoliberal economics. At the same time, we see that the legal tools achieve a completely new status in both political and social life, and on a scale that has never before existed. However, will these tools be sufficient to ensure that indigenous cultures and identities are managed in sustainable ways at the interface with international market actors? Indeed, this is an open-ended question that lacks a clear answer. Indigenous people take and are given the right to participate on an equal basis in the new economic arena, including the right to sell their 'otherness'. The impact that this will have in the long run remains to be seen.

We perhaps need to have competent advisory bodies and actors who can provide advice to companies. In addition, there may also be a need to open up a debate if common guidelines are needed on how to use the traditional *gakti* and other Sami objects, how to manage ritual sites and how to ensure that other elements with references to traditions are both respected for their diversity and not transported into commercial products that are hurtful and invasive to the Sami people. In addition, we need to create the tools for land management that account for TEK and can be implemented in the tourism industry and in the formulation of guidelines for tourists who visit indigenous communities. The use of uncultivated land can mean operational conflicts and pressures on public resources. Sami tourism operators are familiar with this and are striving to maintain a good working relationship with other local interests, such as reindeer herders, local hunters and fishermen, and to take into account the maintenance of population compositions, endangered areas and species. In addition, it is emphasised that ethical judgements must be the basis for deciding which elements of the Sami cultural heritage can be commercialised and offered for sale. At the same time, all the Sami actors in this arena must focus on ensuring cultural sustainability. Not everything can and should be sold.

In addition, tourism operators must account for the needs and preferences of their customers and the arrangements necessary for the introductory sales of the products and the satisfaction with the products. These considerations are not necessarily congruent.

A diversity of interests and assessments are being built into business decisions and development perspectives. This expertise should be made visible and recognised, and should be an important argument for making the local Sami tourism industry responsible for the future development of Sami tourism. In addition, it will be a challenge to renegotiate the forces that pull towards orientalism, towards an exotification of the 'natives' traditional knowledge. The concept of decolonisation, seen as becoming an ongoing process, opens up the possibility of acknowledging the specificities of each locality's work in building respectful and sustainable economic indigenous enterprises. What can innovative programmes, built upon indigenous knowledge look like? And how do we find a balance between the danger of exotification and the quest for autonomy?

In this chapter, I have used postcolonial theory as a possible inroad. Many would claim that colonialism is very much ongoing in Sami areas; others would claim that decolonisation is a better concept. Places are different, as are communities of people and ongoing discourses. Orientalism is ongoing and so is postcolonialism, as framed by Ahmed (2000). Different pasts are in the present and adding to the complexity involved in tourism. Encounters are meetings that are not simply in the present: each encounter reopens past encounters. Achieving the political ambitions of the north requires more systematic work, expertise at all levels and an institutional anchoring. Through this, we can ensure the development of new knowledge and new commercial areas with the capacity to address the challenges faced by the Sami community. In addition, it will be necessary for all Arctic communities to find new and sustainable ways to live their lives in a time of major climatic and geopolitical challenges.

Notes

(1) This early political mobilisation is now celebrated as Sami People's Day.

(2) See https://www.regjeringen.no/globalassets/upload/ud/vedlegg/nordomradene/byggesteiner_nord090323_2.pdf (accessed 2 January 2017).

(3) This was a The Research Council of Norway (NFR) Region project owned by the companies themselves, the Finnmark County Council and the Norwegian Research Council. The management team consisted of associate professor Arvid Viken from the University College of Finnmark, the tourism operator Tore Turi and myself, at that time employed in a research position at Finnmark Research Centre.

(4) A *sieidi* is a sacred stone (in northern Sami language) that indicates a sacred sacrificial place, or *bassi* in the northern Sami language.

(5) The *sáiva* lakes have two bottoms. This meant they functioned as a link between this world and the other, parallel world and served as a place where the shaman, or *noiade*, could go to be advised by the ancestors.

(6) *Árran*, or the fireplace, is just one of the sacred places in a *lavvu*.
(7) See http://www.tv2nyhetene.no/innenriks/sinte-samer-varsler-aksjoner-2231109.html.
(8) 04.11.2008 NRK Sami Radio.
(9) See https://www.rubens.no/kostymer-utkledning/verden-rundt/costume-saami-36-38-s-same-kostyme (accessed 2 October 2016).
(10) Rubens update on Facebook. See https://www.facebook.com/permalink.php?story_fbid=10153354655152559&id=356766977558 (accessed 9 September 2015).
(11) Nordlys 'Det skjærer I øynene å se på dem'. See http://www.nordlys.no/samisk/det-skjarer-i-oynene-a-se-pa-dem/s/5-34-204510 (accessed 9 September 2015).

References

Ahmed, S. (2000) *Strange Encounters: Embodied Others in Post-Coloniality*. London: Routledge.

Åhrén, M. (2010) The Saami traditional dress and beauty pageants: Indigenous peoples' rights of ownership and self-determination over their cultures. Doctoral thesis, University of Tromsø.

Butler, R. and Hinch, T. (eds) (1996) *Tourism and Indigenous Peoples*. London: International Thomson Business Press.

Butler, R. and Hinch, T. (eds) (2007) *Tourism and Indigenous Peoples: Issues and Implications*. Oxford: Elsevier.

Butler, C.F. and Menzier, C.R. (2007) Traditional ecological knowledge and indigenous tourism. In R. Butler and T. Hinch (eds) *Tourism and Indigenous Peoples: Issues and Implications* (pp. 15–28). Oxford: Elsevier.

Certeau, M. (1988) *The Practice of Everyday Life*. Berkley, CA: University of California Press.

Comaroff, J.L. and Comaroff, J. (2009) *Ethnicity, Inc*. London: The University of Chicago Press.

Eidheim, H. (1990 [1971]) Aspects of the Lappish minority situation. *Occasional Papers in Social Anthropology*, 14. Oslo: Department of Social Anthropology.

Graburn, N.H. (1976) Introduction: Arts of the Fourth World. In N.H. Graburn (ed.) *Ethnic and Tourist Arts: Cultural Expressions from the Fourth World* (pp. 1–32). Berkeley, CA: University of California Press.

Halter, M. (2000) *Shopping for Identity: The Marketing of Ethnicity*. New York: Schocken Books.

Hirvonen, V. (2008) *Voices from Sápmi, Sami Women's Path to Authorship*. Guovdageaidnu: DAT.

Kjerland, K.A. and Rio, K.M. (eds) (2009) *Kolonitid Nordmenn på eventyr og big business i Afrika og Stillehavet*. Oslo: Scandinavian Academic Press.

Kramvig, B. (2005) Fleksible kategorier, fleksible liv. *Norsk Antropologisk Tidsskrift* 2–3, 97–108.

Kramvig, B. and Flemmen, A.B. (2008) Møter: Sammenstøt av verdier i samisk-norsk hverdagsliv. In O. Leirvik and Å. Røthing (eds) *Verdier* (pp. 101–120). Oslo: Universitetsforlaget.

Kramvig, B. and Flemmen, A.B. (2010) Mangfold, likhet og likestilling i Sápmi. In A.B. Berg, A.B. Flemmen and B. Gulliksen (eds) *Norske likestillinger* (pp. 167–195). Trondheim: Tapir.

Kramvig, B. and Flemmen, A.B. (2016) What alters when the Sami traditional costume travels? Affective investments in the Sápmi. In J. Fryktman and M. Povrzanovic Fryktman (eds) *Sensitive Objects: Affects and Material Culture* (pp. 179–198). Lund: Academic Press.

MacCannell, D. (1999) *The Tourist. A New Theory of the Leisure Class*. Berkeley, CA: University of California Press.

Mathers, K. and Landau, L. (2007) Natives, tourists, and makwerekwere: Ethical concerns with 'Proudly South African Tourism'. *Development Southern Africa* 24, 523–537.

Minde, H. (ed.) (2008) *Indigenous Peoples: Self-Determination, Knowledge, Indigeneity*. Delf: Eburon Academic Publishers.

Olsen, K. (2002) Authenticity as a concept in tourism research: The social organisation of the experience of authenticity. *Tourist Studies* 2, 159–182.

Olsen, F.E. and Kramvig, B. (2009) Kultur som næring – møter som sammenstøter? *Magma* 9, 22–29.

Saarinen, J. (2001) *The Transformation of a Tourist Destination: Theory and Case Studies on the Production of Local Geographies in Tourism in Finnish Lapland*. Oulu: Oulu University Press.

Said, E. (1985) *Orientalism. Western Conceptions of the Orient*. London: Penguin Books.

Sivertsen, E.A. (2009) *Samisk Ungdom – mellom Tradisjon og Modernitet?* MA thesis, Universitetet i Oslo.

Spivak, G.C. (1999) *A Critique of Postcolonial Reason Toward a History of the Vanishing Present*. London: Harvard University Press.

Star, S.L. (2010) This is not a boundary object: Reflections on the origin of a concept. *Science, Technology and Human Values* 35, 601–617.

Stordahl, V. (1994) *Same i Den Moderne Verden: Ending og Kontinuitet i et Samisk Lokalsamfunn*. PhD thesis, Institutt for Sosialantropologi, Tromsø.

Verran, H. (2002) A postcolonial moment in science studies: Alternative firing regimes of environmental scientists and Aboriginal landowners. *Social Studies of Science* 32, 729–762.

Viken, A. (2006) Tourism and Sami identity: An analysis of the tourism–identity nexus in a Sami community. *Scandinavian Journal of Hospitality and Tourism* 6 (1), 7–24.

Ween, G. and Lien, M. (2012) Decolonialization in the Arctic? Nature practices and land rights in the Norwegian High North. *Journal of Rural and Community Development* 7, 93–109.

Part 2

Arctic Contestations; Resourcification of Indigenous Landscapes

5 Sami Tourism at the Crossroads: Globalization as a Challenge for Business, Environment and Culture in Swedish Sápmi

Dieter K. Müller and Fredrik Hoppstadius

Introduction

Indigenous tourism has often been promoted as a way of balancing economic decline in indigenous industries, and a way of creating understanding of the situation of indigenous peoples (Altman & Finlayson, 1993; Butler & Hinch, 1996, 2007; Ryan & Aicken, 2005). As pointed out previously, however, the situation for indigenous peoples is contingent on geographical and historical circumstances influencing their development to date on the one hand, and their relation to the dominant culture on the other (Müller & Kuoljok Huuva, 2009; Smith, 1996). As a consequence, indigenous tourism is influenced not only by the indigeneity of its proponents and the indigenous theme of their product, but also by its geographical location. This has also been acknowledged by Butler and Hinch (2007), who characterize indigenous tourism as a complex system combining various social, economic, political and environmental dimensions. In an Arctic context, this implies that indigenous tourism cannot be discussed without acknowledging accessibility, seasonality and distance to decision-making power (Müller & Pettersson, 2001). In a Nordic context, it furthermore means that it has to be recognized that indigenous tourism takes place in the setting of a welfare state, leaving individuals belonging to the Sami, northern Europe's indigenous group, with multiple options for professional development and careers, among which tourism is only one.

Hence, though Sami tourism has often been identified as a potential source of income for the indigenous group (Müller & Pettersson, 2006; Palomino, 2012; Pettersson, 2002, 2003), it did not develop as anticipated.

Müller and Kuoljok Huuva (2009) argued that this has been due to cultural constraints identifying tourism as an inferior activity compared to Sami reindeer herding. However, the fact that Sami tourism has to remain a side activity to reindeer herding in order to sustain the authenticity of the product has been mentioned as an explanation for the limited development, as has the Sami perception that tourism development challenges the freedom of the reindeer herders (Viken *et al.*, 1998). Thus, as concluded in a review, '... there was a mismatch between vital values still alive in the Sami society, and the spirit of capitalism' (Pettersson & Viken, 2007: 184).

To what extent this has changed since the 1990s is not yet documented. However, it can be noted that challenges have arisen to the existing Sami tourism industry because of a rejuvenated international interest in northern resources (Müller, 2011). Resources like minerals, timber and wind power entail competing land uses for not only Sami reindeer herding but also Sami tourism. Against this background, the purpose of this chapter is to analyze how Sami tourism entrepreneurs assess these external challenges, also in comparison with other requirements related to the business of tourism.

Conditions for Indigenous Tourism Development

As already mentioned, indigenous tourism does not occur in a societal and environmental vacuum. Research has repeatedly attempted to explain its development in time and space. Often, push factors inherent to the situation of indigenous peoples as marginalized have been employed to explain an engagement in tourism (e.g. Butler & Hinch, 1996, 2007; Buultjens *et al.*, 2005; McIntosh & Ryan, 2007; Ryan & Aicken, 2005). Hence, tourism is considered an exit strategy from a situation in which members of the indigenous community, usually for economic reasons, are forced to find a new and/or complementary livelihood outside traditional trades and industries. Besides economic dimensions, the revitalization and sustenance of cultural practices and traditions are also mentioned as drivers and potential positive outcomes of tourism development (Dowsley, 2009; Müller & Pettersson, 2001).

Still, the development and success of tourism are not only dependent on the motivation of the people involved. Previously, Smith's (1996) 4H approach has been an attempt to analyze indigenous tourism. In accordance, four dimensions – habitat, heritage, handicraft and history – are argued to be decisive in the development of indigenous tourism. In this context, habitat is understood as the landscape that nurtured indigenous peoples, while heritage refers to the accumulated body of knowledge enabling survival in the habitat. Handicraft is an important interface to tourists and provides the base for economic exchange. Finally, history represents the relationship to the dominant culture as a result of historical interactions, which not least influences the willingness to engage in tourism. The 4H approach has been frequently used in research to assess not least the potential and constraints of indigenous tourism (Johansen & Mehmetoglu, 2011; Müller & Pettersson, 2001, 2006; Notzke, 2004; Pratt *et al.*, 2013). While helpful in conceptualizing relations to the surrounding environment and society, the model obviously fails to address, for example, business knowledge and practice as potential explanations for the development of indigenous tourism. This is of interest not least since it has been shown that cultural tourism products with a strong commercial focus and government support are often more successful than those lacking these assets (Hughes & Carlsen, 2010).

Hinch and Butler (2007) try to acknowledge this complex situation by presenting an indigenous tourism system. Besides the general characteristics of the tourism system (Hall, 2005), their model highlights the cross-cultural interaction involved in indigenous tourism and the presence of other major players, namely the often non-indigenous tourism trade, governments and the media. These stakeholders act in multilayered environments covering physical, economic, social and environmental aspects that also constrain potential strategies and actions. The latter is important as indigenous peoples in welfare state settings certainly have different preconditions for leading the lives they desire than people in developing economies, which may include developing tourism and other businesses. The focus on the environmental context also allows for explaining indigenous tourism through generic factors influencing life and tourism in often remote areas in general, rather than employing indigeneity alone as an explanation for the success and failure of indigenous tourism.

In this context it is important to note that, even between welfare states, policies regarding indigenous peoples differ dramatically, as do geographical preconditions (Keskitalo *et al.*, 2013; see Chapter 1, this volume). Hence, the Nordic Sami are neither geographically secluded nor contingent on subsistence policies as the North American Inuit are, but in fact lead modern lives including professions as professors, administrators, craftsmen, etc. Only a minority of the Sami is, in fact, involved in reindeer herding and hence has access to a major albeit non-exclusive asset of Sami tourism, i.e. reindeer (Müller & Pettersson, 2001). Thus, for a majority of the Sami population, tourism is not an important option.

The mentioning of cross-cultural interaction by Hinch and Butler (2007) resembles not least Smith's idea of history as an important factor in the decision to engage in tourism or not. However, it also presupposes that tourists are culturally different from the indigenous populations. Indeed, various studies indicate that domestic demand for indigenous tourism seems to be limited (Pettersson, 2002; Ryan & Huyton, 2000a, 2000b). Instead, it is argued that domestic tourism often implies a limited cultural distance between the tourists and the indigenous hosts and destinations. This is applicable in the setting of European welfare states, where differences between indigenous Sami and dominant ethnicities are in fact marginal (Olsen, 2006). This affects knowledge and constructions of indigeneity. Hence, a certain cultural and probably geographical distance is needed in order to construct and perceive indigenous peoples and their habitats as tourist attractions. As has been pointed out, this is usually perceived as unproblematic as long as the relation refers to Germans and French or Americans and Russians. In relation to indigenous peoples, however, exoticism is combined with ideas of primitiveness and a lack of modernity (Viken & Müller, 2006). This 'othering' may therefore hardly reflect a modern indigenous reality but instead produce a stereotypical and staged image and tourism experience (Müller & Pettersson, 2006; Olsen, 2006; Viken, 2006). Olsen (2003) even claims that this contributes to maintaining ethnic boundaries in a situation that is rather characterized by hybridity, implying that people are both Sami and Norwegian. However, indigenous peoples are not necessarily passive victims only but may in fact contribute to constructing a traditional image of themselves, very much aware of their value in the international touristic marketplace (Amoamo, 2007; Tuulentie, 2006).

Major players

The presence of indigenous peoples and the availability of indigenous culture as a product become a unique selling point in the global marketplace (Figure 5.1). Hence, indigenous peoples are used to sell destinations (Dann, 1996; Hollinshead, 1996; Lew, 1998) and, particularly in an Arctic context, their presence seems to be valuable; Saarinen (1999) argues that Sami were often used in a misleading way to represent the northern destination as exotic. Moreover, Sami can obviously be seen as a viable component of a

Figure 5.1 A historical postcard showing a woman in a traditional Sami dress and a (then) modern hairdo at the Arctic Circle (K. Rune Lundström, AB Publishing, Skellefteå)

touristic wilderness. Hence, the media and non-indigenous stakeholders play important roles in the development of indigenous tourism. However, Hunter (2011) argues that it is certainly not always tourism alone that creates representations and subjective perceptions of them. Hence, though tourism certainly contributes to creating an often stereotypical image, processes are complex and involve both indigenous and non-indigenous stakeholders (Müller & Pettersson, 2006; Tuulentie, 2006). For the example of the Jokkmokk winter festival, Müller and Pettersson (2006) argue that the festival functions as an arena where the creation of touristic image and practice is tested and contested by various Sami and non-Sami stakeholders, and amidst them not least the media critically discusses how Sami should be and what should and should not be allowed in Sami tourism.

Still, as Hollinshead (2007) points out, a great deal has happened in recent decades regarding the representation of indigenous groups in media. Not least indigenous involvement in media production itself has increased, complicating the current situation. This is also applicable to the Sami artists who have gained considerable attention when it comes to film and the arts. While agreeing with Hollinshead's basic notion, Hall (2007) criticizes the lack of awareness regarding the uneven distribution of power within indigenous groups and asks what this means for the creation of image and representation (cf. Müller & Kuoljok Huuva, 2009).

In an Arctic context, indigenous peoples are considered important not only for being part of the imaginary of the north (Ireland, 2003) but also for being partners with non-indigenous enterprises and their influence regarding land use (Notzke, 1999). However, this relationship has not always been smooth (Engström & Boluk, 2012; Müller & Kuoljok Huuva, 2009), and indeed preconditions vary between countries dependent on legal land regimes. In the Nordic case, the right to roam makes it more or less impossible to regulate tourism by means of control of land. Still, there are other reasons for indigenous and non-indigenous stakeholders to cooperate. Notzke (2004) sees a need for cooperation but notes that the lack of business knowledge among the indigenous tourism entrepreneurs is a reason for many missed opportunities, and Getz and Jamieson (1997) list the following reasons for why indigenous tourism has not developed as anticipated: (1) lack of knowledge of tourism potentials; (2) lack of relevant skills; (3) lack of integration in the overall tourism industry and its organizations; (4) lack of market research; and (5) fear of potential loss of cultural identity. Even Fuller *et al.* (2005) argue along the same lines: resources needed for developing tourism are seldom controlled solely by indigenous peoples and hence cooperation is a necessity, though it is not always easy.

Another major stakeholder involved in indigenous tourism is the government (Hinch & Butler, 2007). The government provides infrastructure, public services and legislation, and regulates land use and markets. However, many governments act through economic support, directly sponsoring indigenous tourism initiatives among other things. For the Australian case, Buultjens *et al.* (2005) reveal a number of problems related to this practice. Besides poor coordination between different programs and the poor infrastructure in remote Australia, they also point to the fact that many programs indeed promote the idea of indigenous participation rather than actual participation, and in fact are often focused on start-up phases rather than the actual business operation. Müller and Pettersson (2006) also report on government as a participant in ethnopolitical discourse, stating what indigenous groups should do, not least in order to enable government to cut back on its support.

Methodology

In Sweden, Sami tourism development has a relatively prominent place on the political agenda of Sami business development. In fact, it has been identified as an important Sami industry (Pettersson, 2004). Nevertheless, it has also been noted that Sami tourism did not develop as anticipated for various reasons (Müller & Kuoljok Huuva, 2009). *Samiid Riikaserravi*, The Swedish Reindeer Herders' Union (SSR), initiated a program to develop Sami tourism as an alternative and to supplement reindeer herders' livelihood. This resulted in the establishment of the indigenous destination management organization *Visit Sápmi*, focusing partly on the quality development of existing companies and partly on marketing the Sami tourism products that qualified for the quality label *Sápmi Experience*, developed analogous to the Swedish label for ecotourism, *Nature's Best* (Pettersson, 2006).

This study departs from the work conducted by *Visit Sápmi*. Hence, the definition of Sami tourism enterprises has been established by the Sami community organized in *Visit Sápmi* themselves; only companies acknowledged by *Visit Sápmi* are considered part of Sami tourism. Thus, it can be assumed that the companies are controlled by Sami and also offer products centering on a Sami topic, which is usually seen as essential for being considered part of indigenous tourism (Hinch & Butler, 2007; Zeppel, 2006). However, it is reasonable to acknowledge that even this definition was established in an ethnopolitical context in which definitions of ethnicity and cultural belonging are not self-evident.

The empirical base of the study consists of 37 structured interviews with companies listed by *Visit Sápmi* as Sami tourism companies. Altogether 58 companies were contacted during fall 2012, but not all of them could be reached. Furthermore, three potential respondents did not want to participate in the study, and one did not consider itself a commercial tourism provider. A majority of the non-respondents could be characterized as small-scale entrepreneurs offering cultural activities or Sami craft. Moreover, the connection to reindeer herding was obvious, as 30 of the interviewees were also members of reindeer herding cooperatives. Only 5 of the responding firms had 10 or more employees during peak seasons; 15 firms had 1–5 employees; and the remaining 17 firms had no employees at all. Hence, Sami tourism companies can be characterized as mainly microfirms. Together, they offer a wide variety of activities: 12 companies are mainly accommodation providers, though several of them do so in combination with various activities; 17 mainly provide cultural activities; while the remaining 8 focus on outdoor activities like hunting and fishing. Altogether 20 of the respondents were men and 17 were women. Seventeen of the companies were established in the year 2000 or later, 13 during the 1990s and 1 way back in 1960.

The structured interviews were conducted by telephone. Candidates were first contacted by email in order to set up a time suitable for the respondent. In several cases, this time had to be revised and a new meeting set up. Still, a phone survey was considered suitable in order to reach as many respondents as possible, since the likelihood that respondents would choose to fill in a questionnaire was considered small. Moreover, the survey contained some open-ended questions and it was anticipated that the chance to yield answers from the respondents was higher when talking to them and noting their answers during the interview. Indeed, it turned out that several respondents were rather talkative when finally reached over the phone. Of course, face-to-face interviews would have been preferable, but the fact that the respondents were spread over all of northern Sweden would have made this a challenging logistic exercise.

The questionnaire contained 29 pre-formulated questions. Besides questions mapping company background and motivation, 17 questions referred to the respondents' perception of different challenges. These can be characterized as referring partly to challenges presented by respective changes in the physical, social, political and economic environments, and partly to challenges entailed by the tourism industry and the government. Hence, the questionnaire very much followed the ideas presented in Hinch and Butler's (2007) indigenous tourism system.

The collected data was analyzed using mainly statistical tools. Moreover, open-ended question content and comments by the interviewees were used to exemplify results from the quantitative parts of the survey.

Results

The focus of the study has been to analyze how Sami tourism entrepreneurs assess external challenges, also in comparison with other requirements related to the business of tourism. In a first step, it has been charted how the entrepreneurs perceive threats to their businesses caused by changes in the outside environment (Figure 5.2).

In general, the Sami tourism entrepreneurs were appreciative of their environment and considered it an important asset for their business; in

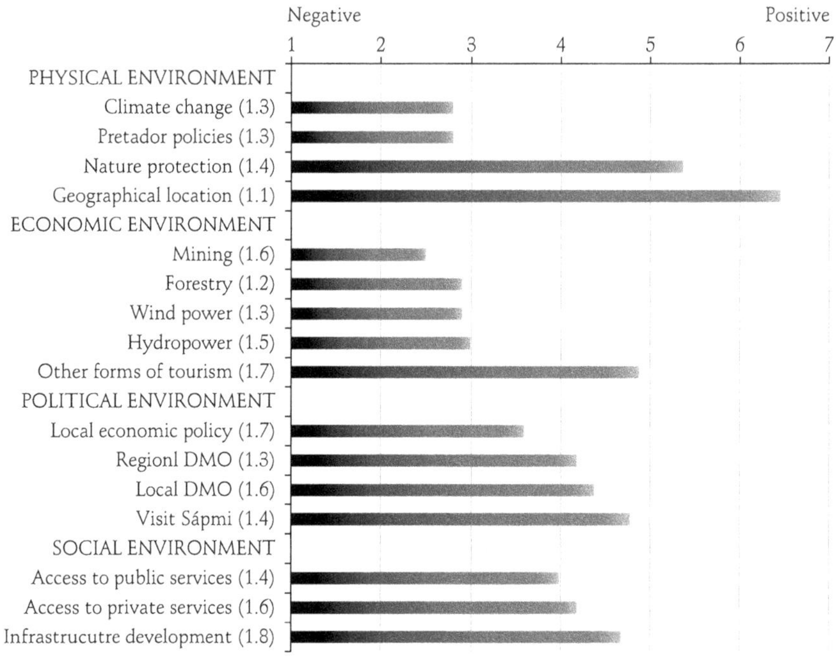

Figure 5.2 Perceptions of the role of various external factors on the Sami respondents' tourism businesses (*n*=37); mean values. Standard deviations in parentheses

answer to the question of how they thought the location in Sápmi affected their business, they considered it extremely positive. State interference with land use, namely the establishment of protected areas, was also seen as an advantage for their tourism business. However, not everything related to the physical environment was considered positive. In fact, even though – somewhat paradoxically – nature protection was embraced, state policy regarding the protection of predators was not, since the latter is seen as a major threat to reindeer herding. Moreover, less surprisingly, climate change was considered a negative change as well. Considering the characteristics of the tourism businesses, this rather unanimous perception is somewhat surprising, since few of the companies would in fact be affected by climate change directly.

Questions regarding another outside-induced change in the environment referred to changing land use owing not least to external economic interests. These are discussed here under the label of an economic environment. Here, opinions regarding other industries were generally rather negative. Particularly mining was seen as a major threat to the tourism business – only two of 37 perceived it positively, and 13 were neutral; but even the presence of forestry, wind power and hydropower was less than appreciated, though a large group was neutral regarding these land uses. Only other, non-indigenous forms of tourism were appreciated, though not by all.

The political environment and the support of government and touristic infrastructure like destination management/marketing organizations (DMO) were seen as rather neutral. The greatest support was given to *Visit Sápmi*, the Sami's own DMO. This is in line with previous research on indigenous entrepreneurship, highlighting trust in business norms developed within the ethnic group (Dana & Remes, 2005). Cooperation with other local stakeholders was seen as neutral as well, but varied notably between locations and enterprises. Even access to public and private services was seen as satisfactory for the respondents' businesses, as was the development of infrastructure.

The latter may be considered surprising, since not least the access to services and the provision of infrastructure are considered major challenges to tourism in remote and peripheral areas (Brown & Hall, 2000; Hall & Boyd, 2005; Müller & Jansson, 2007).

The focus of the second block of questions was on cooperation with the tourism industry and the government. Even here, respondents were asked to agree or disagree with pre-formulated statements (Figure 5.3).

As indicated, earlier cooperation within the Sami group was regarded as certainly important (Dana & Remes, 2005). However, cooperation with other tourism companies and companies within other industries was seen as almost equally important. Furthermore, on average, support from the reindeer herding cooperative was acknowledged, though variations occurred. The same applies for the support from Sami politicians, which

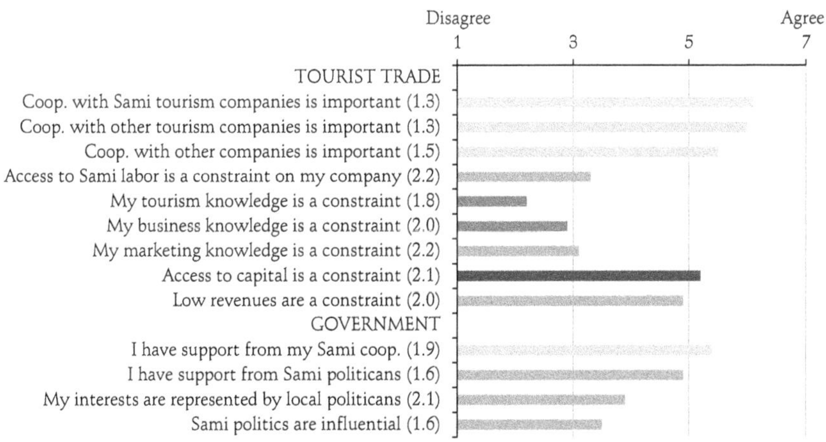

Figure 5.3 Stated relations of Sami tourism entrepreneurs to tourist trade and government (n=37); mean values. Standard deviations in parentheses

was also recognized. Still, many had doubts as to the extent that Sami politicians could influence decision-making.

Another statement that many, but far from all, agreed on was that access to capital is a constraint. The same applied to the low revenues, which were also acknowledged as a limitation. Still, a majority of the responding entrepreneurs did not see any problem with their own knowledge regarding tourism, business and marketing, areas where they displayed great confidence (even though many of them lacked higher education or relevant work experience). It should be mentioned, though, that 9 of the 37 respondents deviated from this pattern and admitted that they saw their own knowledge as a constraint. Answers to an open question indicated that not least marketing knowledge was seen as a major bottleneck to developing the company further. One's own ability to develop the company with regard to time constraints and motivation was seen as hindered, indirectly indicating a lack of capital for employment or a lack of trust.

However, many companies saw changes in the physical environment as a major threat to their businesses, a view not limited to companies within the activity segment. Obviously, a general concern for the image of the destination combined with a concern for the sustenance of reindeer herding and other parts of the Sami culture implied challenges for the respondents. This indicates the strong cultural ethos that characterizes many of the Sami entrepreneurs.

The group of Sami entrepreneurs seemed to be rather coherent in their attitudes and opinions. Factors like the respondents' educational background or the geographical location of the company did not explain variations in the material. An exploratory hierarchical cluster analysis was performed in order to identify cases with similar answers about the stated environmental and business challenges, and altogether four clusters could be distinguished. A comparison of the clusters with regard to the mean values reported for each question showed significant differences (t-test >0.95) between clusters in respect of nature protection, hydropower, local economic policy, promotion by *Visit Sápmi*, infrastructure development, the importance of cooperation with other tourism companies, lacking access to Sami workforce, access to capital and support from local politicians. Hence, significant differences occurred only in approximately a third of all addressed issues.

The analysis further revealed differences between the identified clusters and their characteristics (Figure 5.4). The dominant cluster ('typical companies') represents the most frequent set of attitudes among the Sami tourism firms. Altogether 20 of 37 surveyed companies can be found in this cluster. Closely related to this cluster is a group of three companies that is mainly concerned with access to a Sami workforce, which is seen as a hindrance to further positive development. A third cluster ('economically unconcerned') comprising nine companies does not see access to capital and to a Sami workforce as constraints on their business development. The five companies in the fourth cluster ('economic focus') see nature protection

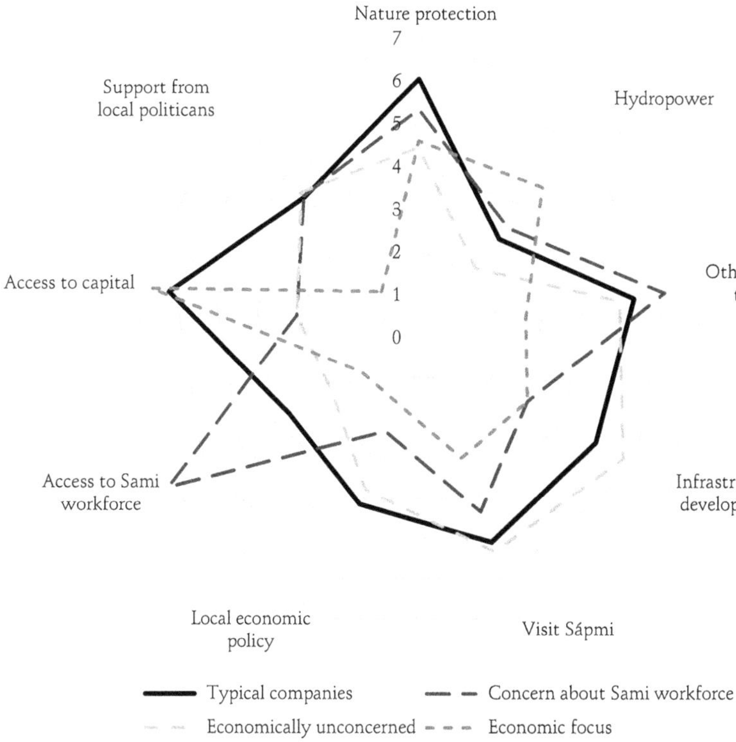

Figure 5.4 Differences between Sami tourism entrepreneurs; mean values

more negatively and hydropower generation more positively than other companies do. At the same time, they seem to be less satisfied with local politics and infrastructure development. A major constraint is access to capital, which is their single most dominant concern.

In contrast to the total surveyed population, in which 17 of 37 entrepreneurs were women, the distribution in the clusters is skewed. While the cluster representing the typical companies was dominated by women, the clusters that were economically unconcerned and concerned about the Sami workforce were dominated by men. Moreover, many of the economically unconcerned companies have been operating for more than 20 years. Companies with an economic focus are mainly among those offering cultural activities.

It is thus important to note that Sami tourism companies are rather united in their assessment of current challenges and changes. Not least, competing land use interests are seen as major challenges to their business development, which is interesting considering the fact that only a minority of the companies actually utilize the environment directly for their own operations. Instead, it seems to be a general concern for the destination that is the main apprehension for the entrepreneurs. Still, it should also be noted that despite these commonalities, differences between companies can be detected.

The future

The question of how the respondents imagined Sami tourism 10 years from now revealed some aspects that were not obvious in the more quantitative part of the study, but are known from previous research efforts (Müller & Kuoljok Huuva, 2009). Several entrepreneurs hoped for a greater number and variety of Sami tourism companies and more acceptance for Sami tourism, obviously referring to criticism from within the Sami community and the Sami reindeer husbandry cooperatives in particular. A desire for even better integration with reindeer husbandry was mentioned, and the fact that economic pressure was likely to limit the number of reindeer herders in the future was seen as a positive development making access to genuine Sami tourism based on reindeer experiences more exclusive. In this context, great hopes were also placed on a Sami DMO and quality labeling scheme.

Only a few entrepreneurs admitted that they do not expect greater positive changes, and some pointed to the need to maintain the place-based and small-scale characteristics of Sami tourism companies.

Most respondents were positive regarding the next 10 years for their own companies, though several pointed to the need to find new owners owing to retirement. Another interesting aspect mentioned by many is a wish to overcome seasonality by, for example, focusing on conferences and workshops. Even a greater focus on the Sami dimension of the company was mentioned as a way to achieve greater competitiveness without the need to grow the company. The latter aspect was mentioned as an aim by several respondents. However, it became obvious that greater income was also high on the agenda of many companies, which desired growth in terms of visitor numbers and employees. International cooperation in Sápmi was mentioned as a way to achieve this, as was improved transportation, not least during the winter. Only one respondent admitted that the company would most likely be discontinued.

Conclusions

Following Butler and Hinch (2007), at the beginning of this text it was argued that indigenous tourism is a complex system involving the activities of indigenous and non-indigenous stakeholders. Thus, preconditions for indigenous tourism are expected to vary dramatically between countries, even though they may contain the same indigenous group, as in the case of the Sami in Norway, Sweden, Finland and Russia. As has been shown elsewhere (Müller & Kouljok Huuva, 2009), tourism has often been discussed as an opportunity or a threat to indigenous peoples. In this study addressing the question of how Sami tourism entrepreneurs assess external challenges, also in comparison with other requirements related to the business of tourism, little of this was visible. Obviously, some Sami tourism entrepreneurs aim to disseminate knowledge about Sami culture, but tourism is in general not seen as a threat. Of course, the study addresses the protagonists of Sami tourism in Sweden directly, and thus negative responses about tourism as such may not be expected, but the general impression is that the respondents see tourism as a self-evident commercial activity that needs no further justification.

The study confirms that a major challenge to the system in fact comes from within the Sami community. A cultural-economic hegemony puts reindeer husbandry at the forefront, and its proponents challenge touristic practices utilizing the reindeer herding culture as a tourist attraction (Müller & Kouljok Huuva, 2009; Pettersson & Viken, 2007). Hence, even today touristic practices need to be negotiated and tested within a complex ethnopolitical framework (Müller & Pettersson, 2006).

Still, a majority of respondents see the major challenges for tourism development in competing land uses. Mining, hydropower, wind power and forestry are the Swedish examples of extractive industries that are perceived as threats to the indigenous tourism industry. This may come as a surprise, since it is not obvious that an indigenous artist or craftsman is affected by the shadow or sound of a windmill, or by a hydro dam nearby. What the answers indicate, however, is that Sami tourism entrepreneurs are comprehensively embedded in their culture and see themselves as representatives of it even within tourism (Dana & Remes, 2005). Consequently, threats to the Sami culture and reindeer herding in particular are believed to be threats to their businesses as well. Thus, even protective measures, for example policies for the protection of predators, are seen as negative and harmful although they hardly seem to affect the tourist companies negatively.

Interestingly, these concerns regarding the physical and economic environment overshadow many of the challenges identified elsewhere as particularly applicable to indigenous tourism. With the possible exception of marketing skills, none of the aspects mentioned by Getz and Jamieson (1997) is considered a challenge; instead, the respondents seem to have confidence in their ability to run a company in cooperation with non-indigenous stakeholders as well, and the risk of a potential loss of cultural identity is not noted at all. This does not mean that they do not suffer from economic concern – in fact, many of the companies desire growth and greater revenues, and see access to capital as an important constraint. Still, the respondents create their own hegemonic worldview whereby the sustenance of their traditional indigenous culture, with reindeer herding at its core, is superior to shortsighted business success. Moreover, socially negotiated interests of an ethnic group override individual desire for economic success. This may be because most of the Sami tourism entrepreneurs are still involved in reindeer herding through reindeer ownership, and thus are at least indirectly dependent on the goodwill of the reindeer herding cooperatives.

However, the stated importance of various threats seems to be an emotional assessment, mainly since there is no evidence that any of the above-mentioned threats has led to the closure of a Sami tourism business. Hence, while they are an integrated part of modern Sweden, Sami tourism entrepreneurs seem to still be caught between traditional ideas about their ethnic group on the one hand and individual economic necessities and desires on the other, a situation that is not uncomplicated for the single entrepreneur struggling for economic survival and ethnic acceptance at the same time.

Acknowledgments

This research has been conducted within the MISTRA Arctic Futures program and the project From Reindeer Herder to Tourism Entrepreneur, funded by the Marianne and Marcus Wallenberg Foundation. The funding from MISTRA, the Swedish Foundation for Strategic Environmental Research and the Wallenberg Foundation is hereby gratefully acknowledged.

References

Altman, J. and Finlayson, J. (1993) Aborigines, tourism and sustainable development. *Journal of Tourism Studies* 4 (1), 38–50.

Amoamo, M. (2007) Māori tourism: Image and identity—a postcolonial perspective. *Annals of Leisure Research* 10 (3–4), 454–474.

Brown, F. and Hall, D. (eds) (2000) *Tourism in Peripheral Areas*. Clevedon: Channel View Publications.

Butler, R. and Hinch, T. (eds) (1996) *Tourism and Indigenous Peoples*. London: International Thomson Business Press.

Butler, R. and Hinch, T. (eds) (2007) *Tourism and Indigenous Peoples: Issues and Implications*. Amsterdam: Butterworth-Heinemann.

Buultjens, J., Waller, I., Graham, S. and Carson, D. (2005) Public sector initiatives for aboriginal small business development in tourism. In C. Ryan and M. Aicken (eds) *Indigenous Tourism: The Commodification and Management of Culture* (pp. 127–147). Amsterdam: Elsevier.

Dann, G.M.S. (1996) Images of destination people in travelogues. In R. Butler and T. Hinch (eds) *Tourism and Indigenous Peoples* (pp. 349–375). London: International Thompson Business Press.

Dana, T.E. and Remes, L. (2005) An ethnographic study of entrepreneurship among the Sami people of Finland. *Journal of Small Business & Entrepreneurship* 18 (2), 189–200.

Dowsley, M. (2009) Inuit-organised polar bear sport hunting in Nunavut territory, Canada. *Journal of Ecotourism* 8 (2), 161–175.

Engström, C. and Boluk, K.A. (2012) The battlefield in the mountain: Exploring the conflict of tourism development on Three Peaks in Idre, Sweden. *Tourism, Planning & Development* 9 (4), 411–427.

Fuller, D., Buultjens, J. and Cummings, E. (2005) Ecotourism and indigenous micro-enterprise formation in northern Australia opportunities and constraints. *Tourism Management* 26 (6), 891–904.

Getz, D. and Jamieson, W. (1997) Rural tourism in Canada: Issues, opportunities and entrepreneurship in aboriginal tourism in Alberta. In S.J. Page and D. Getz (eds) *The Business of Rural Tourism: International Perspectives* (pp. 93–107). London: International Thomson Business Press.

Hall, C.M. (2005) *Tourism: Rethinking the Social Science of Mobility*. Harlow: Pearson.

Hall, C.M. (2007) Politics, power and indigenous tourism. In R. Butler and T. Hinch (eds) *Tourism and Indigenous Peoples: Issues and Implications* (pp. 305–318). Amsterdam: Butterworth-Heinemann.

Hall, C.M. and Boyd, S.W. (eds) (2005) *Nature-Based Tourism in Peripheral Areas: Development or Disaster?* Clevedon: Channel View Publications.

Hinch, T. and Butler, R. (2007) Introduction: Revisiting common ground. In R. Butler and T. Hinch (eds) *Tourism and Indigenous Peoples: Issues and Implications* (pp. 1–12). Amsterdam: Butterworth-Heinemann.

Hollinshead, K. (1996) Marketing and metaphysical realism: The disidentification of Aboriginal life and traditions through tourism. In R. Butler and T. Hinch (eds) *Tourism and Indigenous Peoples* (pp. 208–348). London: International Thompson Business Press.

Hollinshead, K. (2007) Indigenous Australia in the bittersweet world: The power of tourism in the projection of 'old' and 'fresh' visions of Aboriginality. In R. Butler and T. Hinch (eds) *Tourism and Indigenous Peoples: Issues and Implications* (pp. 281–304). Amsterdam: Butterworth-Heinemann.

Hughes, M. and Carlsen, J. (2010) The business of cultural heritage tourism: Critical success factors. *Journal of Heritage Tourism* 5 (1), 17–32.

Hunter, W.C. (2011) Rukai indigenous tourism: Representations, cultural identity and Q method. *Tourism Management* 32 (2), 335–348.

Ireland, M. (2003) Sustaining indigenous peoples in the wilderness areas of Scandinavia and North-West Russia. *Scandinavian Journal of Hospitality and Tourism* 3 (1), 71–81.

Johansen, T.E. and Mehmetoglu, M. (2011) Indigenous tourism from a visitor's perspective: An empirical examination of Valene L. Smith's 4Hs at a Sami festival in Norway. *Journal of Heritage Tourism* 6 (2), 129–141.

Keskitalo, E.C.H., Malmberg, G., Westin, K., Wiberg, U., Müller, D.K. and Pettersson, Ö. (2013) Contrasting Arctic and mainstream Swedish descriptions of Northern Sweden: The view from established domestic research. *Arctic* 66 (3), 351–365.

Lew, A.A. (1998) American Indians in state tourism promotional literature. In A.A. Lew and G.A. Van Otten (eds) *Tourism and Gaming on American Indian Lands* (pp. 15–31). New York: Cognizant Communication Corporation.

McIntosh, A. and Ryan, C. (2007) The market perspective of indigenous tourism: Opportunities for business development. In R. Butler and T. Hinch (eds) *Tourism and Indigenous Peoples: Issues and Implications* (pp. 73–83). Amsterdam: Butterworth-Heinemann.

Müller, D.K. (2011) Tourism development in Europe's 'last wilderness': An assessment of nature-based tourism in Swedish Lapland. In A.A. Grenier and D.K. Müller (eds) *Polar Tourism: A Tool for Regional Development* (pp. 129–153). Montreal: Presses de l'Université du Québec.

Müller, D.K. and Pettersson, R. (2001) Access to Sami tourism in northern Sweden. *Scandinavian Journal of Hospitality and Tourism* 1 (1), 5–18.

Müller, D.K. and Pettersson, R. (2006) Sami heritage at the winter festival in Jokkmokk, Sweden. *Scandinavian Journal of Hospitality and Tourism* 6 (1), 54–69.

Müller, D.K. and Jansson, B. (eds) (2007) *Tourism in Peripheries: Perspectives from the Far North and South*. Wallingford: CABI.
Müller, D.K. and Kuoljok Huuva, S. (2009) Limits to Sami tourism development: The case of Jokkmokk, Sweden. *Journal of Ecotourism* 8 (2), 115–127.
Notzke, C. (1999) Indigenous tourism development in the Arctic. *Annals of Tourism Research* 26 (1), 55–76.
Notzke, C. (2004) Indigenous tourism development in southern Alberta, Canada: Tentative engagement. *Journal of Sustainable Tourism* 12 (1), 29–54.
Olsen, K. (2003) The touristic construction of the 'Emblematic' Sami. *Acta Borealia* 20 (1), 3–20.
Olsen, K. (2006) Making differences in a changing world: The Norwegian Sami in the tourist industry. *Scandinavian Journal of Hospitality and Tourism* 6 (1), 37–53.
Palomino, H.U. (2012) Innovation and entrepreneurship in Swedish Lapland. *International Journal of Tourism Anthropology* 2 (1), 1–15.
Pettersson, R. (2002) Sami tourism in Northern Sweden: Measuring tourists' opinions using stated preference methodology. *Tourism and Hospitality Research* 3 (4), 357–369.
Pettersson, R. (2003) Indigenous cultural events: The development of a Sami winter festival in Northern Sweden. *Tourism* 51 (3), 319–332.
Pettersson, R. (2006) Ecotourism and indigenous people: Positive and negative impacts of Sami tourism. In S. Gössling and J. Hultman (eds) *Ecotourism in Scandinavia: Lessons in Theory and Practice* (pp. 166–177). Wallingford: CABI.
Pettersson, R. and Viken, A. (2007) Sami perspectives indigenous tourism northern Europe: Commerce or cultural development. In R. Butler and T. Hinch (eds) *Tourism and Indigenous Peoples: Issues and Implications* (pp. 176–187). Amsterdam: Butterworth-Heinemann.
Pratt, S., Gibson, D. and Movono, A. (2013) Tribal tourism in Fiji: An application and extension of Smith's 4Hs of indigenous tourism. *Asia Pacific Journal of Tourism Research* 18 (8), 894–912.
Ryan, C. and Huyton, J. (2000a) Aboriginal tourism—A linear structural relations analysis of domestic and international tourist demand. *International Journal of Tourism Research* 2 (1), 15–29.
Ryan, C. and Huyton, J. (2000b) Who is interested in Aboriginal tourism in the Northern Territory, Australia? A cluster analysis. *Journal of Sustainable Tourism* 8 (1), 53–88.
Ryan, C. and Aicken, M. (eds) (2005) *Indigenous Tourism: The Commodification and Management of Culture*. Amsterdam: Elsevier.
Saarinen, J. (1999) Representations of indigeneity. Sami culture in the discourse of tourism. In P.M. Sant and J.N. Brown (eds) *Indigeneity: Constructions and (Re) presentations* (pp. 231–249). New York: Nova Science.
Smith, V.L. (1996) Indigenous tourism: The four Hs. In R. Butler and T. Hinch (eds) *Tourism and Indigenous Peoples* (pp. 283–307). London: International Thomson Business Press.
Tuulentie, S. (2006) The dialectic of identities in the field of tourism: The discourses of the indigenous Sami in defining their own and the tourists' identities. *Scandinavian Journal of Hospitality and Tourism* 6 (1), 25–36.
Viken, A. (2006) Tourism and Sami identity – An analysis of the tourism–identity nexus in a Sami community. *Scandinavian Journal of Hospitality and Tourism* 6 (1), 7–24.
Viken, A. and Müller, D.K. (2006) Introduction: Tourism and the Sami. *Scandinavian Journal of Hospitality and Tourism* 6 (1), 1–6.
Viken, A., Midtgard, M.R., Bakken, T. and Borch, T. (1998) *Innovasjoner i samisk reiseliv*. Alta: Finnmarksforskning.
Zeppel, H. (2006) *Indigenous Ecotourism: Sustainable Development and Management*. Wallingford: CABI.

6 Tourist Hegemonies of Outside Powers: The Case of Salmon Fishing Safari Camps in Territories of Traditional Land Use (Kola Peninsula)

Yulian Konstantinov

Background

Exclusive upmarket salmon fishing camps have a history of a quarter of a century on the Kola Peninsula.[1] They shall be called further 'safari camps' or just 'camps' so that they are distinctly set apart from other tourist ventures. Much of the current low-profile existence of the safari camps I attribute to the lack of such distinction in public and even academic discourse between this very special and highly exclusive tourist form, on the one hand, and all the rest in the way of Kola fishing tourism (cf. Nakhshina, 2012: 233). In tourist information sources all such ventures are lumped together as *turistkie rybolovnye bazy* (tourist fishing bases), and thus it is difficult to get precise numbers (Department of Tourism Development, 2010; Murmansk Region, 2015). It may be taken that they are about 30.

The start of the camps (lagerya, as local people call them) can be situated in the late 1980s (Ogarkova, 2007; Osherenko, 1998). They quickly gained popularity among Western fly-fishing salmon and trout anglers and by the mid-1990s the foreign clientele had stabilized at close to 1000 visitors a year. The camps cater for tourists with means, capable of paying on average some USD1500/day, usually for six-day tours.[2] Or, as a price offer for 2015 says:

> Depending on proximity to peak fishing times, 7 nights and 6 full days at the Ponoi River Company: \$4,390–\$13,900. (Ponoi River Co., 2015b)

Since the 2000s, the number of Russian clients has also grown. The camps attract home anglers from the rapidly and spectacularly emerged class of people with means – the upper crust of what goes for a post-Soviet Russian middle class – in popular speech: the 'New Russians' (*novye russkie*). When talking about clients perceived as Westerners, local people would use the terms *burzhui* (bourgeois), or in an indiscriminate way: 'Americans'.

Construction and Commodization of 'Wilderness'

Since the very beginning of this lucrative business, prospective foreign investors have been handpicking the sites for the camps. Enjoying administrative support at regional and superior levels, they gradually came to possess exclusive rights to exploit the resource of the best salmon rivers that the Kola Peninsula could offer (Osherenko, 1998: 18f). The development seems to have gone in parallel with high rank administrators of regional and even federal scale coming into possession of prime salmon locations as personal residences. This latter is a rather cryptic part of the picture, and whatever information there is, is of an entirely anecdotal nature.[3] Consequently, I cannot dwell any further on this topic here, save for saying that in the hierarchy of wildlife exclusive enclosures, 'residences' would belong to a separate category from 'camps'. The first can be considered to be for the privileged enjoyment of home elites, while the second for that of wealthy foreigners.

Naturally, such wildlife resource locations, currently enclosed for elite enjoyment, have been attracting users since time immemorial. This is well attested in the archaeological record.[4] In this light, I place special attention on a particular type of location. As far back as 7000 BC, Mesolithic settlements of the people inhabiting the peninsula used to be situated at river estuaries. This ensured access to both seawater and freshwater renewable resources, importantly including firewood and drinking water (Gurina, 1982: 27f; cf. Krupnik, 1993). The now uninhabited village of Ponoy at the estuary of the eponymous river is very much a case in point. The village was 'deserted', as safari ads describe its present state, as part of a campaign for 'administrative liquidation' of tundra and coastal villages in the 1960s and 1970s, with the subsequent resettling of their inhabitants into bigger regional towns and settlements (Afanas'yeva, 2013; Gutsol *et al.*, 2007; Konstantinov, 2015). Despite the relative remoteness of this sub-Arctic region, and the exigencies of its climate, it has been a well-recorded site of human presence, stretching roughly as far back as 10,000 years. In particular, its estuary and other lake/river locations have attracted permanent human presences, disrupted by outside intrusions only in recent times. A very traumatic disruption has been the already mentioned liquidation of tundra villages in the 1960s and 1970s. The exclusive safari, or upmarket wildlife tourism, discussed in this chapter, can be considered as a similar disruption vis-à-vis the local, tundra-connected population – on a par with the liquidation of ancient Sami villages in the late 1960s and 1970s. I show further that exclusive safari camps enclosed for their own use the best salmon rivers of the peninsula in a manner very similar to the military enclosing of the Barents Sea coast. My principal argument is that this type of exclusive enclosing of land, and the consequent removal of the local population from it, has created the possibility for an artificial 'rewilding' of an environment inhabited by the Kola indigenous population since time immemorial (cf. Höchtl *et al.*, 2005). An exclusive tourist product has been created in this way, in a tight fit between home and foreign high-power interests. As a way of approaching this post-Soviet development, I first discuss the phenomenon of 'commodization of wilderness' (Øian, 2013: 184) as it has been played out on the local terrain. I then turn to a closer ethnographic look at the camps, their staff and clients. My final conclusion points to the political significance and disruptive implications for indigenous and other tundra-connected local inhabitants of constructing and enclosing 'wild nature' in the sub-Arctic (and anywhere) for exclusive tourist use.

The Significance of 'Red Fish' in Tundra-Based Economies

Salmon and trout –'red fish' (*krasnaya ryba*) as it is generally called in the Russian North – played an important role in the economy of the indigenous Sami people. 'Red fish' was used mainly in bartering for manufactured goods, food products and alcoholic beverages. These economic exchanges between the local Sami population and outside traders happened seasonally at the earliest fortified trading post and town on the peninsula – the town of Kola, known since the mid-16th century. 'Red fish' thus played the role of currency for the Sami people, while 'white fish', in the main grayling (*sig*) and pike (*shchuka*), went for subsistence (Kharuzin, 1890; Luk'yanchenko, 1971; Ushakov, 1998a, 1998b).

'Red fish' is still used as currency in the tundra part of the peninsula. Metonymically referred to in local slang as '(tail-) fins' (Rus. *hvosty*, sg. *hvost*), red fish is used, for instance, for informally paying helicopter pilots or for other services. In one case I know, a visiting dentist to a tundra village would take a 'fin' for pulling out a tooth. Another common use of 'red fish' is for making presents to urban relatives, friends or influential persons, thus belonging to the 'fuzzy' category somewhere between present and bribe. 'Red fish' is, consequently, very much part of the local 'substantive economy' in the classical Polanyian sense (Polanyi, 1957, see also Lajus, 2008 in Nakhshina, 2012: 232). Crucially, 'red fish' (salmon in particular) is here – as other prize species are in many regions across the globe – part and parcel of sophisticated social institutions of sharing (cf. Ziker, 2006). Further, the taking of salmon and its subsequent consumption and sharing (these two being often mixed in the making of memorable social occasions), constitute orienting moments of human and extra-human relatedness. Such relatedness to fish can be seen as key enactments of belonging, and the formation and sustaining of local identities (cf. Dallman *et al.*, 2013; Nakhshina, 2012; Øian, 2013).

In the specific case such a fine and rich texture of an age-long interweaving of social, economic and spiritual currents in the reality of a 'dwelling perspective' (Ingold, 2000) comes to be disrupted through the interests and imaginaries of distant, unseen and powerful agencies. These intrusions have introduced a tension between 'living inside', to extend on Ingold's metaphors, and 'coming from the outside'. Unlike other more benign cases as, for instance in south-eastern Norway (Øian, 2013), the Swiss Alps (Höchtl *et al.*, 2005) and Australia (Burns, 2004), the Kola case presents features of lapsing back into early forms of wildlife enclosing for exclusive enjoyment. Such early impositions of divisions between the 'lived in' as against the 'provisional' (cf. Franklin, 2008, in Øian, 2013: 179), we may trace back along the chain which Eliason (2012) captures in his title 'From the King's deer to a capitalist commodity'.

In the particular case of Kola, the sequence passes through various phases of wildlife enclosing and, in a dialectical way, creates poachers out of the locals, now banished to the wrong side of the pale. This dynamic has produced an extensive literature of the ways in which imposed boundaries have been resisted through the ages (cf. Bell *et al.*, 2007; Essen *et al.*, 2014; Forsyth & Marckese, 1993; Forsyth *et al.*, 1998; Garth, 1978; Muth & Bowe, 1998). This is a topic, however, which cannot be given justice in this short space. Instead, I only note that post-Soviet wildlife enclosing can be easily traced back to Soviet-time practices. This is easily seen in the Kola case. When the greatest 'encloser' of land – the military (Hønneland & Jørgensen, 1999) – temporarily relaxed their hold on coastal and hinterland territory during the early 1990s, safari camps instantly mushroomed. Similar successions from exclusive power enclosures to upmarket hunting/angling camps, as well as new residences for power holders, can be traced in practically all parts of the former Soviet Bloc. The phenomenon of totalitarian wildlife enclosing still awaits fuller investigation. It could be said that at the moment, in cases like Kola, not only 'wilderness' is created to be commodisized, but, in addition, the Soviet era enclosing culture becomes an unspoken but indelible part of the final high-price product. For it is often where formerly the military used to be exclusive users of vast tracts of land, the local sparse population having been evicted, that the safari tourist casts his/her fly today. Functional parts of the whole show – like helicopters from the 1970s or military camp guards (see further) – turn into metaphoric enactments of reinterpretation, rather than of transition. In the ease and smoothness of such an effortless glide between overtly so disparate realities, one can feel an underlying unity of ethical and aesthetical adherences among privileged elites. These render current political divides irrelevant by creating for themselves timelessly hegemonic and exclusive time spaces. A specific sublimation – in the curious retort of fly-fishing – describes provisional placedness in a Bakhtinian chronotope[5] of wildlife enclosure, uniting 'King's deer' forests, Victorian estates and Soviet wildlife residences into a homogenous and unproblematic ethical and aesthetical whole. This I initially read in the advertising texts of exclusive fishing camps, but will flesh out further on by adding from my own chance encounters near such places, from relevant reports and interviews with insiders, as well as from regional media reportage.

Advertising texts

Advertising texts as a first ethnographic stop show an underlying construction of a tangle of bounding exclusions, designed to protect a select group from the 'touch' of others. These separations create the tourist product itself, a 'me-first-in-wilderness!'[6] exclusive package. After removing the locals from the scene, together with their history of habitation since the last Ice Age, an imaginary of buying oneself into a space miraculously devoid of humans is achieved. Indigenous removal – both as a historical and a current presence – is part of the construction, another being the deferral from consciousness of the continuous presence of the exclusive tourists themselves. The working of this wilderness-producing symbolic machinery (in its second part leaping over a schizophrenic abyss) turns out in a type of 'extra-human-but-myself' placedness, a kind of a 'wilderness paradox'. This amounts to the desire (and corresponding tourist product) of experiencing 'untouched wilderness' by paying for a comfortable way of 'touching it'. Registering the occurrence of such a phenomenon in a Swiss Alpine context, Höchtl *et al.* (2005: 92) note that 'some tourists enjoyed the ("wild" – my gloss) landscape, but desired more touristic infrastructure (e.g. accommodation facilities, hiking trails, information points)'. The observation alerts to the large leeway that a 'wilderness'-seeking tourist mentality allows for paradoxes and contradictions.

Here is a typical text, advertising an angling safari camp, situated near the estuary of the Ponoy River, the biggest river on the Kola Peninsula:

> The camp is built into the side of the hill overlooking a wonderful tail-out before the big, un-navigable rapids below. The scenery around *Brevyeni* (camp) is much more dramatic and wild but very beautiful. Here the river has cut a deep canyon hundreds of metres deep. On the top is the tundra plateau where swans and ptarmigan breed in peace, undisturbed except by the occasional helicopter. Brevyeni is on the way to nowhere. (Frontiers, 2015a)

This paradoxical construction, in which a sub-Arctic 'nowhere' is possible by being open only to the select (coming in by helicopter) is captured unintentionally but eloquently in an ad photo of the same camp. In the picture, a proud angler cradles a big glum-looking salmon with a hook in its jaws, against the background of a river. The caption reads:

> (This is) the most distant camp on the Ponoy river, where you will not see any signs of human impact on the environment. (Dudkin & Co., 2015)

The implication that 'human impact on the environment' is about other-than-the-tourist-humans is an entry into numerous and many-layered exclusions. The most general and abstract among them is defined by the much debated 'society–nature' boundary (Bessire & Bond, 2014: 442; Castree & Braun, 2001). The construction firmly separates the world of humans from that of an 'empty-of-humans' space, conceived as 'nature', in the sense of an external, non-human material world.[7] To this, two additional meanings are added. One is that such a 'thing', which is empty of (non-tourist) humans, is 'empty' in the sense of uselessly existing, and thus imperatively available for resource extraction, as otherwise it would be wrong for it to exist just lying about. This ethical implication brings out the colonial impulse of the wrongness of not using, or underusing, constructing the concept of 'wilderness' (Cronon, 1996), or as in the text above: of 'nowhere'. The latter can be literally uninhabited, staying in a raw geological state. When inhabited by people who are not using it in the 'right' way, it is again as good as empty and thus fair game for exploiting by those who know better. From Osherenko's (1998) report, we learn that the net-trap structures that local collective farms used for large-scale trapping of salmon in Soviet times, had subsequently been removed so that abundance in camps' fishing grounds was returned to its pre-local cultural state. Through this 'rewilding' manipulation, the local population lost a vital resource. An ideological logic of the move was that the replacement of industrial river fishing with exclusive angling for a high-paying clientele would create a much more profitable resource base for the locals, with the added value of it being environmentally protective. Today, 20 years after the promotion of such hopes, it can be said with confidence that none of them were destined to be realized. Local administrators have been complaining for a long time that because camp businesses were not registered locally, they did not pay taxes to local municipalities. Furthermore, local labor is hired only as seasonal wage earners at camps, which is another tax-evasive maneuver. All in all, the presence of safari camps, exploiting the greatest renewable riches on the territory of local municipalities, does not contribute in any perceptible way to the improvement of their economic fortunes. Since the concerned municipalities depend on regional and federal funding for over 90% of their budgets, they have little power, if any, to oppose directives coming from superior levels about who should be given a piece of the salmon pie of the peninsula.

This is the invisible side of the local 'wilderness' moon and a matter of hegemonic application of power. To what extent clients are aware of this side of the picture I cannot tell from what I know. The uniform opinion of camp employees with whom I have had the chance to talk has been that tourists are not interested in anything, but their fishing.

Environmentalist sentiments

The point at which some touch with space beyond that of the camp enclosure can be discerned is in the tension between a display of environmental consciousness, on the one hand, and a next-to-mandatory appropriation of a visually recorded trophy, on the other. The tension cannot but arise, as an aesthetic of being skillful enough (in this case: and rich) to cross the line between human and non-human has too much of the associative legacy with the drama and aesthetic of conquest of 'nature'. In the case of exclusive fly-fishing, such an aesthetic of conquest is recreated through the milder and much mediated dramatic tension of hooking a big fish – a triumph over the 'wild' duly photographed. Stress is laid on the use of barbless hooks in a mandatory practice of 'catch-and-release' fishing. A countering aesthetic of triumph, however, seems to pose limits here as, despite recommendations for not taking the fish out of the water lest it be durably harmed, all photographs that we see advertised are with fish cradled by the angler well up in the air. The purpose is to capture the moment in an amazingly same photograph, endlessly reproduced. A picture with a fish indistinctly seen in its own element seemingly does not answer the purpose.

Local attitudes are predictably negative, ranging from dislike to intense hatred. Of the catch-and-release practice, barbless hooks and such refined tools of wildlife torment, I have heard a lot of criticism among herders and other tundra-connected local people. Since the June and September 'runs' are of fish that swim from the sea and into their native rivers to spawn, the local people liken the fish to a pregnant woman, achieving a compelling metaphor. This text of moral opprobrium is at the center of many a heated campfire talk and usually runs like this:

> Can you imagine grabbing a pregnant woman and pushing her underwater to suffocate and then pull her out and dump her? What shall happen to her? What are her chances to live and give birth?[8]

This highly charged and emotionally intense rhetoric questions the claim underpinning the 'catch-and-release' angling practice that it does not harm the environment. Of whether or not it harms a cultural landscape in the sense of a traumatic disruption of an age-long 'political ecology of emotion' (Dallman *et al.*, 2013) there has been, predictably, no pronouncements in texts produced by the exclusive safari companies. The main line is that they take their clients to fish the rivers of the 'last European wilderness', implying that until the advent of the camps it was a kind of *Terra Nullius*, waiting for the taking.

For the local people at the receiving end of a process of losing their accustomed subsistence and bartering/trading resources as well as cultural identities, the predictable form of resistance is to continue fishing their land. Should local people be caught in such fishing grounds, they would be considered trespassers or 'poachers'.[9] To this usually unseen side of the camps, I turn below.

Illegal fishing

The construction of a 'me-first-in-nature!' tourist product as exclusive placedness in an imagined wilderness required the removal of local people, which I have mentioned above. Such people would be let into the grounds only as temporary employees during the angling season. A camp would usually employ, from May–June till September–October, an average contingent of some 30 people. Ogarkova (2007: 25) gives the following list of camp jobs: camp manager, vice-manager, head cook and 2 assistants, 11–12 guides, masseur, 2 chambermaids, 4 mechanics, gas mechanic, 3 general-purpose workers, doctor, helicopter pilot[10] and a fishing inspector. This list gives an idea of the kind of service that safari camps provide for their wilderness-seeking clients. Other jobs are mentioned in the rare case that a glimpse of the camps appears in official media sources. Such are cases, regrettably – and practically solely – in connection with accidents. In the main, these have so far been helicopter crashes. Not a few lives of safari tourists as well as servicing personnel have been lost that way. The most recent of a series of helicopter accidents did not involve safari tourists however, but high-ranking members of the Murmansk government. The crash of an Mi-8 machine in Munozero Lake in the southern part of the peninsula (31 May 2014) took the lives of 18 people, among them a young bar attendant from a safari camp (*Na meste* 21.06, 2014c). The machine took off from camp *'Pyatka'*, located in Ter' District. The visit there, according to a press report, was a working one: 'the passengers were searching for attractive locations for developing tourism in Ter' District' (B-port, 2014a). Whether the trip was for developing tourism in a general sense, or for picking up spots for future safari camps we shall never know. The visit to a safari camp by a high-ranking group of administrators highlights, however, the tight fit between state power and upmarket tourist interests. These both ordinary tundra-connected people and local administrators consider as given – as reflecting some immutable, pre-modern order of power relations in Russia.[11]

When we come to how borders are erected at ground level, we may face the special police force: OMON.[12] The force is known for its deployment in high tension spots, like the North Caucasus, and its Murmansk Region detachment is often praised for its exemplary service in rebel containment in Chechnya and Ingushetiya (B-port, 2014b,d). Such deployment goes by the name of *komandirovki* (business trips), six months at a time. In former years, the Reindeer Husbandry Cooperative '*Tundra*' in Lovozero used to hire soldiers from the force as a deterrent against illegal hunting of their herds (B-port, 2014c). It was thus the case that on several occasions during 2005–2007, I had the chance to meet OMON soldiers at reindeer herding camps and we could talk together in a friendly atmosphere. They shared with me that they looked on those tundra assignments as a kind of holiday after hard and dangerous service in the Caucasus. From time to time, there was ritual spilling of vodka on the floor in memory of fallen comrades. OMON servicemen are also deployed for riot containment – mainly in connection with demonstrations of protesters in Moscow and St Petersburg. The soldiers would refer to these assignments as caused by 'some crazy old women (*babushki*), and down-and-outs (*bychy*)[13] who scream in the streets for a few dollars that the Americans pay them'.

To ensure a trespasser-free stay for these same 'Americans',[14] however, OMON soldiers are known to be hired by safari camps. Of this I know only from persons who would occasionally fall into the category of trespassers, and also from boating tourists' diaries (see further). What I have seen – and not once – is that when local people are on such outings there is a constant radar-like screening of the skies. 'You don't always hear the helicopter', my companions would say, 'and then it's suddenly upon you. The OMON guys first beat you up and then ask questions. It is hard to hide from them in the open tundra'.

A measure of precaution that is taken is to fish for salmon not with a rod, but with an empty tin, big enough to put your fist in. A wooden handle is fixed as a cross-bar inside for holding the tin. On the outside, a cord is wound to the necessary length with a bought or often homemade spoon-bait and hook at the end. When homemade, the spoon-bait is beaten into shape from, literally, an aluminum spoon. In action, one swings out the line by one hand, holding the tin in the other from the inside, so the line pays away without friction off the free outside surface. The attractiveness of the method is that if a helicopter is heard, or guards are seen, one can quickly hide the tackle or if needs be throw it into the water, which would be difficult and wasteful to do with a rod and reel. Should it be impossible to retrieve the tackle later, it is not such a great loss. The difference in price between the trespasser's tackle and the fly-fishing one of the exclusive tourist must correspond, figuratively, to that between a wheelbarrow and a

Ferrari. Concerning fly-fishing tackle and outfit, as fly-fishing itself, there is a whole world of unreachable irrelevancies for the locals (including fly-fishing skills) that separate the tourists from the humble local trespasser, and place them on some incomprehensible orbit of fishing 'for sport'.

Returning to the one between guards and trespassers, it is to be added that this kind of 'cops-and-robbers' chasing and skirmishes in the tundra would not be reported in the media, and access to such knowledge is very much a matter of being together with the insiders themselves. The following rare news item allows, however, some glimpse of this unseen part of the environment in which exclusive tourism happens. I quote part of the text in full, as it highlights important accents of how exclusivity is protected:

> During the past weekend a tragedy happened at a tourist (safari – my gloss) camp on the River Ponoy. A drunken brawl between guards from various camps ended in murder. The killer is suspected to be a 27 year old former OMON serviceman. We have information that he had been fired from the force a long time ago for negative reasons (i.e. incompatibility with the service – my gloss). His victim is a former prisoner,[15] who worked of late in a tourist camp on the river's bank. (B-port, 2013a)

This newspaper item gives an idea of the kind of people camp managers hire as guards: whether of necessity, or as deliberate policy, I cannot say. As the tundra insiders would often say about the guards, 'they are tough guys' (*krutye rebyata*); 'conversations are difficult with them' (*s nimi razgovory trudnye*), the latter meant as a euphemism. I have not yet heard of anyone actually shot at or killed by such safari camp guards, but stories about heavy manhandling are common. On tundra outings to which I have been generously admitted as a trusted observer, the fears and apprehensions of everyone involved have always been palpable. There was an occasion when we – three of us on a small inflatable raft – were half-paddling, half-wading up a small river. What with the difficulties of making headway, the heat of a bright sunny day in August and the torment of clouds of mosquitoes, we did not see two safari anglers till we were almost upon them. They should not have been there in the first place, as we did not know of any camp nearby. The one we knew was much further up. We were under the delusion that the tundra around was still free land. The thought flashed in my head that nothing prevented camp owners from extending their fishing grounds as much as they wanted – the only force that could stop them would be a competitor with higher administrative connections than themselves. In the tension of the moment (for some reason the two human figures appeared abnormally tall to me), I managed to ask in a voice that I found strange to hear: 'Is lake (a name) upstream?'

This I asked in Russian. One of the anglers answered in some pidgin Russian, but with a Finnish accent: 'There is no such lake up there'. We knew pretty well the lake in question was about half a kilometer upstream, but decided that arguing was out of place. We swerved the raft round hard and paddled downstream in great haste until the big lake from which we had started originally came into view. Just then, a helicopter's rumble was heard from the eastern horizon. More mad scrambling downstream until we saw some willow bushes big enough to shove the raft under. Through the overhanging branches, I saw the helicopter in my binoculars: it was the size of a dragonfly and on a disappearing course. We got on the bank and each lit a cigarette with a shaky hand. 'That was a close shave', was the first comment. 'Those Americans have satellite phones on them, with GPS functions and everything, they could have given our bearings to the guards. I thought they had already managed to send a copter (*vertushka*) after us'. After some deliberation, we decided that the *burzhui* must have been good guys after all. They had graciously decided not to phone the guards and report us.

This is how it looks from as close to firsthand as I have come. A look inside the camps is, however, more difficult.

The Clients

When prices are in the range of USD1500 a day, it is not easy for an ethnographer to do participant observation with safari tourists. I have thus not yet seen a safari camp at close range. I have had the occasional chance to see and even talk with the clients. My very first meeting happened to be at Moscow Sheremet'yevo Airport back in the early 2000s. I was waiting for my bags to appear on the flight from Murmansk belt. The man was in his mid-fifties and in an expensive outdoor outfit and fancy hat, which instantly set him apart from the nervously waiting crowd. He was worried about his luggage, as it transpired later on – his angling rods in a fancy 'Simms' bag. But there was something else that was on his mind as we waited for the belt to show signs of life. His trip to Moscow was untimely as, he explained, he had just avoided ending up probably dead in a helicopter crash. It was then that I remembered there had been an accident reported in the press with several casualties – all of them safari tourists. The man – a businessperson from Minnesota – had missed getting on the ill-fated helicopter trip to a safari camp through the incredible luck of being late for his original flight from the USA, hence missing the connection to Murmansk. So now he was travelling back on his own, after his week at the camp had eventually been arranged – what with an investigation of the crash grounding all aircraft for as long it took.

I was impressed by the man's calm taking of all this in his stride, and still having had his salmon experience in a nature which managed to remain wild despite an aircraft crashing into it now and then. I asked him about how it felt. 'Well such things happen back home all the time', he said, 'pity for the guys, but that's part of the game'. So, in a way, the tragedy had sustained an ethos of daring adventure in a frontline kind of place.

My chances to see and even talk with safari tourists have been more frequent on the helicopter pad of Lovozero Airport. With the help of various local connections who had a say about flights, I would be put on board and dropped at my tent camp:[16] it lies midway between Lovozero and the salmon safari camps, which are in the lower reaches of the Iokanga and Ponoy Rivers. To talk in the machine itself is impossible due to the noise, hence whatever there is to say is better said on firm ground. Not that the clients are very prone to enter into conversation with a local, such as they must have thought I was. Camp managers and people from the respective camp office in Murmansk were even less inclined to allow such mixing between locals and the select crowd in their charge. I could almost hear the warnings of Soviet time *INTOURIST* guides[17] to Russians seeking contact with foreigners: 'Do not approach the foreigners! I shall call the Militia!' (*Ne pod"ezzhayte k inostrantsam! Militsiu pozovu!*). Instead, while trying to initiate a conversation with a *burzhuy* on the pad, I heard a young and elegantly dressed woman, looking like a camp official from higher up, say in her cell phone: 'Andrey Mikhaylovich (fictional name) there is some little man here approaching the clients. Did you let him be on this flight?'. As I stand 1.83 m in my socks, the 'little man' (*kakoy to chelovechek*) obviously referred not to my actual physical appearance, but to my status. Luckily, Andrey Mikhaylovich must have told her that there was such permission, so the young lady was pacified and turned to her other duties.

'Wild tourists'[18]

Diaries of 'wild' rafters or paddlers sometimes contain stories of brushing-up with angling *burzhui*, guards or even providing glimpses of the inside of camps. Such is the case, for instance, of a group of rafters from Moscow and other Russian cities of finding themselves out of necessity (inclement weather) in a hospitable safari camp by the Harlovka River estuary in the north-west of the peninsula. That they were let into the camp and hospitably welcomed can be explained by the fact that it happened in mid-August. Between the June and September 'runs' of salmon upriver, the salmon season is slack and, consequently, there were no clients at the camp. In the words of the diary writer:

> We gathered our things under pouring rain and crossed over to the other bank of the river to 'the camp of the American anglers'. By this

> time the camp did not function as such anymore, and those who were there were: the Camp Manager (a rather likable chap called Dmitrich), Fishing-Hunting Inspector Kolya (a somewhat peculiar, but quite friendly guardian of the environment), and, together with them two OMON soldiers with Kalashnikov rifles and full gear (simple guys, with nothing against some cheerful company). Without much fuss we were put up at a luxury cottage. During the time there would be *burzhui* at the camp, the copter pilots lived in it. There was everything in the cottage: beds, fantastic down duvets, besides we were given an electric heater, and when we were all set up, we were asked to dinner, in the course of which we, two by two, bathed in a previously heated up sauna. (Dubovskiy, 1999)

This is the way a safari camp was seen by a boating group of four young men and two women, all in their twenties and thirties, from Moscow and St Petersburg. They mention in a nice way the presence of fully armed OMON guards, possibly patrolling the fishing grounds in preparation for the September salmon run and the next batch of clients. The presence of a fishing-hunting inspector is to be understood in the same sense (Osherenko, 1998). Both OMON and the fishing-hunting inspectorate presence accentuates the well-established status that safari camps have in respect of regional law enforcement specialized units and nature-protective agencies. Against this background, and still on the plane of tourism, the safari tourists, protected by the military and nature-controlling organs of the state, appear in diametrical opposition with the free-roaming group of young people. They are out to enjoy 'wild nature' too, weaving in the interstices still left accessible by powerful stakeholders like the military and the safari businesses. By dint of some miracle, they even happen to be welcomed to a safari camp. The military/safari close assemblage is additionally highlighted by a neighboring military radio-communication unit (*Voennaya chast' po obsluzhivaniyu releynoy svyazi*) with whose assistance they managed to get back to Murmansk in the end. The military coastal settlements are usually the port of call of the only passenger boat along the coast – the steamer 'Klavdiya Elanskaya' – whose service must stretch back to the early 1970s. It is on this boat that 'wild tourists' often return to 'civilization', or are sent there for further investigation when arrested by border-guard troops for lack of proper documentation. The coastal zone – to a depth of 25 km inland (but some say 50) – is an off-limits zone for Russian and, emphatically, foreign citizens (cf. Took, 2004: 172f), a matter arranged in a satisfactory manner for the safari camp clients. It is thus the case – to return to the ad at the beginning – that from a camp like Brevennyy, the road is not to nowhere, but to the military and border-guard settlements and bases along the coast, and from there on, on board 'Klavdia Elanskaya' to Murmansk.

Conclusions: Constructing and Enclosing 'Wilderness'

The impressive, but veiled overdevelopment of exclusive angling tourism in Murmansk Region alerts to the importance of turning analytical attention to two generalized sets of findings. In the first place, these are the political implications of ideological and applied construction of a conservationist, but tourism-related practice of 'rewilding'. As I have shown in this chapter, a policy of 'rewilding', that is of removing by representational, symbolic and physical means human historical and contemporary presence from some part of the planet, dovetails neatly with the colonial practice of human removal and declaring as *Terra Nullius* a territory, targeted for hegemonic appropriation. I have shown how in the given case, a form of tourism enacts such political thinking by a whole set of procedures, ultimately converging on the creation of enclosures for the exclusive enjoyment of 'wild nature'. In the specific case, such a policy and the practical realization of enclosure as a strong boundary building and maintenance, follows almost seamlessly with numerous forms of powered enclosures with which recent Soviet history has been so abundant: military enclosures and elite residences being two prime cases in point. The fact that 'rewilding' carries in the popular consciousness conservationist and environmentally protective associations tends to insidiously remove 'wild nature' enclosures from critical consideration. The ease with which such creations may be keeping robust an age-long practice of hegemonic enactments – uniting King's forests in Medieval Europe, with the exclusivity of game laws, favoring a tiny minority of the landed gentry in Victorian Britain, with exclusive residential estates for the top crust of power in Soviet and Soviet-like societies and, finally, with many forms of elitist wildlife exclusivity in a generalized 'West' – can remain veiled over and unseen. The easy and almost instant reinterpretation of former hegemonic enclosures into present tourist exclusivities, such as I have shown here, has not merited consideration by the general public, save for the immediately concerned – the local population who found themselves on the wrong side of the fence in this process. Neither, as it seems, such historical continuities much trouble exclusive tourists. After all, a commodity is offered, one pays for it and goes back with unforgettable memories and pictures.

Notes

(1) Kola Peninsula is the bigger central and eastern part of what is today Murmansk Region. Since the camps are situated predominantly by the salmon rivers of Lovozero and Ter' Districts and thus in the geographically peninsular part of the region, the designation Kola Peninsula, or simply Kola, shall be used henceforth.

(2) The price is all-inclusive, covering accommodation at a well-furnished wooden cottage or tent camp, meals, helicopter travel from Murmansk to camp and back, boat or helicopter outings to a favorable fishing-spot, and a guide. See Currie

(2000), Dudkin & Co. (2015), Frontiers (2015a), Hunter (2000), Ogarkova (2007), Osherenko (1998: 20f) and Ponoi River Co. (2015a,b).

(3) One thus hears of a former governor of Murmansk Region having a small residence at a coastal village, 'deserted' during the tundra village liquidation of the 1960s and 1970s, of a former prime minister having another at a village with a similar fate, and even of higher-placed persons from among the administrative-cum-business elites enjoying the use of own 'residences'. This is hearsay, however, and I cannot corroborate such evidence, beyond saying that it may very nearly reflect the actual state of affairs.

(4) Particularly illustrative are the rock carvings at Kanozero and other Neolithic sites. See Gurina (1982), Kolpakov and Shumkin (2012) and Likhachev (2011).

(5) Chronotope (time-space) is a key concept in Bakhtin's studies of the poetics of the novel, alluding to the way that time and space are together conceived and represented (Bakhtin, 1981; Dentith, 1995: 50).

(6) Here, I am paraphrasing Yankelowich's (1981, in Suša, 1999: Note 29) 'Me-First!' concept 'denoting the type of a hedonistic consumption attitude spread in modern society'.

(7) That is, as *Terra Nullius* (cf. Connor, 2005). See in sense (iii) of 'nature' as defined by Demeritt (2001: 29).

(8) A variant of this often heard rhetorical text is quoted by Osherenko (1998) in her report.

(9) In agreement with other authors, notably Essen *et al.* (2014: 633), I prefer to use more neutral terms like 'trespassing' or 'unlawful/illegal fishing', rather than 'poaching', to avoid the latter's value-laden pejorative implications.

(10) Camps are usually furnished with a small Mi-2 helicopter, used for taking clients to good fishing spots within, allegedly, the camp's fishing perimeter. Some camps may be acquiring more sophisticated Western makes. A helicopter crash that took the lives of two British tourists and their Russian guide (21 July 2013) was reported to have been with a Eurocopter EC-120 (B-port, 2013b).

(11) For the discursive genre of 'litanies' and 'laments', following popular conviction of a medieval-like immutability of a flawed social order in Russia, see Ries (1997).

(12) *Otdelenie Militsii Osobogo Naznacheniya* (Militia Detachment for Special Purposes).

(13) Rus. sg. *bych* – colloquial abbreviation from *byvshyy chelovek* (former person).

(14) 'Americans' is often used in local speech as a generic term for all Westerners, another one being *burzhuy* (bourgeois), Rus. pl. *burzhui*.

(15) In the original: *zek*, an abbreviation for *zaklyuchenniy* (prisoner). Formerly used for Gulag camp inmates, the appellation is now used for inmates of reformatory colonies (*Ispravitel'nie kolonii, IK*). Such a colony for offenders with long- and maximum-term sentences is located in Revda (*IK* 23), close to the village of Lovozero, the administrative center of Lovozero District. The recruiting of a former prisoner as a camp guard may be connected with this particular colony.

(16) This is basically a polar tent with a stove and some sheds, which has been in use as a tundra research camp since 2000. For its use during the International Polar Year (2007–2008), see Konstantinov and Vladimirova (2008).

(17) State tourist company with a monopoly on foreign tourism (*innostrannyy turizm* – hence the acronym) during Soviet times. Founded in 1929.

(18) A form of tourism in USSR/Russia not organized by state organizations or – in post-Soviet times – by private firms. In reference to the reindeer husbandry part of Lovozero District, this is, in the main, canoeing or rafting down the bigger rivers of the peninsula (Ponoy, Yokanga, Umba, etc.). Canoeists publish reports of their journeys and other relevant information in a number of websites and other electronic fora. See Konstantinov (2015) (*Wild tourism*).

References

Note that [PA] indicates personal archive (contact: yulian.konstantinov@uit.no).

Afanas'yeva, A. (2013) *Forced Relocations of the Kola Sami People: Background and Consequences*. MA thesis, Faculty of Humanities, Social Sciences and Education, University of Tromsø.

Bakhtin, M. (1981) Epic and the novel. In M. Holquist (ed.) *The Dialogic Imagination: Four Essays* (pp. 1–40). Austin, TX: University of Texas.

Bell, S., Hampshire, K. and Topalidou, S. (2007) The political culture of poaching: A case study from northern Greece. *Biodiversity and Conservation* 16, 399–418.

Bessire, L. and Bond, D. (2014) Ontological anthropology and the deferral of critique. *American Ethnologist* 41 (3), 440–456.

B-port (2013a) *Byvshego omonovtsa, podozrevaemogo v ubiystve na reke Ponoy, poymali v tundra*, news article. See www.b-port.com/crime/item/106990.html#ixzz2VyaGBVYQ (accessed 11 June 2013).

B-port (2013b) *Krushenie vertoleta v Murmanskoy oblasti proizoshlo iz-za razrusheniya zadnego uzla krepleniya shassi – MAK*, news article. See http://www.b-port.com/news/item/110379.html#ixzz38vEQyqn0 (accessed 31 July 2013).

B-port (2014a) *2 iyun'a ob"yavleno dnem traura po pogibshim*, news article. See www.b-port.com/news/item/130802.html (accessed 2 June 2014).

B-port (2014b) *Murmanskie politseyskie i OMONovtsy vyekhali v ocherednuyu komandirovku na Severnyy Kavkaz*, news article. See http://www.b-port.com/news/item/121823.html#ixzz2qM4WCIAD (accessed 13 January 2014).

B-port (2014c) *Na meste krusheniya Mi-8 naydeno telo*, news article. See http://www.b-port.com/news/item/131965.html#ixzz35LR1qYxz (accessed 21 June 2014).

B-port (2014d) *OMON vernulsya iz komandirovki na Kavkaz*, news article. See http://www.b-port.com/news/item/133382.html#ixzz37VpwXGRB (accessed 14 July 2014).

Burns, G.L. (2004) The host community and wildlife tourism. In K. Higginbottom (ed.) *Wildlife Tourism: Impacts, Management and Planning* (pp. 125–145). Altona, Vic: Common Ground.

Castree, N. and Braun, B. (eds) (2001) *Social Nature. Theory, Practice, and Politics*. Oxford: Blackwell.

Connor, M. (2005) *The Invention of Terra Nullius: Historical and Legal Fictions on the Foundation of Australia*. Sydney: Macleay Press.

Cronon, W. (ed.) (1996) *Uncommon Ground*. New York: W.W. Norton.

Currie, B. (2000) Kharlovka diary June 26–July 3, 1999. *Atlantic Salmon Review* 5 (1–2), 9–13.

Dallman, S., Ngo, M., Laris, P. and Thien, D. (2013) Political ecology of emotion and sacred space: The Winnemem Wintu struggles with California water policy. *Emotion, Space and Society* 6, 33–43.

Department of Tourism Development (2010) Perechen' turistkikh rybolovnykh baz Kol'skogo poluostrova [tourist information brochure]. Murmansk: Ministry of Industry and Enterprise Development of the Murmansk Region.

Demeritt, D. (2001) Being constructive about nature. In N. Castree and B. Braun (eds) *Social Nature. Theory, Practice, and Politics* (pp. 22–40). Oxford: Blackwell.

Dentith, S. (1995) *Bakhtinian Thought. An Introductory Reader*. London: Routledge.

Dubovskiy, S. (1999) Kharlovka-99. [PA]

Dudkin & Co. (2015) *Lager' 'Brevennyy'. Rybalka i rybolovnye tury na Kol'skiy poluostrov*, product description. See http://dudkin-co.com/?page_id=696 (accessed 13 January 2015).

Eliason, S. (2012) From the King's deer to a capitalist commodity: A social historical analysis of the poaching law. *International Journal of Comparative and Applied Criminal Justice* 36 (2), 133–148.

Essen, E. von, Hansen, H., Nordström Källström, H., Peterson, M. and Peterson, T. (2014) Deconstructing the poaching phenomenon: A review of typologies for understanding illegal hunting. *British Journal of Criminology* 54, 632–651.

Forsyth, C. and Marckese, T. (1993) Folk outlaws: Vocabularies of motives. *International Review of Modern Sociology* 23, 17–31.

Forsyth, C., Gramling, R. and Wooddell, G. (1998) Game of poaching: Folk crimes in Southwest Louisiana. *Society & Natural Resources* 11, 25–38.

Franklin, A. (2008) The 'animal question' and the 'consumption' of wildlife. In B. Lovelock (ed.) *Tourism and the Consumption of Wildlife: Hunting, Shooting* and Sport*fishing* (pp. 31–44). London: Routledge.

Frontiers (2015a) Brevyeni Camp. *Frontiers International Travel*, product description. See http://www.frontierstrvl.co.uk/index3.php?p=the_brevyeni_camp (accessed 11 January 2015).

Garth, C. (ed.) (1978) *James Hawker's Journal: A Victorian Poacher*. Oxford: Oxford University Press.

Gurina, N. (1982). *Vremya, vrezannoe v kamen'. Iz istorii drevnikh laplandtsev*. Murmansk: Murmanskoe knizhnoe izdatel'stvo.

Gutsol, N., Vinogradova, S. and Samorukova, A. (2007) *Pereselennye gruppy kol'skikh saamov*. Apatiti: Kol'skiy nauchniy tsentr.

Höchtl, F., Lehringer, S. and Konold, W. (2005) 'Wilderness': What it means when it becomes a reality: A case study from the southwestern Alps. *Landscape and Urban Planning* 70, 85–95.

Hønneland, G. and Jørgensen, A.K. (1999) *Integration vs. Autonomy: Civil-military Relations on the Kola Peninsula*. Aldershot: Ashgate.

Hunter, B. (2000) Focus on Kola Peninsula. *Atlantic Salmon Review* 5 (1–2), 6–9 and 14–24.

Ingold, T. (2000) *The Perception of the Environment: Essays on Livelihood, Dwelling and Skill*. London: Routledge.

Kharuzin, N. (1890) *Russkie lopari. Ocherki proshlago i sovremennogo byta: Izvestiya imperatorskogo obshchestva lyubiteley estestvoznaniya, antropologii i etnografii. Tom LXVI. Trudi etnograficheskogo otdela. Tom X*, Moscow.

Kolpakov, E.M. and Shumkin V.Y. (2012) *Petroglify Kanozera*. SPb: Iskusstvo Rossii.

Konstantinov, Y. (2015) *Conversations with Power: Soviet and Postsoviet Developments in the Reindeer Husbandry Part of the Kola Peninsula*. Uppsala: Uppsala University.

Krupnik, I. (1993) *Arctic Adaptations: Native Whalers and Reindeer Herders of Northern Eurasia*. Hanover: Dartmouth College.

Lajus, J. (2008) Frozen Frontier in the Russian North in Historical Perspective. Paper presented at BOREAS workshop 'Heading North, Heading South: Arctic Social Sciences Research in Global Dialogue', 6–9 March 2008. Max Planck Institute for Social Anthropology, Halle.

Likhachev, V. (2011) *Risunki Kanozero. Otkrytie, izuchenie, sokhranenie*. Apatiti: IP Likhachev.

Luk'yanchenko, T. (1971) *Material'naya kul'tura saamov (loparey) Kol'skogo poluostrova v kontse XIX-XX vv*. M: Nauka.

Murmansk Region (2015) *Fishing Tours Murmansk Region*, product description. See http://www.murman.ru/guide/tourism/fishing-eng.shtml?print=1 (accessed 11 January 2015).

Muth, R. and Bowe, J. (1998) Illegal harvest of renewable natural resources in North America: Toward a typology of the motivations for poaching. *Society & Natural Resources* 11, 9–24.

Nakhshina, M. (2012) Community interpretations of fishing outside legal regulations: A case study from Northwest Russia. In C. Carothers, K.R. Criddle, C.P. Chambers, J. Fall and P.J. Cullenberg (eds) *Fishing People of the North: Cultures, Economies, and*

Management Responding to Change (pp. 229–242). Fairbanks, AL: University of Alaska Fairbanks.

Ogarkova, M. (2007) Organisatsiya sportivnogo rybolova na territorii Murmanskoy oblasti (Na primere ZAO "Reka Ponoy"), diploma paper. Vologodskiy gosudarstvennyy tekhnicheskiy universitet [unpublished MS, PA].

Osherenko, G. (1998) The Ponoi River Report: Sport Fishing in the Kola Peninsula. A Project of the Circumpolar Conservation Union [unpublished MS, PA].

Øian, H. (2013) Wilderness tourism and the moralities of commitment: Hunting and angling as modes of engaging with the natures and animals of rural landscapes in Norway. *Journal of Rural Studies* 32, 178–185.

Polanyi, K. (1957) The economy as instituted process. In K. Polanyi, C.M. Arensberg and H.W. Pearson (eds) *Trade and Market in the Early Empires* (pp. 243–270). New York: Free Press.

Ponoi River Co. (2015a) *Ponoi River Co. Ryabaga, Brevyeni, Purnache*, product description. See http://www.theflyshop.com/travel/kola/ponoi-river-company.html (accessed 13 January 2015).

Ponoi River Co. (2015b) *Ponoi river*, product description. See http://www.theflyshop.com/travel/kola/ponoi-river-company.html (accessed 13 January 2015).

Ries, N. (1997) *Russian Talk: Culture and Conversation during Perestroyka*. Ithaca, NY: Cornell University.

Suša, O. (1999) The ecological problem in modern society: Solidarity, conflict and human dignity. *Human Dignity: Values and Justice. Czech Philosophical Studies 3,* 111–149.

Took, R. (2004) *Running with Reindeer: Encounters in Russian Lapland*. London: John Murray.

Ushakov, I. (1998a) *Kol'skiy Sever v dosovetskoe vremya. Izbrannye proizvedeniya*, Vol. 2. Murmansk: Murmanskoe knizhnoe izdatel'stvo.

Ushakov, I. (1998b) *Kol'skaya starina. Izbrannye proizvedeniya*, Vol. 3. Murmansk: Murmanskoe knizhnoe izdatel'stvo.

Yankelowich, D. (1981) *The New Rules: Searching for Self-fulfillment in a World Turned Upside Down*. New York: Random House.

Ziker, J. (2006) The social movement of meat in Taimyr, Northern Russia. *Journal Nomadic Peoples* 10 (2), 105–123.

7 Empowering Whom? Politics and Realities of Indigenous Tourism Development in the Russian Arctic

Albina Pashkevich

Introduction

The beginning of the 2000s saw a renewed interest in the Russian Arctic. Not only due to the publicity given to the region by official visits from top Russian politicians (Vladimir Putin and Dimitri Medvedev), but also due to the strategic direction of development of the Arctic areas. The attention that the Russian political elite gave to the Arctic resulted in several initiatives along the coast of the Arctic Ocean and again highlighted the importance of this territory for the national political agenda. Some of these initiatives have focused on the condition of the Arctic environment, which has led to the extension of protected areas and the clean-up of the most remote Arctic territories, and intensified efforts to revive scientific investigations (Pashkevich, 2013; Pashkevich & Stjernström, 2014). These initiatives also resulted in a short period of openness that made it possible to carry out several important research initiatives in the territory of several Arctic regions (International Polar Year 2007–2008, expeditions to Siberia, the Far East and the Kola Peninsula).

In this seemingly open atmosphere, the central and regional authorities of Russian Arctic regions have also become involved in the promotion of tourism. For example, the Russian Geographic Society hosted an international Arctic forum entitled 'The Arctic: A Territory of Dialogue', with several meetings taking place between 2010 and 2013, and with top Russian government officials and international celebrities among the invited guests and keynote speakers. Several regional initiatives focused on renewing attention on tourism in the Arctic, and was followed by three international tourism forums in Arkhangelsk (2011, 2013, 2015), and similar initiatives in other Arctic regions supported by regional governments in the

Republic of Sakha and Komi, the Murmansk Oblast and the Yamal-Nenets Autonomous Okrug.

At the same time, this attention has also highlighted the role of local control, participation and benefit-sharing for all economic activities, including tourism operations in the Arctic. It has also emphasised the need to establish a system of tourism operations built on respect for local indigenous traditions, culture and heritage. There are about 40 different indigenous groups currently living on the territory of the Russian Arctic, which is estimated to include some 244,000 people (Nesterov, 2015). The specifics of indigenous community involvement in tourism development in the Arctic have been the subject of several research efforts (Hendry, 2005; Müller *et al.*, 2013; Pashkevich, 2013). The idea of local communities controlling their own involvement in tourism activities and use of their resources and territories is in line with previous studies in other parts of the world (Saarinen, 2011; Stronza, 2007; Telfer & Sharpley, 2007). However, these research accounts also emphasise the difficulty of achieving these goals of better integration and control by the indigenous communities in relation to tourism (Müller *et al.*, 2013; Saarinen, 2011).

However, in Russia, tourism development is increasingly promoted and discussed; the focus lies more on the potential of this development (Kalashnikova, 2013; Laktionova & Markaryan, 2012; Toskunina & Shpanova, 2009) rather than dealing specifically with the involvement of local communities and especially representatives from indigenous communities (Pashkevich, 2013). It is interesting to note that the Russian situation seems to correspond well to situations in other parts of the world such as Latin America, Africa, Southeast Asia and Australia (Coria & Calfucura, 2012; Fuller *et al.*, 2005; Ruthanen *et al.*, 2015). The institutional setting related to tourism development in Russia is unsettled: it suffers from inadequate expertise in tourism planning, management and marketing, and also from the short planning–decision cycle of the regional administrations (Pashkevich, 2013). Moreover, the lack of political power and effective decision-making by indigenous communities precludes them from realising the benefits brought by tourism, or from mitigating the possible side effects (comparable situations have been described by Nyaupane *et al.* [2006] and Yang *et al.* [2013]). Power relations that have been created by previous modes of thinking and planning traditions continue to influence the interactions between organisations, agencies and individuals (Ruthanen *et al.*, 2015; Yang *et al.*, 2013). Responses to the efforts to create a solid regional–local agenda involving a plurality of interests have shown that underlying power relations in tourism development and planning arrangements are important dimensions influencing the success or failure of this process (Pashkevich, 2013; Pashkevich & Stjernström, 2014).

The present chapter analyses the process of tourism planning, the coordination of efforts concerned with tourism destination development based on the indigenous culture and traditions of one of Russia's Arctic regions – the Nenets Autonomous Okrug (NAO). Several stakeholder groups involved in this process were identified for this study: government agencies, tourism entrepreneurs and representatives from indigenous communities involved in various ways in tourism development in this region. This chapter provides an overview of the recent development of public institutional structures meant to support the overall process of tourism development in the NAO, and indigenous tourism development in particular. In order to illustrate this process, the public organisation of support towards local indigenous communities is examined. Informal views concerning this process are provided by interviews conducted with representatives of indigenous communities on the role of tourism in their everyday lives, combined with the perspectives of the local tourism entrepreneurs acting as mediators between the indigenous communities and their visitors. The concluding remarks are concerned with the general evaluation of the policies and the realities behind the process of empowerment of local communities and considerations for the future.

The chapter is based on a comprehensive analysis of the data collected in the region for the period from 2011 to 2015. This data consists of 50 semi-structured interviews with governmental officials, tourism entrepreneurs, reindeer herders and members of their families, and local inhabitants of urban and rural settlements scattered along the coast of the Barents and Kara Seas. The analysis also includes official documents, electronically published materials, tourist brochures, guidebooks, etc. The author has visited the NAO region during several field trips, covering the northeastern, western and central parts of the region in the winter, autumn and summer months. Field trips included participant observations of touristic experiences based on the indigenous culture, traditions and local folklore of this vast Arctic territory.

Involving Indigenous Communities in Tourism

The role of involving local communities in the process of planning and managing tourism operations has being widely discussed by several research efforts (Ashley, 2000; Ashley & Roe, 2002; Hall & Brown, 2006; Novelli & Gebhardt, 2007; Saarinen, 2011; Saarinen *et al.*, 2009). The culture and traditional way of life of the indigenous communities packaged for tourists has become an important contribution to tourism destination development in the Arctic. However, according to Hendry (2005) concerns have been raised about the authenticity of the roles played by local indigenous people, and that the places recreated for visitors taking part in tourism operations may not be seen as 'the real thing'. Their involvement in tourism operations has allowed for much needed cultural exchange and the elaboration of ancient practices into the realities of modern life (Hendry,

2005). At the same time, having the capacity to undertake the tasks of creating a tourism product could be challenging for a local indigenous community (Fuller *et al.*, 2005). Unresolved issues concerning formal agreements of landownership and problems with the communication and interaction between various stakeholders have often hindered indigenous tourism (Dyer *et al.*, 2003). Thus, the importance of building up social capital and the active involvement of the host community, alongside the roles played by governmental and non-governmental actors, would provide a foundation for the vitality of the tourism operations (Buultjens *et al.*, 2010; Fuller *et al.*, 2005; Nyaupane *et al.*, 2006).

Local indigenous communities become the centre of attention when tourism is promoted as a way to diversify employment opportunities and attract external investment. The ways in which government policy can support this process are not always straightforward and depend on various factors, one of which is institutional settings allowing schooling and training opportunities in order to meet tourists' expectations (Hall, 2008; Novelli & Gebhardt, 2007). The lack of political power, and the absence of effective systems concerning the distribution of income, means land management planning often becomes a barrier to successful indigenous tourism development (Coria & Calfucura, 2012). Once again, the central role played by government agencies in the arrangements for the provision of indigenous tourism experiences is highlighted, and failures to realise adequate promotional and marketing efforts challenge the dynamics of recent efforts in this area, as seen in the Australian example (Ruthanen *et al.*, 2015).

Another interesting aspect of indigenous tourism development relevant to the Russian case is how the centrally determined visions concerning the role of tourism in strengthening the indigenous community can become a source of potential conflict (Yang *et al.*, 2013). Changes in norms, attitudes and the adaptations made by local indigenous people taking part in tourism operations also need addressing in dialogues with governmental and non-governmental organisations (Yang *et al.*, 2013). Finally, the introduction of the indigenous communities to the potentials brought by tourism development, often influenced by 'cosmopolitan locals' and external actors who are more experienced travellers, act as a catalyst for this development (Iorio & Wall, 2012). However, yet again, local institutions continue with the task of setting up a system of respectful and reflective management to involve indigenous people in tourism.

The recognition of tourism as one of the ways to diversify the raw-material-based economies of the Russian Arctic was formalised in a programme issued by the Russian government called 'Strategy of the Development of Arctic areas of Russian Federation until 2020' (Russian Government, 2013). According to this strategy, which was accepted in 2013, tourism development should become part of the overall socio-economic development of the Arctic, with a focus on the development of ecotourism. The conditions for successful tourism development included a further development of the legislative and normative base of tourism operations. 'Regional tourism clusters' became the buzzwords of official tourism development concepts in every Arctic region (for example, the Arkhangelsk and Murmansk Regions and the Sakha Republic). In reality, this meant that regional capitals could become facilitators of tourism development, at least in the initial stage, due to the availability of the necessary transport infrastructure and services that incoming guests need. The idea of a support system based on public–private partnerships has also gained attention in connection with the revitalisation process of tourism in the Arctic. In practice, a seminar series – or several meetings – with private entrepreneurs was organised to inform about the potentials of the scheme and give best practice examples from Nordic countries.

Every Arctic region interpreted the directions of this strategy according to the resources available at the time; regional governments began to work closely not only on the promotion and marketing of tourism opportunities in the Arctic, but also on encouraging the involvement of the local indigenous people in private entrepreneurship. It is interesting to note that, according to the strategy, in areas populated by indigenous people, tourism activities should be developed in environmentally friendly ways to allow the traditional ways of life and areas of traditional land use to be protected. The pressures put on local public authorities by the ambitious Arctic strategy immediately became a topic of discussion in the local media that doubted that the realisation of any of the measures would be executed in the end.

Developing Russian Arctic Tourism: Preconditions and Realities

In recent decades, the process of tourism development in Russia has been based on the predominant development of domestic tourist flows to the largest urban areas (Moscow, St Petersburg, Nizhny Novgorod), as well as outbound international tourism ('sea and sun' destinations) (Laktionova & Markaryan, 2012). Domestic tourism developed at a much slower pace than the flows of Russian tourists to international holiday destinations. However, rising living standards, as well as changes in life considerations, led to local demands for tourism services (Kalashnikova, 2013; Toskunina & Shpanova, 2009). Demands have been supported (even if not fully according to Western standards) by the extensive development of the hospitality and service sector since the beginning of the 2000s.

As noted in Maher *et al.* (2015), tourism in the Russian Arctic is mainly based around hunting and fishing. Adventure tourism, including winter activities such as snowmobile safaris, ice fishing or rafting during the summer months, is gaining popularity (Maher *et al.*, 2015). Ethnographic tours allowing interaction with the indigenous cultures and traditions have also become an important ingredient of the region's tourism offer. The procedures around becoming a tourism entrepreneur involve a certain amount of uncomplicated paperwork, supported by the Federal Tax Authority (FTA, 2015). The application involves a government duty of €10, and the private entrepreneur commitment to pay around €400 in tax annually plus 6% of their profits.

Since the beginning of 2015, fluctuations in the Russian rouble have led to serious changes in the travel patterns of Russian tourists, who had previously opted to spend their vacations abroad. Russia's central and southern regions suddenly became cheaper destinations due to the 20%–30% increase in the cost of international travel. Tourism operators in remote destinations of the Russian Arctic became optimistic about the possibility of redirecting part of this flow to their territories (personal communication, 2015). Thus, it is interesting to see how the current institutional arrangements in one of the Russian Arctic regions are able to facilitate this process, especially with respect to the indigenous Nenets community occupying its territory. Moreover, the author has visited the NAO territory on several field trips since 2011 covering various types of tourism activities available in the region, and was hence able to observe not only the current conditions, but also a certain degree of change.

Public institutional support of tourism development in Nenets Autonomous Okrug

The population of the NAO is 43,400 (2015), the largest proportion of which is below working age. The economic wealth of the region is entirely dependent on natural resource extraction (mainly oil and gas). This industry has gradually expanded since the early 2000s due to the growing volumes of oil production, which peaked in 2009 at 6.5% of Russia's total oil production, and has since progressively declined (Pashkevich, 2013). The area is also home to 7500 indigenous Nenets (Khabarova, 2013). Reindeer husbandry is one of the most important means of existence for the indigenous Nenets people. This industry has shown signs of recovery, from its almost total collapse in 2009, with a growing production of reindeer meat, not least due to state subsidies. Local producers have also noted higher demand for reindeer meat from consumers in the central part of Russia (mainly from restaurants). There are approximately 170,000 reindeers in the NAO across 9,759 hectares of pasturelands, and the industry employs approximately 1,000 reindeer herders (NAO, 2014).

Development of tourism has long been seen as a way to diversify the region's economy. Unofficially, the region holds a very strong position in the domestic adventure tourism market provided in collaboration with private tourism firms from Moscow (personal communication, 2014). The entrepreneurs from the NAO provide the product, mainly consisting of fishing and hunting trips, and their counterparts in the capital city provide the necessary publicity and bookings. However, the local tourism cultural centre – a destination management organisation – which has been operating in the capital of Naryan-Mar since 2011, presents a slightly different view on tourism product development. This view implies that there is much greater potential in the region than just 'hook and bullet' tourism. More specifically, visitors to the region could be introduced to some elements of the indigenous culture and traditions of the Nenets, where the everyday life of the reindeer herders and their families would be part of this experience (including staying with families out on the tundra, winter fishing and ethnic food). Another direction is based on the potential provided by Arctic nature: adventure tourism, including expeditions to remote areas of the Nenets Okrug, white water rafting, snowmobile safaris and birdwatching based in the Nenets Nature State Reserve and other protected territories. The tourism activities in the region are highly seasonal, still mostly confined to winter, which corresponds to the situation in other parts of the Arctic (Müller, 2011).

Political instability and corruption scandals were a reality for the leadership of the NAO during the 2000s. These affected the overall investment climate and seriously damaged the region's position in relation to the other Russian Arctic regions. The nature of the state policy in tourism development was characterised by short-sightedness as it was only ever planned four years in advance (the duration of the appointment of the top positions in the local governmental bodies, and the extent of the mandate for the region's governor). This created many limitations with regard to the establishment of trust and dialogue among the stakeholders involved in the destination development process. Most of the tourism entrepreneurs agree that, in Arctic regions, the role of state support from the regional authorities is essential for their survival (personal communication, 2015).

The high transport tariffs on airfares to the region and within its territory, the need for helicopter transport to remote areas, as well as the absence of a basic road infrastructure undermine the development of a viable tourism sector. This is despite the relative proximity of the region to the largest urban agglomerations of the European part of Russia (which are only a two and a half hour flight from Moscow). Yet, some entrepreneurs mentioned that this proximity might also be key to their problems (personal communication, 2015). They find themselves under constant control by non-professionals who often think they know best about how to develop tourism in the region (personal communication, 2015). This situation is not only damaging for the overall image of the destination, but a clear sign of an ineffective system of governance.

Developing or promoting? Peculiar ways to establish public–private relations within tourism

Several governmental officials interviewed in the NAO mentioned that questions concerning proactive tourism development had arisen as far back as the late 1980s. An understanding that the oil and gas resources would not last forever caused several visionary bureaucrats to suggest measures towards the diversification of NAO's economy. But these ideas were not supported by any financial means during that time. This means that it took almost 30 years before the first attempts towards resolving the issues of further diversification of NAO's economy were made, looking at tourism as one of the possible measures. Local reactions towards the content of the recent strategy for the Arctic development adopted by the Russian government were not specifically enthusiastic, meaning that 'none of these governmental programmes were meant to actually develop tourism, but rather to promote it' (personal communication, 2013).

In 2011, regional authorities finally made it official that tourism development was a necessary element of regional economic development. In the same year, a decision was taken to create a board consisting of private tourism entrepreneurs and representatives of various bodies within the regional administration, although the initiative died out after a few meetings. One year later, a local government passed a law on public support of tourism activities: the Department of Education and Youth Policy (2013) became responsible for the overall governance of tourism development in the NAO under the programme 'Creation of Favourable Conditions for Tourism Development in NAO for the Period 2013–2015'. The Department of Culture, Youth Policy and Sport was in charge of the process, and began to develop plans for local cultural and sport events, and to facilitate a process of establishing a basic infrastructure for the hospitality sector. This department was also responsible for collecting information concerning private entrepreneurs dealing with the flow of visitors to the region. The heads of the local municipalities responsible for tourism planning and event management were advised to follow up on the regional initiative. None of the municipalities in the NAO in the proximity of the region's capital (apart from the largest one located in Zapolarny Okrug in the Poselok Iskateley settlement), had any resources or expertise to follow up on the initiative, and those responsible for the tourism development inventory were unable to do much in the absence of the tourist entrepreneurs.

The national Arctic development strategy promoted by the Russian central government, where tourism was given a vital role, was not followed up by a similar strategy for the NAO. For example, there were no guidelines for developing public–private cooperation, joint ventures with local communities or a system of decision-making involving the use of natural and cultural resources belonging to the indigenous people. There is a lack of statistics showing the number of visitors to the region. This information is collected by the regional branch of the Russian State Committee of Statistics located in Arkhangelsk (the neighbouring region) and delivered on demand to public and private organisations. However, the Committee of Statistics demands payment for this information and this money was not included in the budget of the Department of Education and Youth Policy responsible for regional actions in tourism planning. The rigid and inflexible system of regional administration in Nenets became a serious hindrance to private initiatives connected to tourism. Locally, decisions concerning tourism development came last in a long line of issues that public servants were dealing with daily, such as housing, energy and food supplies, healthcare and high transportation tariffs.

Another important explanation for the rather lengthy process of setting up the regulatory system and realisation of public support was the absence of a coordinating agency responsible for tourism development. A representative from regional administration said, 'Tourism development is rather new for Russia, nothing is clear and sorted; nobody knows how to deal with it, exactly' (personal communication, 2012). This enabled those with personal connections to governmental officials to benefit from the system of state support towards private business. One of the informants in this study mentioned that regional funds were invested in 2009 in the creation of a base camp called 'Forest Saga', established for recreational purposes, located in proximity to the NAO's capital Naryan-Mar (personal communication, 2012). A few years later, it became not only a centre for recreational activities for the local inhabitants in the regional capital, but also a base for the development of adventure tourism in the region.

So far, attempts to develop tourism in the territory of the NAO had largely amounted to individual entrepreneurs using the public support system for their own profits. The programme of grants supporting new tourism business start-ups in the region, with support from the region's administration since 2011, was not based on destination development planning created specifically for this region. The programme aimed at giving incentives to tourism entrepreneurs to acquire technical equipment complying with national safety standards, to develop tours and package tourist products. The programme resulted mainly in the support of existing tourism firms with connections to the regional authorities, investing in their established facilities and thus, not resulting in new business start-ups.

Additionally, in 2011, the complete absence of joint efforts between tourism entrepreneurs and regional administration in destination development in the NAO was illustrated by the creation of a logo promoting the region's territory for investors and positioning it among the other Arctic Russian regions. Local tourism entrepreneurs felt left out by the regional administration, which hired a prestigious public relations agency to develop the NAO's graphic profile (personal communication, 2012). Local community representatives criticised the logotype for being too obvious and implying a disproportionate development of the oil and gas industry. The indigenous community's existence was diminished to their role in the development of extensive reindeer herding. At first, the logotype appeared on promotional products such as calendars, stickers and paper packaging, and eventually, in 2014, it became the background not only for the regional administration's website, but also for all public authorities in the region (NAO, 2014). Later, it turned out that the symbols on the graphic logo were created based on a limited customer survey, which did not include the views of local inhabitants or representatives of the indigenous people.

Overall, this period of public governance development of tourism in the region can be described using the words of one of the informants from local government (in 2012): 'We are the promoters and supporters of tourism development; we are not dealing with the development of tourism. It is a private activity, which follows its own rules of development'. Here, the role taken by the regional administration of the NAO is at least clear, although the same official added, 'Obviously it is a very new system for Russia. A specific organisation responsible for the promotion of tourism in the region still does not exist. Thus there is a lot of uncertainty and personal interpretations of the existing set of rules are widespread'.

Empowering indigenous people – institutions and realities

The NAO has a long tradition of involving its indigenous minority in the decisions made by the legislative powers through the representatives chosen by the indigenous community. In 2009, a Department of Indigenous People of the North (DIPN) was created to implement a 'state policy towards independent, socio-economic and cultural development of the indigenous people, protection of their traditions and livelihood' (Administration of NAO, 2016). On several occasions, members of reindeer herding brigades mentioned the respect that the head of the DIPN holds among the Nenets, mostly for being 'one of us' and travelling long distances to talk to people directly (personal communication, 2014). However, an absence of tourism expertise among planners, decision-makers and the indigenous representatives themselves created a situation where they had to rely on the expertise of regional bureaucrats and non-ethnic tourism entrepreneurs. In 2011–2012, budget allocations were made to build the Tourism Cultural Centre in proximity to the Okrug's capital Naryan-Mar. The centre was created to realise the national policy for tourism development, connected to opportunities to promote tourism products to internal and external markets. The centre was originally supposed to represent the Nenets culture with the help of members of the active 'Nerdenja' family community, led by Kseniya Yavtysaya (personal communication, 2012, 2014).

The role and the position of the Tourism Cultural Centre have been unclear since its creation, not least due to its questionable internal organisation and lack of qualified leadership and staff. Thus, since its creation in 2012, the centre has not been able to plan an organised promotion of tourism activities or play a leading role in benchmarking tourism products in the region. Finally, during 2014, a newly selected governor of the NAO appointed a new director for the centre, Matvey Chyprov, who is also one of the well-known private tourism entrepreneurs promoting the Nenets' culture and traditions. This firm specialises in day tours to the reindeer brigades stationed closer to the Naryan-Mar during the winter months. In reality, there seems to be two organisations located at the same address. The first is the Tourism Cultural Centre supported by the regional budget, and the second is the Centre of Arctic Expedition Tourism, which operates more as Chyprov's own business. Both exist to promote and to develop adventure tourism in the territory of the NAO. The Tourism Cultural Centre advertises its own commercial hunting and fishing tours, and the opportunity to interact with the indigenous culture, but at the same time has established collaborations with eight private tourism firms and the Nenetsky Nature Reserve (VisitNAO, 2015).

The lack of a coherent strategy for tourism development in the region overall, and the absence of clear leadership, with signs of incompetence and a high level of corruption in the public administrative agencies did not allow the creation of an environment which stimulated wider involvement of representatives from indigenous communities into this process. Another example supporting this notion could be mentioned here. The NAO itself, and tourism development in particular, has already been a target of EU-financed initiatives since 2006. During the periods 2006–2008 and 2010–2013, indigenous entrepreneurship in tourism was promoted in cooperation with the Sami Education Institute, Inari, Finland, through the EU's Kolarctic ENPI CBC programme (Neda Ordym, n.a.). The project aimed to develop a platform for involving several public agencies dealing with promotion and practical education specifically targeting young representatives of indigenous people. However, this Finnish–Russian cooperation was unable to challenge the prevalent attitudes towards tourism in the region. In the absence of transparent institutional arrangements for tourism planning and the non-existence of open channels for representatives from Nenets to participate in the overall destination development process, efforts by the Finnish partners failed to contribute to an increase in indigenous tourism entrepreneurship. So did efforts to involve representatives from ordinary Nenets people. Senior Nenets officials took part in the educational activities provided for the students of the local Naryan-Mar Socio-Humanitarian College as part of the training and competence development organised by Finnish partners. According to one of the informants for this study, approximately 60 people took part in the exchanges provided by the programme (both indigenous and non-indigenous students) over the two periods. Of these, only five ended up working in tourism after completing the programme.

Indigenous tourism development in NAO

Indigenous tourism in the NAO is mostly connected with day trips that take visitors to reindeer herders' camps in the vicinity of the region's centre of Naryan-Mar. The situation observed during several field trips into the Nenets' tundra (2012–2015) revealed a mixed picture concerning the indigenous people's perceptions of tourism development. Most of the Nenets were positive about meeting other people, especially if it involved a cultural exchange in which all parties involved learned more about each other. Several Nenets themselves were afraid that their community's consumption of alcohol is on the increase due to tourists bringing alcohol with them as gifts for the indigenous people as signs of friendship. Thus, the Nenets employed by reindeer herding units often find it difficult to decline the offer to drink with guests. Consequently, due to drinking too much, and as a result of failing to carry out their duties, they are suffering economically (personal communication, 2013).

Another common problem mentioned by younger reindeer herders was that tourists are probably more interested in seeing 'real or traditional Nenets people', not those wearing modern clothes and herding their reindeers on snowmobiles. The young men felt torn between their desires for a modern life – including snowmobiles, the newest models of mobile phones, clothes, etc. – and the responsibility of keeping their own indigenous culture alive. In most cases, they did not find it particularly burdensome, rather they enjoyed combining the elements of their traditional living and modern life. Nonetheless, youngsters felt sad that the Nenets had almost totally lost knowledge of their native language, which further translated into losing skills such as constructing the traditional mobile house (*chym*). During one of the trips, a young reindeer herder mentioned: 'Why spend time setting up a *chym* if we can borrow a wooden cabin (built by some of the hunters to the area)?' This meant that the adaptation to a more convenient way of life and the availability of the permanent wooden cabins made it easier for the reindeer herders to set up camps while following their reindeers to the pastures.

Most of the Nenets reindeer herders met by the author during field trips were shy and unprepared for talking to tourists, or at least it took longer for them to get used to having people around watching them. Another group that is positive towards visitors coming from larger urban centres to the region is the Nenets women who take care of the households for the reindeer herders in the tundra. They felt it was a very pleasant way to make the days go by quicker. The economic reward was welcomed, as it could be used for the household. However, they also wondered about the conditions for the visitors. One woman mentioned: 'I am not sure, but I do not think all of the tourists would want to eat from the same plate as we do. They might also feel uncomfortable sleeping in the same *chym* with us' (personal communication, 2014).

Nenets families living in the remote areas of the NAO also enjoy the occasional company of tourists. On tours to the remote areas of the Kara Sea, the tourists practically live with the family of the reindeer herders, drinking the same tea, eating the same food and helping with the household routines. It is a tourist product about the real life of a nomad family, whose members are open towards newcomers and are genuinely interested in meeting visitors from outside. Several Nenets families from the eastern part of the region have got used to having visitors and are interested in receiving more, but they also say: 'We cannot take too many visitors; space is a limitation for us' (personal communication, 2014). The current director of the Tourism Cultural Centre, Matvey Chyprov, agrees, 'Reindeer herders involved in tourism have a limited carrying capacity of maybe five to ten persons 15 times per year. Those people should not be disturbed too much while working' (personal communication, 2015). Furthermore, according to Chyprov, the increased number of visitors to the same brigade would lead to the dilution of the product and increased pressures on the reindeer herders, and receiving visitors would no longer be meaningful and exciting. In this respect, plans to establish closer cooperation with tourist agencies in China are particularly interesting. It shows a real mismatch between the concerns about the quality of the product, and the fact that the opportunity to make a profit always prevails over ethical considerations.

Private entrepreneurs working with Nenets mentioned another issue connected to the recent history of the cultural assimilation of the indigenous people, which had had a profound effect on the nation. One of the entrepreneurs mentioned that sadly, the help provided by the state may also have contributed to the fact that some of them are choosing to wait for subsidies rather than going out to the tundra. It is suggested that 'some Nenets do not know how to take care of the reindeer, which may lead to the death of many animals, but nobody from outside can say anything because Nenets are supposed to know reindeers best' (personal communication, 2015). There are still a few enclaves of private reindeer herders, far from the larger settlements, that are less affected by the influences of civilisation. They say, 'one of the advantages of being far away from the control of the state is living in a harmony with nature' (personal communication, 2014). At the same time, one of the leaders of the reindeer herding family told the author, 'the further into the tundra you are, the more you have to discipline yourself and rely on your own skills. You cannot sit and wait for the state to come and rescue you' (personal communication, 2014).

Concluding Remarks: Empowering Whom?

The author's travels and meetings with the indigenous people of the NAO revealed the raw reality of the modern-day development of this region in the Russian Arctic. The administrative system created in the region is unable to provide support for the local indigenous communities that wish to share the benefits brought by tourism activities. The politics of empowering the indigenous people have so far suited the regional administration's own agendas. The reality of this development empowers only those who are appointed to hold this power as a part of or closely related to the regional or local administrative elite. The inability of governmental agencies to form close alliances with local non-governmental organisations, in order to create channels of inclusion for the Nenets into planning and monitoring current and future tourism development, is mainly due to a lack of trust and transparency in this process.

The reality of the situation in Nenets Okrug has been formed by incompetence in tourism planning and management, and a lack of collaboration and long-term planning. Research by Buultjens *et al.* (2010) suggests closer collaboration with other powerful sectors of the economy (oil and gas sector) would greatly benefit the indigenous communities and tourism entrepreneurs if properly arranged. The Nenets do not hold any control over tourism activities and can only sense the scale of the challenges that await them if the volume of visitors increases. Apparently, tourism entrepreneurs are taking advantage of the access to the remote areas of the NAO's territory in order to gain access to reindeer herders maintaining their traditional nomadic way of life. In general, there is a positive attitude towards tourism; however, issues such as hospitality skills and basic infrastructural needs, such as water and sewage treatment, represent some of the unresolved issues. At the same time, similar to the notions described by Hendry (2005), there are fears that the Nenets could lose their attractiveness as a tourism product as they do not look like stereotypical reindeer herders in their modern clothes.

The system of public support towards tourism is a major impediment for the interplay between local tourism entrepreneurs and state authorities. Furthermore, the fragmented character of current developments has resulted in uncertainties about how the Nenets communities are being involved in tourism. Tourism entrepreneurs themselves are beginning to question the existing order and are trying to influence the initial stages of the fundamental restructuring that is necessary in order fully exploit the region's features as a destination. However, they lack the formal mechanisms to make lasting changes without proper support from the public institutions. Thus, the region's tourism expansion is proceeding without specified goals; this chaotic development will eventually lead to the overexploitation of sensitive territories. The present governance system of tourism development has led to the formation of double standards. Those who should benefit most from the renewed attention to tourism are the indigenous Nenets communities. However, they are not being included as one of the major stakeholders and are therefore unable to fully participate in this process. This situation is not unique: the gap created by varying views on policies and planning supporting indigenous tourism development elsewhere has been well documented (Higgins-Desbiolles *et al.*, 2014; Ruhanen *et al.*, 2015; Saarinen, 2011; Telfer & Sharpley, 2007; Yang *et al.*, 2013). Recent developments have not corresponded to the terms provided by the indigenous communities in a dialog with the public authorities and local tourism entrepreneurs. The platform for the establishment of this dialog has not been created so far. Nonetheless, the ability of local tourism entrepreneurs to survive through the leadership of regional governments holds the key to collaboration with representatives from the Nenets and is a first step towards the Arctic becoming a territory of dialogue and collaboration that will actually flourish. These alliances are already acting as a catalyst for the recent efforts in tourism development.

References

Ashley, C. (2000) *The Impacts of Tourism to Rural Livelihoods: Namibia's Experience* (ODI Working Paper 128). London: Chameleon Press.

Ashley, C. and Roe, D. (2002) Making tourism work for the poor: Strategies and challenges in Southern Africa. *Development Southern Africa* 19 (1), 61–82.

Buultjens, J., Brereton, D., Memmott, P., Reser, J., Thomson, L. and O'Rourke, T. (2010) The mining sector and indigenous tourism development in Weipa, Queensland. *Tourism Management* 31 (5), 597–606.

Coria, J. and Calfucura, E. (2012) Ecotourism and the development of indigenous communities: The good, the bad, and the ugly. *Ecological Economics* 73, 47–55.

Department of Education and Youth Policy (2013) Department of Education and Youth Policy NAO, the results of 2012. *Naryana-Vynder* 21(19933). See http://nvinder.ru/article/vypusk-no-21-19933-ot-28-fevralya-2013-g/167-otchet-upravleniya-obrazovaniya-i-molodezhnoy (accessed 15 May 2015).

Dyer, P., Aberdeen, L. and Schuler, S. (2003) Tourism impacts on an Australian indigenous community: A Djabugay case study. *Tourism Management* 24 (1), 83–95.

Federal Tax Authority (FTA) (2015) How to start your own business? (In Russian). See http://www.nalog.ru/create_business/ip/creation/registration/ (accessed 8 September 2015).

Fuller, D., Buultjens, J. and Cummings, E. (2005) Ecotourism and indigenous micro-enterprise formation in northern Australia opportunities and constraints. *Tourism Management* 26 (6), 891–904.

Hall, D. and Brown, F. (eds) (2006) *Tourism and Welfare: Ethics, Responsibility and Sustained Well-Being*. Wallingford: CABI.

Hendry, J. (2005) *Reclaiming Culture: Indigenous People and Self Representation*. Gordonsville, VA: Palgrave Macmillan.

Higgins-Desbiolles, F., Trevorrow, G. and Sparrow, S. (2014) The Coorong Wilderness Lodge: A case study of planning failures in indigenous tourism. *Tourism Management* 44, 46–57.

Iorio, M. and Wall, G. (2012) Behind the masks: Tourism and community in Sardinia. *Tourism Management* 33 (6), 1440–1449.

Kalashnikova, N. (2013) Problems and perspectives of the investments projects on the territory of Arkhangelsk region. In G.A. Kainova (ed) *Proceedings of III International Scientific Conference 'Problems of Modern Economy'* (pp. 40–46). Chelyabinsk, Russia.

Khabarova, N. (2013) Number of Nenets people is growing in Nenets Autonomous Okrug. *DvinaInform Agency*, 24 October. See http://www.dvinainform.ru/society/2013/10/24/18613.html (accessed 14 February 2015).

Laktionova, U.V. and Markaryan, I.N. (2012) Current state and prospects for development of tourism in Russia. *European Researcher* 22 (5), 783–786.

Maher, P.T., Gelter, H., Hillmer-Pegram, K., Hovgaard, G., Hull, J., Þór Jóhannesson, G., Karlsdóttir, A., Rantala, O. and Pashkevich, A. (2014) Arctic tourism: Realities and possibilities. In A. Pelyasov (ed.) *Arctic Year Book 2014*. NRF, University of the Arctic.

Müller, D.K. (2011) Tourism development in Europe's 'last wilderness': An assessment of nature-based tourism in Swedish Lapland. In A.A. Grenier and D.K. Müller (eds) *Polar Tourism: A Tool for Regional Development* (pp. 129–153). Montréal, QC: Presses de l'Université du Québec.

Müller, D.K., Lundmark, L. and Lemelin, R.H. (eds) (2013) *New Issues in Polar Tourism: Communities, Environments, Politics*. Dordrecht: Springer.

Neda Ordym (n.a.) Culture tourism project of the indigenous peoples of the North. See http://www.nedaordym.fi/index.php/en/neda (accessed 11 March 2014).

Nenets Autonomous Okrug (2014) *News*. See http://adm-nao.ru/press/ (accessed 9 April 2014).

Nesterov, A. (2015) *Indigenous People of the North*. See http://www.arctic-info.ru/Encyclopedia/Rubric/korennie-narodi (accessed 4 September 2015).

Novelli, M. and Gebhardt, K. (2007) Community based tourism in Namibia: 'Reality Show' or 'Window Dressing?' *Current Issues in Tourism* 10 (5), 443–479.

Nyaupane, G.P., Morais, D.B. and Dowler, L. (2006) The role of community involvement and number/type of visitors on tourism impacts: A controlled comparison of Annapurna, Nepal and Northwest Yunnan, China. *Tourism Management* 27 (6), 1373–1385.

Pashkevich, A. (2013) Tourism development planning and product development in the context of Russian Arctic territories. In R.H. Lemelin, P. Maher and D. Liggett (eds) *From Talk to Action: How Tourism is Changing the Polar Regions* (pp. 41–60). Thunder Bay: Centre for Northern Studies Press, Lakehead University.

Pashkevich, A. and Stjernström, O. (2014) Making Russian Arctic accessible for tourists: Analysis of the institutional barriers. *Polar Geography* 37 (2), 137–156.

Russian Government (2013) *Strategy of the Development of Arctic Areas of Russian Federation until 2020*. See http://government.ru/docs/1869/ (accessed 8 September 2015).

Ruhanen, L., Whitford, M. and McLennan, C. (2015) Indigenous tourism in Australia: Time for a reality check. *Tourism Management* 48, 73–83.

Saarinen, J. (2011) Tourism, indigenous people, and the challenge of development: The representations of Ovahnimbas in tourism promotion and community perceptions toward tourism. *Tourism Analysis* 16, 31–42.

Saarinen, J., Becker, F., Manwa, H. and Wilson, D. (2009) Introduction: Call for sustainability. In J. Saarinen, F. Becker, H. Manwa and D. Wilson (eds) *Sustainable Tourism in Southern Africa: Local Communities and Natural Resources in Transition* (pp. 3–19). Bristol: Channel View Publications.

Stronza, A. (2007) The economic promise of ecotourism for conservation. *Journal of Ecotourism* 6 (3), 210–230.

Telfer, D. and Sharpley, R. (2007) *Tourism and Development in the Developing World*. London: Routledge.

Toskunina, V. and Shpanova, N. (2009) Problems of tourism development in Arkhangelsk region. *Regional Economy* 1 (5), 66–73.
VisitNAO (2015) Tourism companies in the Nenets Autonomous Okrug. See http://visitnao.ru/tourist-resources%20 (accessed 8 September 2015).
Yang, J., Ryan, C. and Zhang, L. (2013) Social conflict in communities impacted by tourism. *Tourism Management* 35, 82–93.

8 Destination Development in the Middle of the Sápmi: Whose Voice is Heard and How?

Seija Tuulentie

Introduction

The Finnish village Kilpisjärvi is located in the middle of the Sami region (Sápmi), in the borderland where Finland, Norway and Sweden meet. However, in Finnish discussion, Kilpisjärvi does not have an image as a Sami village. Its long history in hiking and cross-country skiing tourism is emphasized as well as the importance of the Saana Fell beside the village as a national symbol. In current development plans of the municipality of Enontekiö, it is described as a village which is clearly a colony-like enclave in the middle of a wilderness with a multicultural population coming from different parts of Finland. In addition, it has turned into a second-home village for Norwegians, and the municipality has put hopes into strong tourism development of the village.

The Sami Parliament in Finland has claimed that Sami culture, nature and reindeer herding are not taken into account in the development plans. However, nature conservation and reindeer herding are also on a collision course as reindeer herding is forbidden in the nearby strict nature reserve of Malla.

Kilpisjärvi serves also as a good example of the notion that 'local community' is not one but consists of a multiplicity of voices; the issue which has been recognized e.g. in relation to the mapping of indigenous areas (Roth, 2009) or to community-based tourism (Blackstock, 2005). However, more often, community has been understood as one entity. Blackstock (2005) states that this view of one community assumes shared interests and a consensus on the preferred tourism outcomes, which is not possible in reality. Parallel to the case of Kilpisjärvi, conflictual issues can be found in discussions of the role of the Sami people in regard to nature conservation and tourism development in Swedish mountains (Engström & Boluk, 2012; Müller, 2013).

Thus, issues related to Sami culture, recreation, tourism, reindeer herding and nature conservation are intertwined in the discussion on the development of Kilpisjärvi. The discussion is especially interesting from the point of view of power relations. Despite being a small village, power structures are extremely complicated because of the many stakeholders and interested parties. Researchers, authorities and regular tourists from the south as well as local authorities, decision-makers and local residents have dissenting views of how to develop Kilpisjärvi. Thus, the voice of the indigenous people is not at the forefront. Still, the role and meaning of Sami people is a focal issue as the village is located in the Sami domicile region and its indigenous image is much used in the destination marketing.

The development plans for Kilpisjärvi as a tourism destination are here examined through planning documents and related discussion. Strategies and planning documents set guidelines for the future tourism industry and, thus, are an important part of governance. They channel different lines of societal development and are aimed at changing things, but often do not reveal the choices that have been made in the formulation of the text. The issues are expressed as self-evident facts, but the choices behind the factual statements in the documents can be examined via rhetorical analysis (Kietäväinen & Tuulentie, 2013). This chapter asks who has a say in the development of Kilpisjärvi and its surroundings. Is there a specific Sami perspective to the development plans, and if so, how does it differ from the majority populations' view? And how has the official planning and management system reacted to the Sami issues?

Governance Practices, Power and Knowledge

Theoretically, this chapter is about power relations and governance. The concept of governance represents a departure from the more formal studies into the roles, responsibilities and activities of governments, to a more fluid sphere of interaction between the state, civil society and business interests emphasizing participation and deliberative processes (Dredge *et al.*, 2011; Hall, 2011). A governance perspective includes both structural and agency aspects and encompasses the full-range of emerging and self-organizing state–society interactions rather than just the purposeful efforts directed at controlling and steering society and societal sectors. It denotes that the public sector may continuously, but need not, occupy an important role in governing activities (Löf, 2014).

The question of what role indigenous people are given, or which role they take, in the processes of planning and how their role is understood in relation to tourism are at the core. Especially winter tourism has developed rapidly during the last two decades in Finnish Lapland. Sami culture has been used as a part of the marketing. Ethnically Finnish tourism workers have used Sami costumes, which has created controversy (Suomi, 2011), but also Sami people themselves have profited from tourism by selling their handicrafts and arranging reindeer-related activities (Tuulentie, 2006). The division between Sami culture and majority culture in Finland differs from the situation in Norway and Sweden as reindeer herding has traditionally not only been a Sami but also a Finnish livelihood. Thus, the mix of cultural features is more understandable in Finnish Lapland than in neighboring countries, but this also makes them more complicated (Viken & Müller, 2006). In the tourism sector, the problem is that these cultural issues are not always expressed to the visitors, and they are not touched upon in tourism planning and decision-making.

There are important power relations around tourism governance, with some groups in society, for example, having relatively more influence than others on the governance processes affecting tourism. There can be significant conflicts around tourism governance as groups seek to secure their favored policy decisions (Bramwell & Lane, 2011). The main actors seeking power to define the issues at stake in the Kilpisjärvi case are especially the municipality; the state through Metsähallitus, which is the manager of most of the land area; local entrepreneurs and residents; and the Sami political organ, the Sami Parliament. Governing from a governance perspective is always an interactive process because no single actor, public or private, has the knowledge or resource capacity to tackle problems unilaterally (Kooiman, 1993; Stoker, 1998). Knowledge and power are, however, closely linked to each other and, as Foucault (1979) says, there is no power relation without the correlative constitution of a field of knowledge.

Tourism can be seen as an interesting context in which to study governance as it lies at the intersection of the public, private and community sectors (Ruhanen *et al.*, 2010). The question of 'top-down' or 'bottom-up' governance has been to the front in recent discussion. However, this division simplifies the issues since the communities are heterogenous and multicultural as the case of Kilpisjärvi shows.

It is difficult to give a clear definition of the concept of governance but it is as difficult to say what a destination is. Here, it is understood – from the governance point of view – as a cluster of interrelated stakeholders embedded in a social network (Scott *et al.*, 2008). Such a network of stakeholders produces the experience that the travelers consume. These destination stakeholders include accommodation businesses, attractions, tour companies and others providing commercial services; government agencies and tourism offices as well as representatives of the local community (Baggio *et al.*, 2010).

Akwé: Kon guidelines as a new governance mechanism

A new and emerging mechanism in the governance of the Sami domicile region in Finland is the Akwé: Kon procedure. Akwé: Kon guidelines are created as a part of the Convention on Biological Diversity and are defined as 'Voluntary guidelines for the conduct of cultural, environmental and social impact assessments regarding developments proposed to take place on, or which are likely to impact on, sacred sites and on lands and waters traditionally occupied or used by indigenous and local communities' (Secretariat of the Convention on Biological Diversity, 2004). In Finland, the Ministry of the Environment has decided that the guidelines are intended to be taken into account in the preparation of projects and plans, and in the assessment of their impacts, which are to be implemented on lands traditionally occupied by indigenous peoples, such as the Sami homeland in Finland, when the projects and plans affect the Sami culture, livelihoods and cultural heritage (Ministry of the Environment, 2011). If the project is seen to have negative impacts on indigenous peoples, those responsible for the project and the representatives of indigenous peoples should agree on mitigating measures to minimize the negative impacts.

Finland has been the first country to apply the Akwé: Kon guidelines, and until 2013 no other country had implemented them (Lakkala, 2013) although they were already adopted in 2004 as part of the Convention on Biological Diversity by the United Nations (Ministry of the Environment, 2011).

In Finland, the Akwé: Kon voluntary guidelines are intended for the authorities responsible for land use planning and environmental impact assessments in the Sami homeland, legislators, those who carry out assessments, those responsible for EIA projects, planners, Metsähallitus, permit authorities, the Sami Parliament and village meetings of the Skolt and the Sami community (Ministry of the Environment, 2011). Metsähallitus (2013) points out that since the Akwé: Kon guidelines have never before been applied to land use planning anywhere in the world, there was, consequently, no prior experience to draw on with regard to the application of the guidelines to land use planning. They also state that the guidelines were drawn up to apply to quite diverse legislative and cultural conditions from the Nordic situation. The name of the guidelines comes from the Mohawk language; the word Akwé: Kon means 'everything in creation' (Metsähallitus, 2013).

The Akwé: Kon guidelines are applied in some of the processes dealt with in this chapter while some of the processes have not taken them into account. Thus, it is possible to compare the situation with or without the application of the guidelines. In practice, the application of the guidelines means that the Sami Parliament is able to nominate specific group to examine the issues at stake from the Sami point of view.

Kilpisjärvi and its Tourism

The village with 100 permanent residents is part of the municipality of Enontekiö, 180 kilometers away from the municipal center. The Norwegian city of Tromsø with 75,000 inhabitants is closer to Kilpisjärvi than the municipal center or any other town in Finland. Kilpisjärvi is a fairly typical destination for tourists seeking nature and wilderness experiences resembling e.g. Abisko in Sweden with its research station and few tourist services. Kilpisjärvi is located in an extremely sparsely populated area that suffers from migration to the population centers in the southern part of Finland. The land is almost entirely owned by the state, but the indigenous Sami people are making claims for land rights. For tourists, the main attractions are activity holidays in nature, such as hiking or cross-country skiing. Also, some enthusiasts of more extreme sports such as telemark-skiing have discovered Kilpisjärvi in recent years. Snowmobiling – especially by Norwegians – has been a contested issue since reindeer herders and some villagers complain of the wild traffic while the entrepreneurs emphasize that the snowmobile tourists bring a lot of money to the village (Hannonen *et al.*, 2015; Tuulentie, 2005).

Earlier, the most common Kilpisjärvi tourist was one who had been there before. For the high season in winter, people had been known to make bookings with the hotel as much as five years in advance. Nowadays, emblematic of the place are the growing numbers of Norwegians, who stay in their caravans or cottages. In regard to second homes, private construction was not possible until the 1980s as all land was owned by the state. Only at the end of the 1990s were the first plots for second homes sold, mainly to Norwegians. To date, in total 100 plots have been planned with the majority having built a second home are owned by Norwegians. In addition, there are 250 places for caravans in the village, and most of them have a Norwegian 'nail tent' (spikertelt, a semi-permanent building that supplements the living space of a caravan) beside them (Hannonen *et al.*, 2015).

The village of Kilpisjärvi serves as an interesting and specific example of power relations in tourism for several reasons. First, the origins of the Kilpisjärvi village, as it is now known, are in administration and tourism. It is a young village, and thus very few can claim to be natives of the area. Second, the village belongs to the Sami domicile region, and it has been part of the Sami reindeer herding area for centuries. Third, the geographical position of the village is a special one as it is located at the meeting point of three countries, Finland, Sweden and Norway and, thus, is in the middle of the Sápmi, the land of the indigenous Sami people (Tuulentie, 2005).

As said, from the Finnish point of view, Kilpisjärvi is a very remote place (Figure 8.1), but since the biggest fells in Finland are situated in the area, it is an important landscape from the national perspective. The nature of the area differs a lot from the rest of Finland and Finnish Lapland. Tourism has been important for the village since the 1930s. At that time, the fells were 'discovered' as providing excellent hiking terrain. Before that, only the Sami reindeer herders had lived in these fells. Nature conservation has

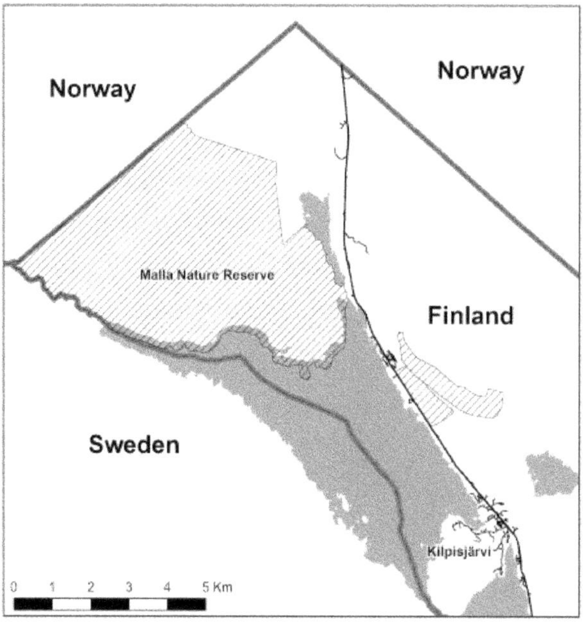

Figure 8.1 A map of Kilpisjärvi village and its surroundings

been a contested issue in the area as well, similarly to the whole of Lapland. There were plans to construct a road to Saana Fell and ski slopes to other parts of the municipality of Enontekiö, but these plans miscarried because of conservation regulations. More than 75% of the municipality's area has been set aside to serve different kinds of nature reserves.

Data and Method: Qualitative Content Analysis

The data in this study is qualitative by nature and consists of documents created by various actors in planning processes (Table 8.1). The municipality of Enontekiö has produced a Kilpisjärvi development plan to the year 2020 and a general land use plan for the village is in the making. In this process, several reports and accounts have been produced and they have been commented on by different stakeholders. In addition to these municipality-related documents, I analyze the documents produced by Metsähallitus, which is the state land owner and, thus, the manager of the surrounding wilderness area. Since 2012, Metsähallitus has been preparing a management plan for the area. This process has taken a long time because in the middle of the planning an initiative to establish a national park in Kilpisjärvi was introduced. Metsähallitus was also commissioned to produce an account for the Ministry of the Environment concerning the plans to have more national parks in Finland, Kilpisjärvi area being one

Table 8.1 Analyzed documents

	Organization in charge	*Executor*	*Year*	*Focus*	*Size*
Kilpisjärvi 2020 (K2020)	Municipality of Enontekiö	Consulting company Seitap	2010	Tourism development	26 pages
General land use plan (GLUP)	Municipality of Enontekiö	Consulting company Seitap	2012–2013	To direct construction and tourism activities	11 pages (participation and evaluation plan); 63 pages (account)
National park account (NP)	Ministry of Environment	Metsähallitus	2013	To find out the possibilities to establish a new national park	
Malla management and use plan (Malla)	Metsähallitus	Metsähallitus	2014	To make the plan especially because of increasing use pressures	79 pages
Internet discussion on the development of Kilpisjärvi and Käsivarsi area	Metsähallitus	Metsähallitus, participants	2012	To get opinions from a wide audience for the plans	100 messages, 33 users, 14 topics
Statement of the Sami Parliament: Kilpisjärvi development plan (SP1)	Sami Parliament	Sami Parliament	2011	Comments on the plan	22 pages
Statement of the Sami Parliament: National park (SP2)	Sami Parliament	Sami Parliament	2013	Comments on the plan	6 pages
Statement of the Sami Parliament: Malla (SP3)	Sami Parliament	Sami Parliament	2014	Comments on the plan	10 pages

Note: In the analysis the documents are referred to by an abbreviation which is in parenthesis after the name of the document.

of the most important candidates. Here, the management plan for Malla nature reserve and the national park report are analyzed since they are the public documents produced by Metsähallitus to date. The others are in the making but the internet discussion concerning the whole planning process will be taken into account. It was arranged by Metsähallitus in order to have a more interactive planning process. The participants in this internet discussion used pseudonyms and nothing can be said about who participated. Only the participant who gave the answers on behalf of Metsähallitus can be identified. I also analyze the statements by the Sami parliament given to all these plans.

The method of reading these documents can be described as qualitative content analysis which pays attention from a selected theoretical point of view – here governance – to communication structures (Mayring, 2004) which are regarded as important both because of the context and of the typologies in governance literature.

Based on the content analysis, three dimensions of governance have been taken as a point of departure to classify the important elements in the discussion. These dimensions are (1) participation and power-sharing, (2) flexibility and (3) knowledge and learning, and they have been selected on the basis of the conceptual reviews and typologies of governance by de Loë *et al.* (2009), Hall (2011) and Stoker (1998).

de Loë *et al.* (2009) note in their review on the environmental governance literature that the academic literature offers useful insights on six key concerns that must be considered in pursuing innovative approaches to environmental governance. These relate to accountability and legitimacy; actors and roles; fit, interplay and scale; adaptiveness, flexibility and learning; evaluation; and knowledge. Ruhanen *et al.* (2010) make a slightly different list, which includes six most frequently identified variables: accountability, transparency, involvement, structure, effectiveness and power. Naturally, many other elements have been listed in the vast governance literature and wordings vary but the critical issues of involvement or participation, flexibility and learning and the use of knowledge come to the forefront in the Kilpisjärvi data.

Findings

Participation and power-sharing

The analyzed governance processes were initiated either by the state (Metsähallitus) or by the municipality of Enontekiö (development plan and general land use plan). Involving all stakeholders has been mentioned, however, as one of the most important starting points in each plan. Although top-down processes at the start, the involvement of all stakeholders has been regarded as important.

In the Malla management and use plan by Metsähallitus, a comprehensive steering group of stakeholders was nominated. It was the same group that continued its work with the national park plan – a planning process which came up suddenly in the middle of the original process concerning protected areas around Kilpisjärvi. In addition, an internet discussion forum was created in order to gather the comments of a wider audience.

In the municipal process of the general land use plan, the legal demands of the Land Use and Building Act were followed but not voluntarily exceeded. However, the account for the plan included a large interview with the Sami steering group representative. He considered that the relationship between tourism and reindeer herding has improved but that Sami culture, handicraft and reindeer herding should be more available to the visitors.

The Kilpisjärvi 2020 development plan was carried out as a thoroughly participatory process. Surveys, workshops and an internet discussion were used in order to collect various views about destination development. On the basis of the planning process, the emphasis was on entrepreneurs' views. Based on this data, the plan came to the conclusion that the most important parties in cooperation are local enterprises, the municipality, Metsähallitus and the actors from Norway (K2020). Local residents together with the biological station of the University of Helsinki were the least important actors in the network. Sami people or reindeer herders were not mentioned separately in the development plan which had a strong focus on tourism. Thus, indigeneity is not emphasized in Kilpisjärvi tourism development from the municipal point of view.

When it came to participation, the Sami Parliament was very critical in its comments on the Kilpisjärvi 2020 plan. They saw that in the Kilpisjärvi 2020 plan their comments had not been taken into account at all (SP1). They had suggested that other kinds of alternatives than the one based on a mass tourism type of development should also be considered. The Sami Parliament had suggested that the tourism profile of Kilpisjärvi should be more clearly based on nature values and reindeer herding culture but this had not been discussed.

The Sami Parliament's criticism toward the national park account was that the tourism entrepreneurs' voice was heard but negotiations with reindeer herders and the Sami Parliament had not been carried out. In relation to the Malla process, the Sami Parliament was disappointed that the Akwé: Kon guidelines were not applied although they were already applied in other management and use plans in the Sami domicile region.

Power relations were most explicitly discussed in the internet discussion. The power of money and the business people coming from outside were referred to. The power was said to be concentrated on a few entrepreneurs. From the Sami perspective especially topics related to dog sleds and building in the wilderness were raised. From the point of view of participation, the issue of 'who is a local' was interesting. One writer claimed that the Sami are the original population and their occupation has been reindeer herding although their livelihoods have changed:

> New business people have mainly come from outside and they now have the power and they enlarge their ownership to new sectors. Isn't anything enough? If there will be base camps for motorized vehicles, nothing holds them. More traffic, nature suffers. (Comment in the internet discussion, 11/05/2012)

Flexibility

The question of flexibility is related especially to the application of the Akwé: Kon guidelines which came as a new element to the planning processes. The guidelines were new and no examples of their application existed in Finland or in other parts of the world. The state organization Metsähallitus had first applied the Akwé: Kon procedure in relation to the planning of Hammastunturi wilderness area in the eastern part of the Sami homeland but they did not establish a separate Akwé: Kon group for the Malla Management and Use plan. This was criticized by the Sami Parliament although they were able to nominate four representatives to the steering group because of the guidelines (SP4). This possibility to have four representatives came to the process after the Sami critique that the entrepreneurs were overrepresented in the group. This can be seen as an emblem of flexibility on Metsähallitus' part.

When preparing the general land use plan for Kilpisjärvi, the Akwé: Kon guidelines were said to be applied. This process was still ongoing as of October 2016 and thus no statements by the Sami organizations exist yet.

The use of the Akwé: Kon guidelines shows increased interest in taking the indigenous voice into account. The experiences of that kind of governance practice cannot be evaluated yet. In any case, it means that more indigenous representatives are involved in planning.

Other emblems of flexibility in the planning processes can be seen in the result of the Kilpisjärvi 2020 plan: such constructions as a lift to the Saana Fell or a specific arena for all terrain vehicles were removed from the plan because of local and Sami resistance. However, the planning process was quite traditional with its participation events, which were claimed to concentrate on hearing mostly the entrepreneurs' views and leaving the Sami views unnoticed.

Knowledge and learning

The use of indigenous and traditional knowledge in the tourism development of Kilpisjärvi was lacking in the analyzed processes. Since reindeer herding is the most important livelihood of the Sami in the area and tourism is run by entrepreneurs mainly belonging to the majority population, the experiences of these two livelihoods did not meet in planning.

In the internet forum, the discussion about the Sami was scarce, mainly related to the relationships between reindeer herding and tourism disturbance. The disturbance was seen as coming from dog sleds and snowmobiles, and the debate concentrated on the issue of if they disturb the reindeer or not.

However, what was common to the internet discussion and to the statements by the Sami Parliament was that both emphasized the possibilities of more ecological and non-motorized tourism development. The Sami Parliament favored the establishment of the national park since it was regarded as enhancing Sami culture and conducting tourism in a more nature-based direction (SP2). Nevertheless, the conclusion was that the national park was not established because of 'local resistance' (Karjalainen, 2014) which again raises the question of who are the locals and whose knowledge and experiences are taken into account.

Traditions were discussed especially in relation to dog sleds: many of the commentators in the internet forum favored dog sleds as an environmentally benign mode of tourism while the statements by the Sami claimed that the activity does not belong to the traditional activities of the Sami homeland but has been adopted from other parts of the Arctic area. Thus, the Sami Parliament stated (SP2) that it should be replaced by reindeer safaris and other more traditional tourism services.

From the Sami point of view there was also a lack of knowledge concerning the consequences of the municipal plans to increase the number of tourists. The Sami Parliament demanded more research of the consequences on nature, the landscape and Sami culture (SP1). In addition, a fundamental question of land ownership was seen as an important point to be clarified before the state sells its land property to private purposes (SP1).

As a conclusion to the learning and knowledge aspects, it can be said that although a vast body of knowledge was collected as a basis for the plans, nothing really new came up in regard to the use of knowledge. The application of the Akwé: Kon guidelines was the only attempt to use new modes of governance.

Discussion

The overall view of the role of the Sami in the discussion of the development of Kilpisjärvi as a tourism destination is that Sami culture is taken into account especially from the reindeer herding perspective but that it is totally separate when the discussion deals with tourism business and tourism development. This may be due to the fact that the Sami people in the region are mainly reindeer herders and that the Sami culture is often defined through reindeer herding and not related to the traditional hiking tourism or the emerging safari tourism. Tourism traditions in the region are more related to admiration of the landscapes and nature-based activities than e.g. the local culture. Of course, as observed for example in the Kilpisjärvi development plan, 'multiculturalism' is an advantage for the marketing of Kilpisjärvi, and the common view is that Sáminess and reindeer herding culture should be somehow visible in the village.

However, there is no unwillingness among the Sami to participate in tourism business. In its statement on the Lapland Tourism Strategy 2015–2018 (Regional Council of Lapland, 2014), the Sami Parliament notes that none of the strategic goals are is related to the Sami culture or sustainable development. The Sami Parliament suggested that one strategic goal should be to develop ecological tourism entrepreneurship of the Sami and to support the networking and marketing of the Sami tourism entrepreneurs. According to the statement, one of the actions should be the creation of a Sami tourism brand. This shows that the Sami in general are ready to develop indigenous tourism.

Similar tones are in the Sami visions of the tourism development in Kilpisjärvi. The Sami Parliament has stated that tourism should be developed on a small-scale and non-motorized basis, and by appreciating natural values. The Sami were one of the few parties favoring the establishment of the national park. The decision not to establish the park was made by the Ministry of the Environment. Earlier, the Minister of the Environment had stated that the national park could be established only if it had strong local support. Thus, the support by reindeer herders and the Sami population was not enough since the tourist entrepreneurs, municipal decision-makers and the Regional Council of Lapland were against the park. This discussion on the establishment of the national park parallels the situations in Sweden. Müller (2013) has described how tourist entrepreneurs' resistance hindered the establishment of a national park in the Vindelfjällen area, which gives support to the idea that national parks are good for tourism does not always work (see also Puhakka *et al.*, 2009). Müller (2013) describes this as a core–periphery conflict and as a transfer of power over land use from the periphery to core regions. This is very much the case in Kilpisjärvi, as well, what comes to the opinions of the opponents of the national park who are mainly tourist entrepreneurs and other non-Sami locals. However, the indigenous people's view makes it more complicated: they seem to trust more the state-led management of the lands than the local private actors.

A plethora of laws and acts exist which demand that the indigenous people as well as reindeer herders are heard on issues that affect their culture and the possibility to practice their traditional livelihoods. However, when it comes to such livelihoods as tourism, which is not regarded as a traditional Sami livelihood, the voice of the Sami is not among the most powerful in decision-making. As Engström and Boluk (2012) have noted in the context of the Swedish mountain case, traditional claims to land are not considered as superior to economically driven plans.

Conclusion

Governance is a concept which directs attention from top-down government processes to more deliberative practices and to the participation of all stakeholders in governing processes. Still, the planning and destination development processes and the initial understanding of the development needs are often initiated by the state or municipal authorities and carried out by similar procedures regardless of the local conditions. Problems arise from defining the local community which usually consists of a multiplicity of voices.

Kilpisjärvi is an example of a small destination with complex land use and management issues. The state has had a dominant position in tourism as well as other development in the region because of the dominance of the state-owned land and conservation areas. Fairly recently, private tourism business has obtained a footing in the village. Its location at the corner of three nation states has been both a strength and a weakness for the place: it has been used in tourism marketing, but remoteness from the Finnish point of view has made it less accessible than many other destinations in Finnish Lapland. Conflictual issues such as who is regarded as a real Sami and who should own the land in the Sami homeland remain in the background of all development processes in the Sami region.

Attempts toward more bottom-up type of planning processes have been made but as the discussion around the development plans shows they have not been totally successful. From the indigenous point of view, the barriers between various land-use modes, such as tourism and reindeer herding, should be dismantled and more open discussion between various actors created. Also, what the statements, especially for the Kilpisjärvi 2020 plan, make clear is that few alternatives for future development are given as a basis for discussion. Innovations of truly ecological and culturally benign tourism development and noticing the possibilities of indigenous culture in tourism are still to come. From a new governance point of view, it remains to be seen whether the Akwé: Kon guidelines have any effect on the hearing and influence of the indigenous voice. In the current situation, they are anyway something to which the Sami people can appeal to when they feel that their voice is not heard enough.

References

Baggio, R., Scott, N. and Cooper, C. (2010) Improving tourism destination governance: A complexity science approach. *Tourism Review* 65 (4), 51–60.

Blackstock, K. (2005) A critical look at community-based tourism. *Community Development Journal* 40 (1), 39–49.

Bramwell, B. and Lane, B. (2011) Critical research on the governance of tourism and sustainability. *Journal of Sustainable Tourism* 19 (4–5), 411–421.

de Loë, R.C., Armitage, D., Plummer, R., Davidson, S. and Moraru, L. (2009) *From Government to Governance: A State-of-the-Art Review of Environmental Governance.* Final Report. Prepared for Alberta Environment, Environmental Stewardship, Environmental Relations. Guelph, ON: Rob de Loë Consulting Services.

Dredge, D., Jenkins, J. and Whitford, M. (2011) Tourism planning and policy: Historical development and contemporary challenges. In D. Dredge and J. Jenkins (eds) *Stories of Practice: Tourism Policy and Planning* (pp. 13–34). Farnham: Ashgate.

Engström, C. and Boluk, K.A. (2012) The battlefield of the mountain: Exploring the conflict of tourism development on the Three Peaks in Idre, Sweden. *Tourism Planning & Development* 9 (4), 411–427.

Foucault, M. (1979) *Discipline and Punish: The Birth of the Prison.* New York: Vintage.

Hall, C.M. (2011) A typology of governance and its implications for tourism policy analysis, *Journal of Sustainable Tourism* 19 (4), 437–457.

Hannonen, O., Tuulentie, S. and Pitkänen, K. (2015) Borders and second home tourism: Norwegian and Russian second home owners in Finnish border areas. *Journal of Borderlands Studies* 30, 53–67.

Karjalainen, S. (2014) Tyytyväistä hyrinää ja pettymystä. Kansallispuistopäätös: Puolentoista vuoden mittainen odottelu päättyi [Decision on national park: One and half years waiting ended]. *Lapin Kansa* 1.2.2014.

Kietäväinen, A. and Tuulentie, S. (2013) Tourism strategies and climate change: Rhetoric at both strategic and grassroots levels about growth and sustainable development in Finland. *Journal of Sustainable Tourism* 6, 845–861.

Kooiman, J. (1993) *Modern Governance.* London: Sage.

Lakkala, A. (2013) Saamelaiskäräjät sai tunnustusta artikla 8(j):n toimeenpanosta. *Yle*, news program. See http://yle.fi/uutiset/saamelaiskarajat_sai_tunnustusta_artikla_8jn_toimeenpanosta/6877045 (accessed 15 September 2015).

Löf, A. (2014) *Challenging Adaptability. Analysing the Governance of Reindeer Husbandry in Sweden.* Umeå: Department of Political Science, Umeå University.

Mayring, P. (2004) Qualitative content analysis. In U. Flick, E. von Kardorff and I. Steinke (eds) *A Companion to Qualitative Research* (pp. 266–269). London: Sage.

Metsähallitus (2013) Application of Akwé: Kon Guidelines in the Management and Land Use Plan for the Hammastunturi Wilderness Area. Natural Heritage Services, policy document. See http://julkaisut.metsa.fi/assets/pdf/lp/Muut/AkweKonraportti2013.pdf (accessed 15 September 2015).

Ministry of the Environment (2011) Akwé: Kon Voluntary Guidelines. *Ympäristöhallinnon ohjeita 1/2011*, Edita Ltd. Helsinki.

Müller, D.K. (2013) National parks for tourism development in sub-Arctic areas: Curse or blessing? In D.K. Müller, L. Lundmark and R.H. Lemelin (eds) *New Issues in Polar Tourism. Communities, Environments, Politics* (pp. 189–204). Amsterdam: Springer.

Puhakka, R., Sarkki, S., Cottrell, S.P. and Siikamäki, P. (2009) Local discourses and international initiatives: sociocultural sustainability of tourism in Oulanka National Park, Finland. *Journal of Sustainable Tourism* 17 (5), 529–549.

Regional Council of Lapland (2014) *Lapland Tourism Strategy 2015–2018*, policy document. See http://www.lappi.fi/lapinliitto/c/document_library/get_file?folderId=1252845&name=DLFE-25241.pdf (accessed 19 May 2015).

Roth, R. (2009) The challenges of mapping complex indigenous spatiality: From abstract space to dwelling space. *Cultural Geographies* 16, 207–227.

Ruhanen, L., Scott, N., Ritchie, B. and Tkaczynski, T. (2010) Governance: A review and synthesis of the literature. *Tourism Review* 65 (4), 4–16.

Scott, N., Cooper, C. and Baggio, R. (2008) Destination networks: Four Australian cases. *Annals of Tourism Research* 35 (1), 169–188.

Secretariat of the Convention on Biological Diversity (2004) *Akwé: Kon Guidelines*, guidelines. See https://www.cbd.int/doc/publications/akwe-brochure-en.pdf (accessed 18 May 2015).

Stoker, G. (1998) Governance as theory: Five propositions. *International Social Science Journal* 50, 17–28.

Suomi, K. (2011) *Quest for Authenticity – Representation of the Sami Image in Christmas Tourism*. Master's thesis, University of Helsinki.

Tuulentie, S. (2005) The power of nature-based tourism: Tensions between different understandings of nature at the peripheral tourist resort of Kilpisjärvi. In J. Saarinen and C.M. Hall (eds) *Nature-Based Tourism Research in Finland: Local Contexts, Global Issues*. Research Papers 916 (pp. 47–62). Rovaniemi: Finnish Forest Research Institute.

Tuulentie, S. (2006) The dialectic of identities in the field of tourism: Discourses of the indigenous Sami in defining their own and the tourists' identities. *Scandinavian Journal of Hospitality and Tourism* 6 (1), 25–36.

Viken, A. and Müller, D.K. (2006) Introduction: Tourism and the Sami. *Scandinavian Journal of Hospitality and Tourism* 6 (1), 1–6.

9 Culture in Nature: Exploring the Role of 'Culture' in the Destination of Ilulissat, Greenland

Karina M. Smed

Introduction

Ilulissat is a polar tourism destination dominated by its proximity to the Ilulissat Icefjord, which is a United Nations Educational, Scientific and Cultural Organisation (UNESCO) World Heritage Site under the category of natural properties. The promotion of Ilulissat is managed by the national destination marketing organisation (DMO) of Greenland, Visit Greenland, which launched a national brand in 2010 named 'the pioneering nation'. The brand was created to stress the special relationship between nature and people as the brand identity of Greenland: 'The core story of Greenland is about the relationship between nature and people. The key concepts in the nation brand are "Powerful & Pioneering" – nature's might and man's pioneering spirit and the relationship between the two' (Visit Greenland, 2014). The significance of the Icefjord to Ilulissat as a tourist destination is quite obvious when inspecting promotional materials from local tour operators and administrative documents, and the use of 'nature' for promotional purposes is therefore also quite clear. On the other hand, the dimension pertaining to 'people' in the Greenlandic brand is less obvious and straightforward; therefore, this chapter attempts to explore the role of 'culture' in the tourism product promoted and sold in Ilulissat in light of the dynamic nature–culture relationship that Visit Greenland emphasises as a unique selling point for Greenland. Indigenous tourism, therefore, also seems to be a somewhat secondary and implicit factor in a Greenlandic context, although it is occasionally emphasised as a feature of Greenlandic identity (Greenland.com, 2015). Because nature is the number one attraction in Greenland, indigeneity is less obviously used as an attraction value, although there is an aim to focus on and promote nature and people.

The question is how a local tourism environment facilitates and uses this strategic proposition of a dynamic nature–culture relationship in practice to promote and sell its products, if at all? And also, how this manifests itself in the local tourism environment? Indigeneity therefore comes into this study as part of the exploration of the culture dimension of the nature–culture relationship in Ilulissat.

Throughout this chapter, a point of departure is taken in the local tourism industry in Ilulissat in order to explore what is being sold and promoted to tourists in this respect. Since the contention is that the Icefjord, its UNESCO World Heritage status, as well as the general context of the Arctic as nature driven, present an emphasis on nature as a distinct feature of attraction, cultural dimensions relating to the indigenous population are less evident and will therefore be explored further. The theoretical foundation rests on a central discussion of the nature–culture relationship, particularly pertaining to both polar tourism and heritage, which may reinforce certain perceptions of nature and culture. In addition, the concept of visitability will be used to explore the process of inviting tourists in, and subsequent influences on the destination and its core product. According to Dicks (2004: 8), visitability can be defined as '… the production or makeover of various kinds of spaces (physical or virtual), so that they actively call out to and invite the attention of visitors'. This will affect the destination in various ways, which has been examined throughout this study and points towards ways in which culture and nature are addressed in Ilulissat for the purpose of tourism. The basis for this study is therefore made up of a combination of a generic theoretical discussion of the nature–culture relationship and visitability as an analytical tool. The aim is then to create a direct relation between the overall perspective on nature and culture distinctions in tourism and actual tourism practices in Ilulissat. This chapter thus begins by establishing the context of the study, including the design, data and case presentation on which the study rests. The context of tourism in Greenland and Ilulissat in particular has been paid some attention due to its significance and impact on the issue at hand. Following the contextual dimensions of the study, some theoretical concepts and perspectives are introduced, and subsequently, the analysis of visitability is presented by applying it to the case material and evaluating it in relation to the theoretical framework. Lastly, the chapter is concluded with a summary of the findings.

Methods

For this study, the following triangulation of methods will be used: (1) documents such as strategy papers, government reports, administration plans and commercial websites; (2) interviews with local tourism actors, specifically tour operators and administrators; and (3) participant observations of the tourism products. The first set of data has formed the initial access to this destination's tourism environment whereby an insight into tourism in Ilulissat was established. This data will primarily be used indirectly, as background information to inform the study, but also directly when it carries analytical relevance. The second set of data consists of 10 interviews with local tourism operators and administrators in Ilulissat where the interviews also took place in June 2013. The third set of data consists of participant observations of various representations of the tourism product in Ilulissat; for example, tours provided by tour operators, the physical environment as it presents itself to tourists and local museums were observational focal points. The interviews as well as participant observations are primary data used in the analysis to understand local approaches to uses of nature and culture in the Ilulissat tourism product.

This case study is focused on Ilulissat as a tourist destination and different dimensions of the tourism environment that frame tourism in Ilulissat, as illustrated in Figure 9.1. UNESCO and Visit Greenland play the roles of external influences having internal impacts on the tourism environment in Ilulissat, which is why they will be paid some attention. As mentioned, the nature–culture relationship is discussed by applying a theoretical foundation that combines polar tourism, heritage and visitability. This study is therefore built up of these different perspectives, which are a direct extension of other texts that the author has contributed to this topic (e.g. Smed, 2014).

This study presents an insight into the nature–culture relationship in Ilulissat from a relatively commercial, business-oriented perspective, which may seem odd for a topic like this. However, this is done to emphasise a less explored view of the locality in which tourism exists as a business interest, and it is therefore the aim to explore and discuss a potential impact of tourism

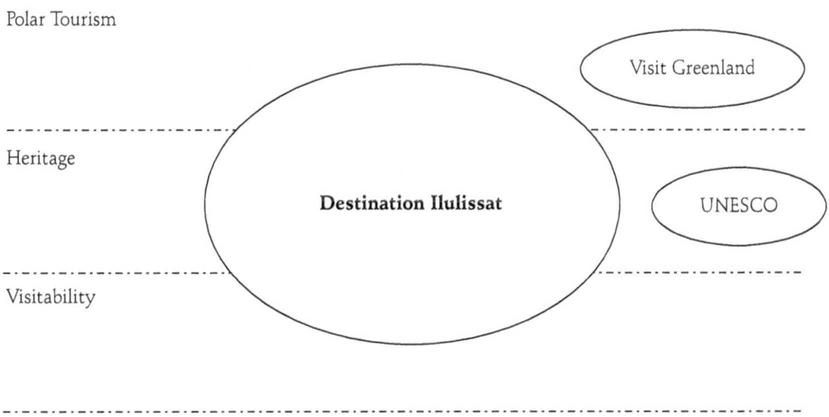

Figure 9.1 Approach to the case: Destination Ilulissat (based on Smed, 2014)

on a broad social scale. This perspective also adds to the understanding of the role of tourism primarily as a business to the community around it, and it will thus be highlighted that indigenous people in Greenland are perhaps a very small part of the local tourism environment, which also explains the somewhat secondary role of culture in this locality.

The Case of Ilulissat, Greenland

Tourism has been part of a national development strategy for Greenland since 1991 (Kaae, 2002), and it is expected to offer an alternative to declining income in the fishing industry (Government of Greenland, 2011). By 2014, the total number of tourists to Greenland had reached approximately 210,000 (Statistics Greenland, 2014), which should be viewed as relative to a total population of only 56,370 in the world's largest island of 2,166,086 km^2 (Statistics Greenland, 2013). Tourism in Greenland is still under development while also facing great changes in terms of developing infrastructure, new markets and a focus on developing the tourism product to meet the needs of new tourists (Visit Greenland, 2012). As with many other destinations in the Arctic, the dominant product that tourists come to Greenland for is nature. In a visitor survey from 2013, it was established that tourists visiting Greenland had one favourite experience, the natural phenomenon of ice (Visit Greenland, 2013). In the same survey, it was added that *'Nature and wildlife are also highly favoured experiences, more than Greenlandic culture or interactions with locals'* (Visit Greenland, 2013). This confirms the contention that nature is dominant and a primary factor in tourism in the Arctic, and culture only a secondary interest. Indigenous tourism in this context is therefore also less obvious as an attraction value or as an actual practice, because culture-related products are secondary to nature, and therefore not emphasised or developed extensively at this point. In addition, Visit Greenland speaks of the *'indigenous nation with an indigenous people'* (Greenland.com, 2015) as a feature of the Greenlandic community that has been carried into modern living. However, the question is if this feature is merely a passive historical feature that could potentially be used more actively. This contention is part of the basis for this chapter.

When it comes to the demographic profile of tourists to Greenland, the average tourist to Greenland is 55 years old, resides in Western Europe, visits only once, for five to eight nights and travels in pairs (Visit Greenland, 2013). The Visit Greenland (2013) visitor survey makes a distinction between cruise tourists and land-based tourists. Due to the geography and infrastructure of Greenland, this distinction seems logical, as there is a great difference between a cruise, where several locations are visited briefly and from the seaside, and a land-based visit, where you tend to stay longer in a relatively small community, due to the relative difficulty of moving from one place to another.[1] Findings from the Visit Greenland survey show that the majority of these land-based tourists tend to stick to mainly one but at times two regions while visiting Greenland. The number one destination for these tourists is Disko Bay where Ilulissat is situated. Moreover, repeat visitors tend to be land-based tourists, mostly Scandinavians visiting friends and family (VFF) or coming for business or educational purposes (Visit Greenland, 2013). Not surprisingly, the land-based tourists stay longer in one location which also reflects them staying in fewer places while visiting Greenland.

Visit Greenland also distinguishes between spring and summer tourists, and this is based on prevalent perceptions of extremities in climate, which is assumed to appeal to different types of tourists (Visit Greenland, 2013). However, in the context of this study, there is very little difference in their touristic behaviour in Ilulissat, and it needs to be stressed that although different regions are popular at different times of the year, Disko Bay, home to Ilulissat and the Ilulissat Icefjord, is equally popular all-year round (Visit Greenland, 2013).

Specifically in Ilulissat, tourism provides an important source of income and has done so for the last couple of decades (Qaasuitsup Municipality, 2014), and it is proposed that approximately one-third of all tourists coming to Greenland will come to Ilulissat, which would make it the most visited of all destinations in Greenland (CRT, 2013; Kaae & Råhede, 2011; Statistics Greenland, 2012). Statistics states that in 2012 around 56,000 tourists stayed in Ilulissat (Statistics Greenland, 2012), but because of the relatively high number of tourists that are not staying overnight at the hotels, as cruise tourists and VFF tourists who are not part of these statistics, the number might be even higher. Receiving a relatively large number of tourists for a town of only 4558 inhabitants (Qaasuitsup Municipality, 2014), tourism becomes quite a noticeable industry in Ilulissat, but challenges also exist mainly in relation to a relatively short peak season in July and August, short stays and lack of local involvement (CRT, 2013). Particularly the last issue could indicate a lack of connection between the tourism industry and the indigenous population caused by some sort of detachment. This may be a reason why culture is secondary in a tourism context, or maybe reversely, because there isn't a great cultural perspective, there is a lack of involvement.

Visit Greenland, which markets Ilulissat and the rest of Greenland, zooms in on the Icefjord and the spectacular natural landscape that surrounds Ilulissat, while less attention is paid to its explicit cultural attraction value (Greenland.com, 2015). This obviously also reflects the profile of the visitors as described above, although there is no indication of whether a cultural profile could be strengthened to benefit the tourism industry. The data collected for this study demonstrates the same tendencies as already identified. This is exemplified by one of the local tour operators in Ilulissat:

> This [the Icefjord] is why people come here, it gets sold. The vast majority of our tours have something to do with the Icefjord. If you sail, hike, or fly, the Icefjord is involved. Then we have a bit of a different product up north, which means a lot to us, but overall for Ilulissat it is the Icefjord that attracts, no doubt about that. (Tour operator, Ilulissat, 12 June 2013, author's translation)

This gives a first indication of what is perceived important and valuable in a tourism context, which will be explored throughout this study, and which is built on the notion of natural attraction value as a label for tourism in Ilulissat and cultural attraction value as a less evident label for tourism in Ilulissat, although it certainly carries potential value for the destination.

Nature–Culture Conceptualisation in Polar Tourism

Polar tourism, that is tourism to the Arctic and Antarctica, is increasingly becoming an academic point of attention (Hall & Saarinen, 2010), and various definitional points of polar tourism have emerged from research. One definitional point pertains to nature as the main attraction of polar tourism (Hall & Johnston, 1995; Lemelin *et al.*, 2012; Maher *et al.*, 2011; Müller *et al.*, 2013; Olwig & Lowenthal, 2006; Snyder & Stonehouse, 2007). It seems that tourists to polar regions strive to obtain unique nature experiences which confirm perceptions of human-free, untouched wilderness, which has become rather difficult to encounter in 'older' more established destinations (Saarinen, 2005). In addition, it is well established that culture is secondary to tourists to polar regions (CRT, 2013; Maher *et al.*, 2011; Müller *et al.*, 2013), and moreover, culture takes a secondary position to nature and is mostly explained in combination with nature (Johnston, 1995, 2011). Nature as it characterises and defines polar tourism is hereby stressed, while cultural dimensions entangled in polar tourism often relate to more dominant, tangible aspects of polar destinations.

Different conceptualisations of nature and culture exist, many of which put the two in opposition. This is also a tendency detected in relation to heritage, which is central to Ilulissat tourism due to the fact that the Icefjord is a World Heritage Site. Heritage is very much focused around categorisation and, according to Smith (2013), a rather distinct categorisation of heritage sites exists where architecture, monuments and historical buildings are perceived as culture and wilderness and physical landscapes as nature. Oftentimes, the contention of human-made versus nature-made defines culture and nature in this respect, which is very much a Western and largely European construct. Due to the Western dominance of the global tourism industry – by way of the West being a large generating region, and therefore shaping many tourism businesses' outlook (Smed, 2011; Weaver & Lawton, 2006) – many aspects of tourism are heavily influenced by these Western constructs. Therefore, such constructs also inadvertently influence tourists' expectations. In this way, the tourism industry itself as well as the expectations, motivations and preferences that it creates come to support such distinct nature–culture categories (Dicks, 2004). For instance, in the process of determining primary motivations within a market segment, such as nature, secondary motivations, such as culture, are very often ignored and become almost invisible (Pearce & Lee, 2005). Moreover, certain ideas are commoditised through tourism, for example 'wilderness', as Saarinen (2005) points out. He claims that wilderness has been reinvented by tourism due to its marketing efforts of an 'untouched and human-free' landscape. The heavy mediation, however, points towards a cultural construct rather than a natural one in this respect (Saarinen, 2005). Destinations can thereby appear rather one-dimensional as in the case of Ilulissat as a nature destination.

At a destination like Ilulissat where human life is necessarily adapted to the natural environment (Olwig & Lowenthal, 2006), it becomes rather complex to operate with such distinctions of nature and culture, and one might wonder if it is at all relevant or fruitful to do so. However, tourism clearly reinforces categorisations of nature and culture, thus such a distinction works against the complex intertwining of the two concepts that characterises the tourism product of Ilulissat.

It is argued that this prevailing ethnocentrism 'positions tourism in a frame of reference that expects and promotes categories such as cultural versus natural tourism products' (Smed, 2014: 288), which does not necessarily translate well from a European context to an Arctic one, such as Ilulissat, because of the significance of nature in shaping polar communities (Hansson, 2012; Olwig & Lowenthal, 2006). An objective of this study is therefore to explore ways in which this may be addressed differently in the tourism context that Ilulissat presents. It is also the intention that this will bring a new level of understanding to tourism in Ilulissat and the challenges that Arctic destinations in general are facing.

Heritage as Culture and Nature

Heritage is placed at the centre of tourism in Ilulissat due to the proximity and status of the Ilulissat Icefjord. However, the question is whether or not heritage has any practical impact on tourism practices in Ilulissat. Heritage and tourism are linked in a business relationship, since elements of conservation, protection and understanding of a given site or object, as well as catering for visitor experiences are combined (Dicks, 2004; Smith, 2003; Timothy & Boyd, 2003). Heritage may thus play different roles in a given community, to local tourism businesses and incoming tourists. The International Council on Monuments and Sites (ICOMOS) defines heritage as follows:

> a broad concept that includes the natural as well as the cultural environment. It encompasses landscapes, historic places, sites and built environments, as well as biodiversity, collections, past and continuing cultural practice, knowledge and living experiences. It records and expresses the long processes of historic development, forming the essence of diverse national, regional, indigenous and local identities and is an integral part of modern life. It is a dynamic reference point and positive instrument for growth and change. The particular heritage and collective memory of each locality or community is irreplaceable and an important foundation for development both now and in the future. (ICOMOS, 1999)

The meaning and significance of heritage is hereby underlined. This definition includes the identification of distinctively different natural and cultural environments. Considering the case of Ilulissat, this is a noticeable observation due to the close connection between nature and culture that characterises Greenlandic identity. Particularly in terms of cultural practice, which is intangible culture (Smith, 2003), it becomes difficult to distinguish between natural and cultural heritage, and therefore a broader understanding of what heritage could be is required. In light of this, it is striking that indigenous tourism is also difficult to pinpoint as part of the tourism context, because it is closely related to the concept of cultural heritage and definitely part of this intangible, anthropological culture. However, part of the problem seems to be that obvious connections are not addressed due to the fact that natural heritage takes priority, and therefore indigeneity in general is a neglected issue in the tourism environment. Moreover, because of the complexity of the nature–culture relationship that is at the core of indigenous Greenland, it has also become more of an implicit factor not addressed directly, perhaps because it does not translate as nature or culture in a tourism environment, thereby making it difficult to promote and sell.

The nature–culture relationship is already conceptualised as rather complex, which is also reflected in definitions and perceptions thereof, and the general impression is that it all depends on the context in which the nature–culture relationship is explored (Heyd, 2003; Olwig & Lowenthal, 2006; Smith, 2003). It is, however, quite evident that the interconnectedness of nature and culture has dominated recent discussions to a greater extent than the aforementioned opposition of the two concepts (Heyd, 2003). This is directly linked to the prevailing perception that everything is 'cultured' one way or another (Dicks, 2004; Saarinen, 2005). This perception relates directly to the contention that because nature is heavily mediated, for example, through the positioning of polar regions as remote wilderness, peripheral destinations and last frontiers (Müller & Jansson, 2007), a cultural relation will always be present in any encounter with nature (Smith, 2003;

Wilson, 1992), and exactly because of this, nature and culture only exist as a pair, opposing each other or not.

The impact of the UNESCO World Heritage Committee on these perceptions is somewhat implicit due to the fact that the World Heritage List has become a global phenomenon (Smith, 2003), while it has very local effects at the sites at hand, which are often quite small and unique and therefore cannot be standardised through a common list. Nonetheless, UNESCO does provide definitions of cultural and natural heritage as separate, distinctive types of sites, and moreover, as particular understandings of what cultural heritage entails:

> Cultural heritage refers to monuments, groups of buildings and sites with historical, aesthetic, archaeological, scientific, ethnological or anthropological value. Natural heritage refers to outstanding physical, biological and geological formations, habitats of threatened species of animals and plants and areas with scientific, conservation or aesthetic value. (www.whc.unesco.org)

This reflects an understanding of culture as material culture, which means that this is the only view of culture that is presented and promoted. The fact that the heritage list has been divided into the categories of 'cultural', 'natural' and 'mixed' heritage sites (or cultural landscapes as stated by Smith [2003]), suggests a separation nonetheless. These categories are thus somewhat paradoxical because they on the one hand emphasise difference, while simultaneously holding on to an overlap of categories. It is even stated that:

> The most significant feature of the 1972 World Heritage Convention is that it links together in a single document the concepts of nature conservation and the preservation of cultural properties. The Convention recognizes the way in which people interact with nature, and the fundamental need to preserve the balance between the two. (www.whc.unesco.org)

The idea behind this may very well be found in the above-mentioned lack of acknowledgement of distinctly different types of culture. It seems in this quote that there is some recognition of the overlap between the concepts but only to the extent that interaction, the anthropological, non-material type of culture, is overlapping with nature, while the first UNESCO quote indicates a contradicting proposition in which hierarchical, material culture seemed to be underlined as distinctly different to nature. Over the years, Smith (2003) claims that rethinking the different types of sites on the list has become necessary, because of new perceptions of heritage and what constitutes heritage. In tourism, non-material culture is gaining ground

(Dicks, 2004), and as such a broader understanding of culture in tourism is required in order to challenge the perception of Ilulissat primarily as a nature destination only entailing natural heritage.

The mere existence of nature–culture categories as proposed by UNESCO emphasises differences and opposition, which directly reflects prevailing categorisations in the tourism industry in general, such as ethnocentric perceptions of nature versus culture destinations and products, or the World Heritage List as a global standard. These categories will thus be used analytically as a frame of reference for tourism in Ilulissat, and the dynamic nature–culture relationship will also be analysed in order to generate an understanding of culture as it exists in Ilulissat tourism although primarily perceived as a nature destination.

Employing Visitability

The concept of visitability is introduced by Dicks (2004) and serves the purpose in this study of concretising the complex nature–culture relationship in the tourism product of Ilulissat. Dicks (2004) defines visitability as a calling out to visitors, and this relates directly to the process of inviting tourists to consume place and subsequently influence the destination and its core product. It is within this process that the nature–culture relationship is positioned by the tourism actors in one way or another as part of the tourism product and the tourism environment in which it exists. Therefore, understanding the extent to which visitability is employed in Ilulissat may shed light on the nature–culture relationship. The central contention in visitability is that a process of standardisation is bound to take place due to the market orientation that visitor environments rely on in order to be competitive places of consumption (Dicks, 2004).

Dicks (2004) further zooms in on visitors' consumption of cultural display as a representation of so-called 'actual' culture at the destination. What is 'actual' in this context is here defined by the interviewees, as the product being sold and promoted, the cultural display, is also defined by them. However, Dicks (2004) further presents as a key point in her argument that the cultural display is not entirely self-imposed, as the expectation of a certain level of standardisation by visitors puts the receiving environment under pressure to display culture as expected and as similar to other, Western, destinations. This contention resembles Boorstin's (1964) pseudo-events which he saw as a necessary outcome of tourists' encounters with culture at a destination due to the fact that staging is a given within tourism, a cultural display, as opposed to 'actual' cultural practices as they exist outside a tourism context. Obviously, this contention rests on a very particular understanding of culture, which sheds light on the fact that the cultural display that tourism is, is not simple nor unaffected by its surroundings. This means that even though the tourism actors in Ilulissat display culture – and nature – in the ways that they choose, external influences on the environment also affect what and how cultural display is in Ilulissat. Then, the question remains how this process is undertaken as well as what is left 'undisplayed' and ignored within the 'actual' culture of Ilulissat. It thus logically follows that visitability in practice, including tourist expectations as a central component, presents certain challenges to the unique attraction values of a destination.

The role of nature in relation to visitability is that nature is increasingly viewed as an extension of culture, because it is now increasingly viewed as visitable space (Dicks, 2004). In the case of Ilulissat, it is now evident that people are venturing into areas that were previously perceived to be wild and untouched by humans. This means that such areas are becoming increasingly cultured and objects of cultural display. As an example from Ilulissat can be mentioned the number of helicopter tours in the Ilulissat area, particularly in the protected areas, that has increased due to a growing tourist interest (Qaasuitsup Municipality and the Greenland Home Rule, 2009). This may have some paradoxical environmental impacts, but it may also have some impact on the tourist experience in terms of more visible signs of human presence (e.g. traffic signs and noise). This interferes with expectations of a unique, nature experience that should not be obviously shared with or disturbed by other people's presence. This is one of the rather simplistic or naïve visions of a grand nature experience where culture is left out, but nonetheless expectations that exist among modern tourists. Although perhaps rather one-sided and ignoring the fact that culture has always been affected by the nature that surrounds it, the frame of reference in which tourism exists may thus be said to comply with this narrow distinction between nature and culture.

The assessment of visitability rests on a number of indicators that Dicks (2004) presents, and the analysis of this study is built around an exploration of these indicators as they may be detected in Ilulissat, and this will be presented subsequently. The study thus seeks to understand how visitability is employed and consequently how the nature–culture relationship is practiced in tourism at a destination level in Ilulissat.

Nature and Culture in Ilulissat

According to Dicks (2004), visitability can be assessed on the basis of some indicators which are legibility, standardisation of cultural display and cultural display of nature. Legibility entails an aspect of 'talking environments' as an image of spaces being easily accessible, comprehensible and welcoming of the visitor and inviting people in by the way in which it is managed and organised for the accommodation of tourists. In this regard, the data material showed that perhaps legibility is not present to the fullest extent in Ilulissat. A lot of information relevant and useful for tourists is not obviously presented to them, or that is, only when they actively seek out information themselves, for example when asking for activities to engage in or events that take place in the area. The data indicates that there may be a challenge in the different interests of local actors in this respect, some of which are directly related to the business of providing for tourists, and some of which are not. A local tour operator articulates these conflicting interests:

> We really want these cultural experiences here in Greenland as well ... If it was possible to plan ahead and tell people, this and that weekend we expect a dog sled race, or this is the weekend we'll host the Greenlandic championships, and this is what is going to happen on national day—I expect to get the programme the day before, but that's not good enough. (Tour operator, Ilulissat, 12 June 2013, author's translation)

The level of legibility is thus affected by these conflicting interests between those concerned with tourism income, in this case this local tour operator, and those who are not, in this case the local organisations in charge of the events taking place. Consequently, legibility is perhaps present yet challenged by some of the underlying structures present in Ilulissat. This is not to say that visitability and therefore legibility is the aim for tourism in Ilulissat, but merely that this could be an indicator of where the destination of Ilulissat is headed, that is, whether or not it is moving towards greater visitability. Subsequently, the consequences of employed visitability need to be assessed.

The data also revealed that legibility is not very obvious in other ways. The observations showed that the physical environment also presents challenges to legibility for example in terms of scarce signing. Participant observation in Ilulissat clearly showed some difficulty in finding one's way around town, which at times requires a level of knowledge that seems very 'local' and implicit and not readily available to tourists. There obviously is a lack of legibility.

Another indicator of visitability is the extent to which standardisation of cultural displays is evident at the destination. Standardisation in this respect is thought of as cultural displays being overly market produced and therefore overly commoditised to expected tourist needs (Dicks, 2004). As mentioned, standardisation in Ilulissat is expected as a result of the Icefjord's status as a World Heritage Site, which entails some degree of global recognition. Standardisation is indicated by a rather negative perception that local tour operators hold of tourists' expectations of local culture:

> They visited a young, local family, she is an accountant, and he works at the airport, and in their imagination they probably imagined that they were to visit a housewife sitting on the kitchen floor with a seal,

> and a hunter coming home with a few seals, but that is not the real Greenland. (Local tour operator, 12 June 2013, author's translation)

The complexity of what these mutual expectations might entail is crucial for understanding the underlying thinking behind prioritising certain products over others. Clearly, a standardised way of thinking about tourists is also at play here. Particularly in terms of what tourists would expect Greenlandic culture to be and what stereotypes they might therefore hold. These negative perceptions of tourist expectations thus impact the product development quite significantly. There is a good chance that they have imposed some level of standardisation on the product as tourism businesses succumb to what they perceive as tourists expectations. However, whether or not these are correct is not really addressed, and as such it may very well be the tourism businesses that carry false expectations to tourists instead.

The third indicator of visitability addresses the cultural display of nature or what Dicks (2004) describes as giving nature 'a human face', which presents a way of interpreting and mediating nature for tourists. The human face of nature is where culture and nature are combined to cater for tourists. For example, a means to interpret the heritage site of the Icefjord is presented by the fact that directions are provided at the entrance to the Icefjord. It is pointed out where to walk, where to stop and where to enjoy the view, which assists tourists in understanding this activity or experience in a very particular way. This nature experience thus becomes a culturally mediated experience as well, as it is not completely value-free or open to tourists' own interpretations. Another example of this was evident at a museum exhibit, where local citizens were asked to express their views of the effects that climate change might have on local life in the area. This was exhibited as portraits and statements by these local people. Many of these statements pointed towards specific consequences such as changed behaviour or outlook on life in Ilulissat (Ilulissat Museum, climate change exhibition, June 2013), which then projects into a particular interpretation of nature from a local perspective, hence representing the human face of nature.

When it comes to the expectation of cultural experiences in the area, nature is very often drawn into the discussion of what is important or not for the tourism industry in the area. This may very well be due to the fact that nature and culture are so closely connected in this area, but it may also be a very direct expression of what is perceived the sellable attraction of Ilulissat:

> That thing up in Eqi, I would definitely call that a cultural experience. The Icefjord is not the primary issue, the settlement is [visited on the same trip]. That the Icefjord is close by is of course a factor in attracting people to exactly this place, of course. (Tour operator, Ilulissat, 12 June 2013, author's translation)

What constitutes a cultural experience is obviously a question here, and one which is not explicit in this statement. However, the fact that the Icefjord is continuously mentioned in relation to nature as well as culture experiences may be the direct result of visitability and trying to see nature in a human light. In addition, it is worth noticing that mainly non-material forms of culture are emphasised in this respect, which is perhaps not surprising. This shows that the type of culture perceived as valuable for tourist consumption needs to be nuanced to some extent in order to expose what Ilulissat has to offer in terms of culture and indigenous tourism. A statement from one of the local tour operators underlines this exact point by being rather abstract in explaining what he perceives culture to be: 'Culture is and will always be letting Greenlandic people live their daily lives, and us tagging along on the sideline' (Tour operator, Ilulissat, 13 June 2013, author's translation). This has a strong relation to the intangible form of cultural practice that could be displayed in Ilulissat while emphasising a close connection to nature.

These three indicators show that some level of visitability is present in the tourism product of Ilulissat, while at the same time there are elements indicating that perhaps a different direction is underway and could perhaps be strengthened to avoid a high level of standardisation. There is no question that the tourism product in Ilulissat is based on its natural attraction value, which is quite logical considering its uniqueness in this respect. At the same time, it is quite evident that certain perceptions of how nature is to be presented to tourists dominate the product at the destination in general. In other words, visitability is employed to a certain degree, although not opening up for indigeneity in its various forms. In addition, the nature–culture relationship that Visit Greenland proposes and which seems an obvious and unique feature of Ilulissat is less evident, perhaps due to the implicit understanding of culture as part of nature that may prevail in the local community, and which indigenous people may take for granted.

Conclusion: Potentials for Visitable Indigeneity in Ilulissat

It is hereby evident that existing tourism practices reinforce the strong emphasis on nature rather than the proposed unique nature–culture relationship that is also a feature of identity in indigenous people in Greenland. Therefore, nature tends to dominate the Ilulissat tourism product, and as a consequence, culture is less exposed to tourists. This has become difficult to change due to reinforcing powers that exist locally and globally, and which tend to constrain the development of the Ilulissat tourism product from becoming more balanced in its representation of this nature–culture relationship. It is quite clear that there are implicit indications of a more nuanced understanding of this relationship, including indigenous people's role in it, but there are also indications that these tend to be undermined by what is perceived more suitable and visitable for contemporary tourists, and therefore the dynamic nature–culture relationship as well as the core of Greenlandic indigenous communities is not used to the extent that it possibly could have been.

Some paradoxes exist in this nature–culture relationship though. Dynamics in this relationship are recognised, but as demonstrated, categorisations are also present at the same time. Obviously, nature is put forth as the most important attraction value of Ilulissat, and in doing so, culture is undermined as a value in itself. Due to the implicitness of the nature–culture relationship, alignment between internal understandings of the product and external understandings and expectations presents a challenge. It is therefore suggested that the imbalance within this relationship is addressed through a more nuanced view of not only the nature–culture relationship, but also the tourists who are to visit Ilulissat and what their expectations might be. Rethinking the core values of the Ilulissat product is thus required in order to develop and strengthen new dimensions within the products with a slightly different perspective and aim in mind. This would also entail changing direction from going towards visitability and thereby the level of standardisation that visitability presents, and towards a more unique approach. This would require determining the unique features of Ilulissat that may be used as attraction values to tourists, which could very well entail features of indigenous living, historically speaking, which already play a minor role in the tourism product, but also in relation to modern living. In the long run, this could potentially also provide a stronger basis for Ilulissat as a competitive tourist destination to stand out from the competition that may rely more directly on standardised, visitable products. From this perspective, product development provides a basis for more marketing-related initiatives, which then link up to Visit Greenland's brand proposition. In addition, this type of development could also support a stronger anchoring of the tourism product more directly in the local community due to the fact that 'culture' and 'people' are put more directly on the tourism agenda, and this would also present possibilities for strengthening Ilulissat's position as a competitive and sustainable tourist destination in years to come.

The culture–nature relationship has already been pointed out by Visit Greenland as a particular focal point for tourism to Greenland, but its use at a significant Greenlandic destination like Ilulissat is less evident in tourism practices explored throughout this study, and there could be more potential in this unique selling point than what is currently being exposed. However, some questions also remain unanswered. For example, are there any differences between Ilulissat and other destinations in Greenland that need to be considered when speaking of local culture and indigeneity? Moreover, could it be that this is not just an unexploited possibility that is not being used in Ilulissat, but that the 'nature' label is a necessity for Ilulissat with the status that it currently possesses as a strong natural heritage site? These are all questions that deserve attention in order to understand this tourism environment to the fullest extent, and hence to propose a strategic approach to tourism in the Ilulissat area.

Note

(1) Because of the icecap covering about 80% of the island of Greenland, the land is only inhabited at the coastline, and only in the best-suited places. Due to the very low population density, communities are quite far from each other and quite isolated, and it is therefore difficult, not to say impossible, to move from one to another by land for the regular tourist, and therefore air or sea routes are the most viable options for tourists.

References

Boorstin, D.J. (1964) *The Image: A Guide to Pseudo-Events in America*. New York: Atheneum.

CRT (2013) *Turisme i Ilulissat. Fakta og Potentialer* [In Danish]. *Tourism in Ilulissat. Facts and Potential*. Nexø: Centre for Regional & Tourism Research.

Dicks, B. (2004) *Culture on Display: The Production of Contemporary Visitability*. Maidenhead: McGraw-Hill Education.

Greenland.com (2015) Tourist promotion. See http://www.greenland.com/en/ (accessed 24 September 2015).

Government of Greenland (2011) *Redegørelse om regional udviklingsstrategi* [In Danish]. *Declaration on Regional Development Strategy*. Nuuk: Government of Greenland.

Hall, C.M. and Johnston, M.E. (eds) (1995) *Polar Tourism. Tourism in the Arctic and Antarctic Regions*. Chichester: Wiley.

Hall, C.M. and Saarinen, J. (eds) (2010) *Tourism and Change in Polar Regions. Climate, Environments and Experiences*. London: Routledge.

Hansson, H. (2012) An Arctic Eden: Alexander Hutchinson's Try Lapland and The Hospitable North. *Northern Review* 35, 147–165.

Heyd, T. (2003) *Encountering Nature*. Aldershot: Ashgate.

ICOMOS (1999) *International Cultural Tourism Charter: Managing Tourism at Places of Heritage Significance*. 8th Draft, XIIth General Assembly. Charenton-le-Pont: ICOMOS.

Johnston, M.E. (1995) Patterns and issues in Arctic and sub-Arctic tourism. In C.M. Hall and M.E. Johnston (eds) *Polar Tourism: Tourism in the Arctic and Antarctic Regions* (pp. 27–42). Chichester: Wiley.

Johnston, M.E. (2011) Arctic tourism: Arctic tourism introduction. In P.T. Maher, E.J. Stewart and M. Lück (eds) *Polar Tourism: Human, Environmental and Governance Dimensions* (pp. 17–32). New York: Cognizant.

Kaae, B.C. (2002) Nature and tourism in Greenland. In A.E. Watson, L. Alessa and J. Sproull (eds) *Wilderness in the Circumpolar North: Searching for Compatibility in Ecological, Traditional, and Ecotourism Values* (pp. 43–53). Proceedings RMRS-P-26. Ogden, UT: US Department of Agriculture, Forest Service, Rocky Mountain Research Station.

Kaae, B.C. and Råhede, M.K. (2011) Arctic governance dimension: Tourism strategies and environmental labeling in Greenland. In P.T. Maher, E.J. Stewart and M. Lück (eds) *Polar Tourism: Human, Environmental and Governance Dimensions* (pp. 80–101). New York: Cognizant.

Lemelin, H., Dawson, J. and Stewart, E.J. (eds) (2012) *Last Chance Tourism: Adapting Tourism Opportunities in a Changing World*. New York: Routledge.
Maher, P.T., Stewart, E.J. and Lück, M. (eds) (2011) *Polar Tourism. Human, Environmental and Governance Dimensions*. New York: Cognizant.
Müller, D.K. and Jansson, B. (eds) (2007) *Tourism in Peripheries: Perspectives from the Far North and South*. Wallingford: CABI.
Müller, D.K., Lundmark, L. and Lemelin, R.H. (eds) (2013) *New Issues in Polar Tourism: Communities, Environments, Politics*. Dordrecht: Springer.
Olwig, K.R. and Lowenthal, D. (eds) (2006) *The Nature of Cultural Heritage and the Culture of Natural Heritage. Northern Perspectives on a Contested Patrimony*. London: Routledge.
Pearce, P.L. and Lee, U. (2005) Developing the travel career approach to tourist motivation. *Journal of Travel Research* 43, 226–237.
Qaasuitsup Municipality (2014) Municipal information. See http://www.qaasuitsup.gl/da-DK/Inoqarfiit/Ilulissat (accessed 5 May 2014).
Qaasuitsup Municipality and the Greenland Home Rule (2009) *Forvaltningsplan for Verdensarvsområdet Ilulissat isfjord* [In Danish]. Administration Plan for the World Heritage Area Ilulissat Icefjord.
Saarinen, J. (2005) Tourism in the northern wildernesses: Wilderness discourses and the development of nature-based tourism in northern Finland. In C.M. Hall and S. Boyd (eds) *Nature-Based Tourism in Peripheral Areas: Development or Disaster?* (pp. 36–49). Clevedon: Channel View Publications.
Smed, K.M. (2011) Tales of tourism – Global changes and tourism discourse. *Academic Quarter* 2, 237–249.
Smed, K.M. (2014) The culture of nature: Destination visitability in Ilulissat, Greenland. *The Northern Review* 38, 269–293.
Smith, M.K. (2003) *Issues in Cultural Tourism Studies*. London: Routledge.
Snyder, J.M. and Stonehouse, B. (eds) (2007) *Prospects for Polar Tourism*. Wallingford: CABI.
Statistics Greenland (2012) *Turismen i tal* [In Danish], statistics. See http://www.stat.gl/dialog/topmain.asp?lang=da&subject=Turisme&sc=TU (accessed 5 May 2014).
Statistics Greenland (2013) *Greenland in Figures*, statistics. See http://www.stat.gl/publ/kl/GF/2013/pdf/Greenland%20in%20Figures%202013.pdf (accessed 24 September 2015).
Statistics Greenland (2014) *Turismen i tal* [In Danish], statistics. See http://www.stat.gl/dialog/main.asp?lang=da&sc=TU&version=201501 (accessed 23 September 2015).
Timothy, D.J. and Boyd, S.W. (2003) *Heritage Tourism*. Harlow: Pearson Education.
UNESCO (2014) *UNESCO World Heritage List*. See http://whc.unesco.org/en/list/1149 (accessed 5 May 2014).
Visit Greenland (2012) *Tourists in Greenland*, statistics. See http://corporate.greenland.com/en/tools/tourism-statistics/ (accessed 23 September 2015).
Visit Greenland (2013) *Spring 2013 Visitor Survey Report: Greenland*, statistics. See http://corporate.greenland.com/en/tools/tourism-statistics/ (accessed 23 September 2015).
Visit Greenland (2014) *Branding Tool kit vol. 1*, strategy document. See http://www.visitgreenland.dk/ (accessed 5 May 2014).
Weaver, D. and Lawton, L. (2006) *Tourism Management* (3rd edn). Brisbane: Wiley.
Wilson, A. (1992) *The Culture of Nature: North American Landscape from Disney to the Exxon Valdez*. Oxford: Blackwell.

Part 3

Touristification of the Arctic – Indigenous Wrapping

10 Peripheral Geographies of Creativity: The Case for Aboriginal Tourism in Canada's Yukon Territory

John S. Hull, Suzanne de la Barre and Patrick T. Maher

Introduction

The Canadian Arctic is a region experiencing rapid social, economic and environmental change as a result of globalization, the growth of Northern governments and institutions and from the impacts and challenges of climate change (Maher *et al.*, 2014; Southcott, 2013). In the context of these changes, the role that endogenous communities and economic development opportunities present in the Canadian periphery are recognized. Music, dance, visual arts, storytelling, ceremonies, rituals and folklore are all traditions of communities that enhance place-making and place-marketing through creative tourism development (Gibson, 2012a). Creative tourism is defined as travel directed toward an engaged and authentic experience, with participative learning in the arts, heritage or special character of a place, providing a connection with those who reside in a place and create its living culture (UNESCO, 2006). The growing trend of creative development is significant given the call for increased mandates for economic diversification and the sustainable benefits this can bring; these mandates are reflected in the discourse of governments at all levels, and through the recent efforts of Aboriginal communities across the Canadian Arctic. Though not without its challenges, a major theme for community, industry and governments in Canada's North has focused on Aboriginal tourism as a tool for sustainable community and economic development (Butler & Hinch, 2007; Colton, 2005; Johnston & Hall, 1995; Lemelin & Blangy, 2009; Notzke, 1999; Stewart & Draper, 2009; Williams & O'Neill, 2007).

Over the last two decades, global economic restructuring has resulted in the need to offset declines in traditional resource-based economies. As part of the knowledge economy, the creative and cultural sector has the potential to impact many areas of destination development (OECD, 2014). Gibson (2012b) argues that the 'creative' in small, remote and rural places is where the connections between the knowledge economy, social inclusion agendas and cultural policy come together to support the creative economy. These creative industries are increasingly recognized as important resources appropriate for reimagining the economic and cultural life of Aboriginal communities in the Arctic. How the creative sector engages with these opportunities, how they operate and are supported and how they engage with other diversification strategies, for instance, with tourism, are increasingly important questions. Richards (2011) proposes that the functional development of creativity for tourism must consider the '4Ps' of creativity: the creative person, the creative process, the creative product and the creative environment; the latter requiring a holistic, multisectoral approach to development. The growth of the creative economy involves local cultural and natural amenities support, employment opportunities for residents and host–guest encounters that generate benefits and that foster the development of social and cultural capital, economic diversification, community, economic and sustainable development.

This chapter aims to identify key opportunities and challenges for strengthening the representation of the Aboriginal creative tourism product in the Yukon's tourism marketing efforts. To achieve this objective, the authors conducted an inventory of Aboriginal tourism experiences shown on tourism websites in the Yukon Territory. A tourism experience co-creation model was used to analyze the 'creative product' in relation to empowerment aspects of First Nations representations as communicated and represented to tourists. Seminal research conducted in this area over the last three decades formulated the argument that, when people travel, exposure to place representations shapes people's perceptions of host cultures and also directs their gaze (Bruner, 1991; Urry, 1990).

The first section of the chapter offers a review of the literature to provide a context for understanding the Aboriginal tourism product in the Yukon Territory in the context of the creative and cultural industries. The second section outlines the methodology, where we employ a content analysis of images and text from the websites of Aboriginal and territorial organizations to gain a greater understanding of Aboriginal creative tourism products, how they are being marketed and the nature of the images that are being used. A discussion and analysis section summarizes the results of the research and identifies opportunities and challenges for strengthening the representation of the Aboriginal creative tourism product in marketing materials for the territory.

The Yukon Territory

The population of the Yukon is 37,183 people, as reported in December 2014 and Aboriginal people, more commonly referred to as 'First Nations' in the Yukon, form an estimated 20.9% of that number (or 7,650 people) (Yukon Government, 2015b). There are 14 First Nations in the territory, and 8 language groups (Yukon Government, 2014). Land claim settlements in the Yukon between 1993 and 2006 have provided recognition and authority for 11 of the 14 First Nations communities to self-govern and direct their own future development (Council of Yukon First Nations, 2015) (see Figure 10.1).

Similar to other peripheral areas in Canada, the region has been historically defined in colonial and nation-building terms as an area characterized as being an 'empty wilderness' (Coates & Morrison, 2005;

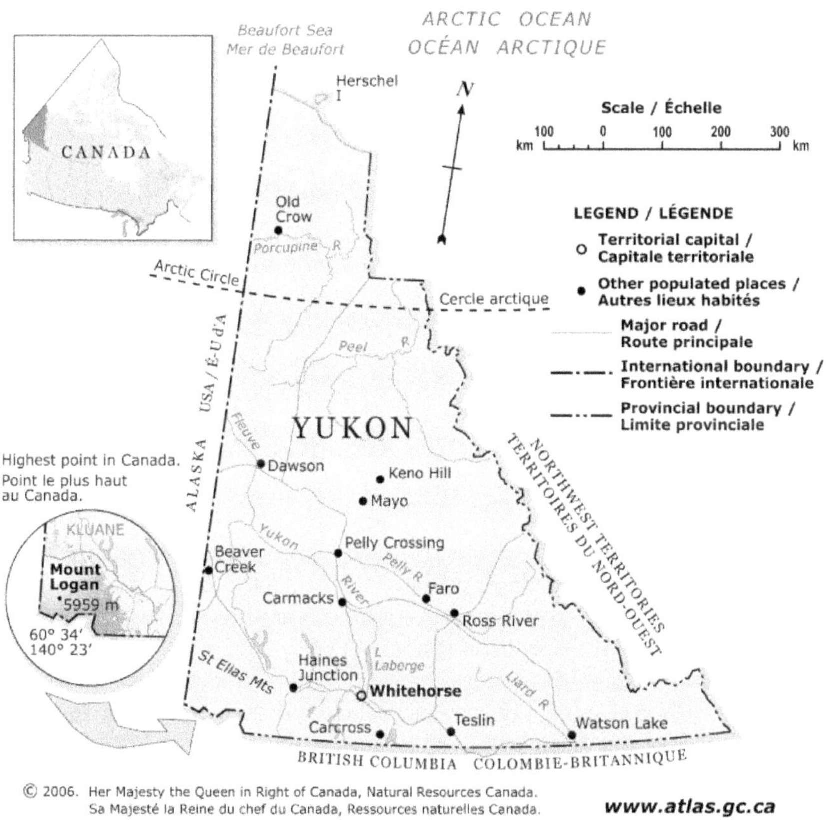

Figure 10.1 The Yukon Territory, Canada (Source: Natural Resources Canada http://open.canada.ca/en/open-government-licence-canada)

Shields, 1991) or as an area for natural resource extraction. Cooke (2004) argues that colonization of the Yukon was unique and happened in two concerted phases: the first entailed tens of thousands of gold seekers and other settlers who arrived in the late 1880s. This early settlement led to the establishment of a Canadian-owned and controlled territory. The second rush of settlers was linked to the arrival of 34,000 American soldiers in 1942–1943 with orders from Washington, DC, to build the Alaska Highway as a way to improve national security (Coates & Morrison, 1998). Cooke (2004) discusses the many reminders of the devastating legacy of the gold rush and the development of the Alaska Highway for Yukon First Nations people. These two settlement periods have had far-reaching implications for Yukon First Nations people (Coates & Morrison, 1998; Cruikshank, 1998; Cooke, 2004). Postcolonial theorists argue that colonialism still has influence on the West's interpretation and interactions with different cultures (Echtner & Prasad, 2003). Researchers point out that there is still recognition that there are economic and ideological power imbalances that legitimize particular racial and cultural differences (Caton & Santos, 2008; Hall & Tucker, 2004).

Aboriginal Tourism in Canada and the Yukon

Aboriginal tourism in Canada is recognized as a source of economic growth and independence for Aboriginal communities (Butler & Hinch, 2007; Notzke, 2006; Williams & O'Neill, 2007). The Aboriginal Tourism Association of Canada (ATAC) is a non-profit organization founded in 2014. It is a consortium of over 20 Aboriginal tourism industry organizations and government representatives from across Canada and serves as the main coordinating and advocacy group committed to growing and promoting a sustainable, culturally rich Aboriginal tourism industry in Canada. The ATAC defines Aboriginal tourism as:

> ... all tourism businesses majority-owned, operated and/or controlled by First Nations, Metis or Inuit peoples that can demonstrate a connection and responsibility to the local Aboriginal community and traditional territory where the operation resides. (ATAC, 2015b: xi)

At the national level, public policy and legal decisions since the first *National Study of the Aboriginal Tourism Industry in Canada* in 2002 (ATAC, 2015a) have resulted in the creation of new organizations and provisions aimed at improving the socio-economic situation of Aboriginal people. Of note, strategies have been operationalized that provide collective support, promotion and marketing of authentic Aboriginal cultural tourism businesses in a respecting protocol (ATAC, 2015b; Williams & O'Neill, 2007).

Provincial and territorial governments have collected data on the value of cultural sectors in their jurisdictions, which is complemented by the research executed by not-for-profit and industry sectors (see Creative Newfoundland and Labrador, 2003; Hills Strategies Research, 2010; Hume, 2009; Yukon Arts Centre Corporation, 2004). Results from the ATAC (2015b) national economic impact study point to the continuing rapid growth of Aboriginal tourism in Canada from 2002 to 2014 with the number of Aboriginal businesses increasing from 892 to 1,500; the number of direct employment opportunities increasing from 12,566 to 33,100; and the gross domestic product (GDP) contribution from Aboriginal tourism increasing from CAD $596 million to CAD $1.4 billion.

ATAC promotes the Yukon Territory as Canada's westernmost and smallest northern territory and highlights its diverse mix of history and culture, its different ecosystems and its abundant wildlife (ATAC, 2015a). Aboriginal culture and heritage is promoted through the eight tourism regions of Travel Yukon (2015a) with the territory's First Nations people contributing their rich tapestry of dialects, arts, crafts, cuisines and cultural practices.

The Yukon is the only Canadian territory with its own Aboriginal tourism organization, the Yukon First Nations Culture and Tourism Association (YFNCTA), which is a non-profit stakeholder-based organization committed to growing and promoting vibrant and sustainable arts/culture and tourism economies. The association acknowledges the symbiotic relationship between the two sectors through organized festivals and events as well as tourism itineraries and packages (ATAC, 2015a; Yukon First Nations Culture and Tourism Association, 2015). Launched in 2011, the Adäka Cultural Festival is produced by the YFNCTA and has become a cornerstone arts and cultural festival showcasing Yukon First Nations arts and culture to Yukon residents and visitors alike (Adäka Cultural Festival, 2015). The Yukon government's travel website offers visitors the potential to create their own self-guided First Nations community tour. The Yukon Culture Quest tour is an example of the territory's efforts to promote First Nations, their communities and cultural centers, and help share with visitors the creative and cultural industries by viewing artifacts and handcrafts, meeting elders and artists, touring local galleries and attending celebrations of music and art (Travel Yukon, 2015c).

In the Canadian Arctic, the Yukon Territory is the most visited tourism region. In 2014, it attracted 443,300 visitors, generated CAD $250 million in gross revenue for Yukon businesses, contributed to just over 4% of Yukon's GDP and comprised over 11% of Yukon employment. In the last few years, the Yukon has had the highest growth in tourism jobs when compared to any other province or territory in Canada (Tourism Industry Association of the Yukon, 2015).

Theoretical Background: Creative and Cultural Industries

The following literature review documents the evolution of the creative and cultural economy and traces it as an area of enquiry from an urban to a peripheral context. Scholars have noted that research on creative and cultural economies has largely focused on urban areas (Gibson, 2012a, 2012b; Petrov, 2008) as well as post-industrial cities (Margulies Breitbart, 2013). In the mid-1990s, the phrase 'cultural economy' emerged in the social sciences and the humanities and became a subject of scholarly investigation for two main reasons: first, as a result of interest in the culturalization of the economy, and second, to address the commodification and materialization of cultural consumption (Gershuny & Miles, 1983; Lash & Urry, 1994; Pratt, 2008). These new industries incorporate activities linked to hobbies, traditions, popular culture, art and new media, and are recognized as drivers of economic growth that coincided with the rise of the creative class in urban areas (Florida, 2002; Scott, 1999, 2000, 2010). Addressing less dominant or accepted versions of what creative is, Gibson (2012b: 6) considered: 'What counts as creative in small, remote, rural places – those places assumed by others to be "uncreative" because of the histories of farming or manufacturing'. He defines the creative economy as activities, processes and purposes, and examines them for how they are differently deployed in non-urban areas. This wave of research has emerged exploring creativity through ethnographic approaches that explore working-class contexts linked to debates about crafting, place identities, emotion, social displacement, precarious labor and cultures of exploitation (Barnes & Hutton, 2009; Christopherson, 2008; Gibson, 2003). Kong *et al.* (2006) argue that the creative economy comes into being as a site of knowledge and policymaking as creative ideas contribute to the economy at specific locations. This focus on 'sense of place' requires an understanding of how local, regional, urban or national economies are put together and the relationships, decisions and principles that govern them (Gibson, 2012a).

There is growing evidence that creative economies exist for different reasons in non-urban areas (Gibson, 2012b; Kneafsey, 2001). Scholars interested in peripheral, remote and rural areas have questioned perspectives that position proximity to urban areas as vital for the creative economy's development and success (Brouder, 2012; Petrov, 2007, 2008). The production of culture is increasingly linked to place, defined as a spatial aggregation of industrial capabilities and skills, in addition to a stockpile of knowledge, traditions, memories and images (Richards, 2011). 'Place' functions as a source of inspiration for artists, designers and craftworkers whose skills and talents are attuned to place-specific energies, traditions and values. The creative industries are increasingly recognized as important resources appropriate for reimagining the economic and cultural life in peripheral and rural regions suffering from economic dislocation and decline (Cloke, 2007; Fleming, 2009; Gibson, 2012b; Huggins & Clifton, 2011; Leriche & Daviet, 2010). These place-based factors highlight different kinds of relationships and approaches to tourism development for communities on the margin that are worth investigating. For instance, creativity promotes economies based on both consumption and production, encouraging a 'small is beautiful' approach to economic development where benefits are also understood in a relative sense. Moreover, Nelson *et al.* (2012) suggest that the cultural and creative economy can be used as a strategy for community transformation. In light of the above geographical positioning, there is a need to better understand how the creative sector is engaging, portraying and representing First Nations, as a significant people inhabiting peripheral space on a global worldwide scale, and more specifically in a context of historical Western colonial development in the circumpolar Arctic. In the case of Canada, these enquiries are increasingly significant given land claims, settlements and self-government agreements, which have supported Aboriginal initiatives toward increased economic and cultural self-sufficiency.

Caton and Santos (2008: 8) argue that tourism often 'contributes to a social construction of visited cultures as exotic, primitive, sensual, servile and dependent on visitors for advancement and modernization'. Commenting on the growing tourism sector, Alaska Native American scholar Bunten (2011) describes how the influx of visitors to destinations in the Arctic northwest presents challenges to creating development and positive change strategies that contest exploitation and commoditization. Colonial-influenced ideas specific to the Yukon as a destination include that it has an increasing allure for travellers offering romantic perceptions of wilderness and its Aboriginal people, and that it is located far away from 'civilization' and the maladies of modernity (de la Barre, 2012). Nonetheless, and alongside colonial legacies, the Yukon is also recognized as a cultural landscape of Aboriginal and settler populations (new and old) with diverse, dynamic and resilient communities that 'refuse to fold in the face of strong exogenous influences and challenging socio-economic circumstances' (de la Barre & Brouder, 2013: 214). In consideration of the above, Butler and Hinch (2007) are among those who point out that there is a need for greater empowerment, involvement and control in the way in which Aboriginal people are portrayed in the media, the presentation of their cultures and their resources.

Methodology

Tourism is recognized as an information-intensive industry. The increasing use of the internet has played a central role in the marketing and promotion of the sectors, businesses and destinations through the adoption of websites containing imagery and words (Garín-Muñoz & Pérez-Amaral,

2011; Sun & Lück, 2015). Tourism websites provide images of place, culture and ethnicity that can also reflect the desires and interests of the culture in which they are created; they also contribute to shaping tourists' experiences of place and people (Rakić & Chambers, 2011). The research presented in this chapter adopts 'critical multiplism', where research findings are based on multiple sources of data, making it less likely that there will be distorted interpretations (Jamal & Everett, 2007). The data gathered for this research comes from a secondary review of the literature as well as from two significant Yukon tourism websites, which were selected for comparative purposes. The first is Travel Yukon (2015b), which is the main government website for the territory, and is controlled and managed by a government agency. The second is the Yukon First Nations Culture and Tourism Association (2015), which is a website owned by the territory's First Nations non-profit stakeholder-based Aboriginal tourism organization that promotes and markets First Nations tourism experiences.

In conducting the research on these websites, the researchers applied the use of a content analysis of website images and a textual analysis similar to methods used in other tourism research (see Albers & James, 1988; Caton & Santos, 2008; Echtner & Prasad, 2003). Content analysis is defined as an observational research method used to make valid inferences from the actual and symbolic content of all forms of recorded communication in a systematic and replicable manner (Bryman, 2012; Hall & Valentin, 2005; Weber, 1990). These inferences are generated from the senders of the message, the message itself or the audience of the message (Weber, 1990). Content analysis has been recognized as a tool for examining social, policy and conceptual-oriented issues of tourism (Garrod, 2003; Jamal & Hollinshead, 2001; Padgett & Hall, 2001; Sirikaya *et al.*, 1999), and in analyses of aspects of marketing (Echtner & Prasad, 2003; Hall & Valentin, 2005; Pritchard & Morgan, 2001).

The first phase of data collection focused on a review of the secondary literature on Aboriginal tourism and the creative and cultural industries. The review aimed to understand and identify important focal themes and content categories and provide a context for website analysis. The second phase involved a content and textual analysis, which focused on investigating the actual representation of images used in marketing the Aboriginal creative tourism product in the Yukon on the two chosen websites.

The second phase analysis reflects the interpretations of the researchers who conducted the coding of the images to identify specific patterns and frequency of specific images and subjects. Content analysis categories identifying three 'focal themes' specific to the Aboriginal creative tourism product were identified: (1) who the people are in the images; (2) what the cultural or creative activities are that people are engaged in; and (3) what the built environment and landscape features present in the image are and how they appear to support the tourism experience. These themes were subsequently reorganized into four new thematic categories: (1) the presence of hosts and guests; (2) the appearance of people and their representation in terms of Aboriginal regalia and dress; (3) the activities people are engaged in; and (4) the specific aspects of the creative and cultural product represented; specifically, storytelling, visual arts, music, dance, ceremonies (e.g. cultural practices). The categories assisted with creating an inventory of co-created cultural interests; that is, interests that form part of the culture in which they are created, as well as interests that relate to how the Aboriginal creative/cultural tourism product is engaging potential customers because of their interest.

In the final phase of the research, a content analysis of the text on the websites was used to identify emergent themes that were supported by the literature on tourism and indigenous communities (Butler & Hinch, 2007). The themes from the analysis included, but were not limited to, host–guest interaction, control/ownership, education/self-determination.

Table 10.1 Research steps

Step	*Application*
(1) Familiarization with the content to be examined and identification of focal themes and content variables for research.	• Representation of Aboriginal tourism from Travel Yukon and Yukon First Nations Culture and Tourism Association websites.
	• Generated from themes identified in a literature review linked to three themes of Aboriginal tourism in the Yukon, creative/cultural industries and postcolonial theory.
(2) Determining the sample and coding variables	• Quantitative analysis (frequency analysis) of images from Travel Yukon's heritage and culture and cultural events web pages (77 images). • Quantitative analysis (frequency analysis) of images and text from Yukon First Nations Culture and Tourism Association web pages (73 images). • Human coding conducted independently during the week of 6 July 2015 by two researchers – then compared for validity and reliability.
(3) Tabulation and reporting	• Qualitative (content analysis) of website text (identification of emergent themes). Prioritization and comparative assessment of results with literature review and results of frequency analysis.

Adapted from Ahuvia (2001), Caton and Santos (2008), Echtner and Prasad (2003), Hall and Valentin (2005), Neuendorf (2002).

The results of the final analysis assisted in interpreting and summarizing the way in which the Aboriginal creative tourism product is portrayed and represented on territorial Aboriginal tourism websites, as well as how these representations challenge or support goals for Aboriginal empowerment and self-determination. The research steps in this study are adapted from previous research and are outlined in Table 10.1.

Image Analysis

The results and analysis are summarized in two sections based on the methodological approach. In the first section, the quantitative analysis of images based on the content variables is summarized. The second section delves deeper into a number of key emergent themes from the website narratives and the secondary literature review. This deeper probing allows for additional website evaluation and the identification of

Table 10.2 Themed content analysis of website images

	Travel Yukon (n=*77*)	*YFNCTA* (n=*73*)
Host/guests		
Hosts	22	24
Guests	7	3
Both	24	13
None	24	33
Appearance		
Aboriginal regalia/dress	23	27
No Aboriginal regalia/dress	52	34
Both	2	12
	Travel Yukon (n=*163*)	*YFNCTA* (n=*141*)
Activities engaged in		
Creative	21	10
Cultural	58	68
Recreational	22	9
Event	41	24
Nature based	21	30
	Travel Yukon (n=*87*)	*YFNCTA* (n=*133*)
Creative/cultural industries		
Storytelling	33	43
Visual arts	20	38
Music	17	17
Ceremonies/cultural practices	14	20
Dance	13	15

key representational issues for the Aboriginal creative tourism product. A summary of the results of the image analysis is presented in Table 10.2.

Host/guest

The summary of hosts and guests in the images revealed that for the Travel Yukon website the highest representation was with both hosts and guests present (*n*=24) or with neither present (or difficult to determine if one or the other was present) (*n*=24). The home page with the link to heritage and culture does provide an excellent example of the co-creation of a tourist experience between a group of Aboriginal dancers and a visitor. For the YFNCTA website, the highest representation was with neither present (*n*=33) or with only the host present (*n*=24). The lowest number of images was with only guests present for both websites.

Appearance

When evaluating the appearance of people on the websites, the Travel Yukon website had twice as many images of people not wearing traditional Aboriginal regalia/dress (*n*=52) compared to people who were wearing Aboriginal regalia/dress (*n*=23). On the YFNCTA website, the images were almost equally divided between those individuals with Aboriginal regalia/dress (*n*=34) and those without (*n*=27). On the YFNCTA website, approximately 16% of images with First Nations people included their depiction in both Aboriginal regalia/dress and those without in the same image (*n*=12).

Activities engaged in

In evaluating the activities engaged in on the Travel Yukon website, the majority of images were of people engaged in cultural activities (*n*=58) and events which included parades, festivals, performances and ceremonies (*n*=41). On the YFNCTA website, the majority of people were engaged in cultural (*n*=68) or nature-based activities (*n*=30). There was overlap in summarizing activities, as a festival, for example, could be categorized as both 'cultural' and as an 'event'.

Creative/cultural industries

In specifically categorizing the creative and cultural industries represented on the two websites, the Travel Yukon website and the YFNCTA website both scored highest on representing storytelling (TY *n*=33, YFNCTA *n*=43) and the visual arts (TY *n*=20, YFNCTA *n*=38). The Travel Yukon website represented music (*n*=17), ceremonies/cultural practices (*n*=14) and dance (*n*=13) and the YFNCTA represented ceremonies/cultural practices (*n*=20), music (*n*=17) and dance (*n*=15).

Textual Analysis and Discussion

The concept of creativity is understood as a practice that unites consumers and producers in the construction of space and sense of place (Aoyama, 2009; Richards, 2011). Creativity provides activity, content and atmosphere for tourism, while tourism supports creative activities (Richards, 2010). Creative tourist encounters are based on experience 'co-creation' between consumers and producers who employ local creative resources to generate distinctive community identities (Richards, 2011; Richards & Wilson, 2006). Bruner (2005) refers to cultural tourism (which includes creative tourism) as a 'borderzone' between consumers and producers. He defines the borderzone as a 'performance space' where tourists and locals meet. 'Sense of place' is recognized as a key factor in understanding people and place relationships, and in promoting intercultural communication and interaction in the Canadian Arctic generally (Grekin & Milne, 1996; Milne *et al.*, 1998; Notzke, 1999) and in the Yukon specifically (de la Barre, 2013). Sennett (2008) argues that tourists are increasingly viewed as crafters of experiences. Viewed in this light, tourists contribute as 'placemakers' adding to the vitality and livelihood of destinations, and support cross-cultural communication and creativity (Edwards *et al.*, 2008; Maitland & Newman, 2009; Richards, 2011).

Based on a further review of the images and textual analysis of the websites, three key emergent themes were prioritized for further investigation. The additional theming allowed for a deeper understanding of the representations of Aboriginal and creative and cultural images. They derive from a textual analysis of the websites, and are supported by the content analysis results. The themes are (1) creative tourism in the Yukon; (2) control and ownership; and (3) education and self-determination.

Creative tourism in the Yukon

Findings from the content analysis for the creative/cultural industries highlight that a majority of the images on the two websites depict cultural activities that are performed by artists and/or craftspeople (TY=58 and YFNCTA=68). However, these images do not incorporate the presence of visitors. Visitor presence is implied; performers and artists look to the camera as if the camera is the audience. Only 31% of the images from the Travel Yukon website and 33% of the images from the YFNCTA website include images that represent both hosts and guests actively 'co-producing' a creative experience. Both websites also had a large percentage of images with no host or guest present (TY=24 and YFNCTA=33). On both websites, the host and guest images are supplemented with accompanying text detailing the many opportunities for the co-production of a creative/cultural experience in the territory (see Table 10.3).

Table 10.3 Narrative text suggesting experience co-production

Travel Yukon	*YFNCTA*
...celebrate Yukon's diverse culture and heritage with us year-round	*...sharing unique tradition*
	...you'll always be welcome
...meet elders	
	...a culture that's kept alive by sharing
...meet talented Yukon artists	
	...we welcome you to our gatherings and festivals
	...everyone is invited to be a part of this celebration
	...to share and celebrate Han traditions and culture

In a study that examined similar phenomena two decades ago, Zukin (1995) also explored the importance of creating a cultural space that connects tourism, consumption and host/guests. These types of connections are evident on the YFNCTA website through the use of the word 'sharing', which is mentioned multiple times and is a central experiential theme. It is also used as part of an approach to promote the experience of unique traditions and to support keeping a culture alive: everyone is invited to be a part of the celebration, and visitors can participate in traditions and culture alongside those who live them. These specific word choices suggest an ability to directly link visitor, resident and cultural traditions through the act of engaging in activities together. The Travel Yukon site underlines co-production in the context of celebration; that is ceremonies and events, for instance. By suggesting ways for visitors to meet Yukon First Nations people, the YFNCTA website uses the language of engagement and is grounded in processes that focus on creative activities and the ways they are embedded in the experiences that are produced and consumed by residents and visitors alike.

Richards (2011) argues that one of the problems in developing active involvement of tourists in the everyday creative life of a destination is the extent to which their involvement supports the commodification of those activities. In the context of Aboriginal tourism, visitor involvement in everyday activities may perpetuate a colonization process, which places everyday activities in the marketplace. In light of this, understanding the control and ownership of the Aboriginal tourism experiences on offer is a significant feature that will be explored in more detail in the next section.

Control and ownership

The Canadian Arctic is recognized as a home to Aboriginal people and also as a frontier land that can provide epic adventures for explorers. The two websites explored in this study both address the issue of control and ownership of the Aboriginal creative tourism product in the Yukon using different representational approaches. Bunten (2011) argues that destinations that host cross-cultural tourism often witness 'Western' desire for 'the Other' which is facilitated through colonialism, popular imagery and tourism. In relation to Yukon tourism, de la Barre (2013) proposes that the value of the Yukon as a 'place' is defined by the special qualities recognized and attributed to the territory by residents, settlers, newcomers and First Nations. As a place, the territory is increasingly distinguished from its colonial past by recent changes that result from land claims settlements, Aboriginal self-government and the devolution of powers from the federal government to the territorial government. Though not without colonial legacies still to be resolved, these changes have resulted in increasing power, control and ownership for Aboriginal peoples in the region. In a study comparing Aboriginal tourism development in the Yukon and in China, de la Barre (2005) found that the success of tourism to foster Aboriginal empowerment depends upon: self-government initiatives that devolve rights and responsibilities to Aboriginal groups; a functional and operational level of democratic infrastructure; non-Aboriginal government support for tourism; the presence of sustainable tourism development principles; and economic development strategies that allow Aboriginal groups to be involved in promotional and marketing strategies.

How is the control and ownership of Aboriginal tourism in the Yukon represented on the two websites? The text on the Travel Yukon website states that the website is administered by the Yukon government's Department of Tourism and Culture, and that the website's objective is to promote modern and historic communities, proud First Peoples' cultures, World Heritage Sites, the Klondike gold rush, wilderness and northern hospitality. The promotional tag line, *Larger than Life*, sets the Aboriginal tourism experience within a vast romantic wilderness where the visitor can experience the Northern Lights, the midnight sun and wildlife. Like many Arctic tourism destinations, tourism is socially constructed as a destination primarily built on biophysical features and remoteness (Müller *et al.*, 2013). In this context, Aboriginal tourism is recognized as one of many 'uniquely' Yukon attractions on the website (Travel Yukon, 2015b). Of the images on the heritage and culture web pages, 42% (n=31) represent Aboriginal tourism experiences with approximately one-third of the images (n=25) depicting Aboriginal peoples in Aboriginal regalia/dress. Aboriginal businesses and cultural interpretation centers are also directly involved in the promotion of the website through the listing of operators and itineraries (Travel Yukon, 2015a). Aboriginal control is demonstrated on the Travel Yukon website mainly through government support for Aboriginal participation in the marketing of the territorial tourism product, and through the use of Aboriginal images elaborated with text.

In the case of the YFNCTA website, it features a translation of the phrase, 'We are happy to see you' in Yukon's eight Aboriginal languages. Images are linked to five main subject areas for tourists: (1) cultural and interpretation centers; (2) festivals and gatherings; (3) attractions and experiences; (4) places to see and buy art; and (5) our communities. Of the images, 92% (*n*=67) represent Aboriginal tourism experiences with 53% of persons in the image depicted in Aboriginal regalia/dress (*n*=39) and 47% not wearing Aboriginal regalia/dress (*n*=34). The sharing of traditional language and the images of individuals in Aboriginal regalia/dress demonstrate the important fact that even though Aboriginal people may share aspects of a common ethnic identity, they are not a homogenous group with differing heritages, backgrounds and life experiences (Bunten, 2011).

Aboriginal control and empowerment, in the case of the YFNCTA website, is characterized by the active participation of many different members of the Aboriginal community and their direct involvement in Aboriginal tourism. The YFNCTA has direct control over the text and images used in the marketing of First Nations people and their culture, and the tourism experiences on offer. The vision for the YFNCTA also demonstrates the importance of a development process that acknowledges 'generative' versus 'distributive' power. Generative power assumes that everyone has skills and capabilities, and through collective action, all can benefit from the common good as part of an inclusionary and collaborative process (Bramwell & Sharman, 1999; Jamal & Getz, 1995; and see de la Barre, 2005 for the application of this framework in relation to Yukon's Aboriginal tourism). It is worth noting that the Yukon government (2016) increased operational funding to the YFNCTA from 60K to 160K in March 2016.

Education and self determination

One of the important trends in the field of creativity and consumption is based on the increased desire of visitors for skilled consumption (Richards & Wilson, 2006). Studies examining the sociocultural impacts of tourism in the Arctic emphasize the importance of tourism as a tool for educating tourists about their role in Aboriginal people's lives. In addition, tourism can also be viewed as a source of building creative clusters or enclaves that foster pride in community through a process of learning, creativity and innovation (Butler & Hinch, 2007; Notzke, 1999; Richards, 2011; Scott, 2010; Smith, 1989; Stewart *et al.*, 2005; Wolfe-Keddie & Reid, 1993).

Moscardo (2010) explains that the growing importance of creative consumption is linked to the shift from a service economy to an experience economy, as well as to an increased focus on relationship marketing as part of brand development. An earlier contention by Richards and

Wilson (2006: 1215) proposed that 'creative consumers are increasingly looking for more engaging, interactive experiences which can help them in their personal development and identity creation, by increasing their creative capital'. For tourists specifically, Prentice (2004) associates the co-production of tourism experiences as part of the growing interest in experiential aspects of a lifestyle formation paradigm in tourism.

The most frequently represented creative activities for tourists on both the Travel Yukon and YFNCTA websites are storytelling (TY *n*=33, YFNCTA *n*=43) and the visual arts (TY *n*=20, YFNCTA *n*=38). Tourism scholars have proposed that stories provide tourists with interactive, participatory and engaging experiences that are personally relevant and educational (Beckendorff *et al.*, 2006; Moscardo, 2010). The key role of a story is to pass on information about a theme of value ensuring ongoing social success and survival for the individual (Moscardo, 2010). Cross-cultural studies point out that there are a number of recurring and universal themes that are addressed through stories and their telling, among them: survival, reproduction and social success as a way to communicate a teaching or 'moral of the story' that individuals in all cultures have the ability to communicate based on their own experiences (Davis & McLeod, 2003; Moscardo, 2010).

On the Travel Yukon website, stories are positioned as key elements in the brand development of the Yukon. For example, their significance is found in the themes on the heritage and culture web pages, as well as on the web page that provides tourist itineraries:

> Yukoners celebrate a varied history and a dynamic arts culture, and our vast northern landscapes figure prominently in art and stories.
>
> From festivals to galleries to dozens of museums, historic sites, and interpretive and cultural centres, Yukon's story is brought to life for visitors in so many ways.
>
> This meandering journey connects several First Nation communities and six impressive cultural centres. View artifacts and handicrafts, meet elders, and learn the story of Wolf and Crow.

The significance of stories for educating and increasing visitor awareness of Aboriginal heritage on the YFNCTA website is especially clear in relation to specific cultural events where interactive engagement with Aboriginal communities is facilitated and encouraged. Two celebrations that feature 'storytelling' are the *Hà Kus Teyea* (Teslin Tlingit Heritage Centre, 2015) and the Adäka Cultural Festival (2015). *Hà Kus Teyea* occurs in the community of Teslin, which is situated along the Alaska Highway two hours south of Whitehorse. The biennial summer gathering brings together families, elders, artists, friends and visitors to share and celebrate Tlingit culture on the shores of Teslin Lake. The Adäka Cultural Festival

(2015) has occurred annually in Whitehorse every summer since 2011. It brings together over 200 artists and performers from all 14 Yukon First Nations, and also showcases Aboriginal artists from across Canada and internationally.

The visual arts and crafts sector is recognized across the Canadian north as a key component of cultural identity, way of life and the economy (Yukon Government, 2013). Travel Yukon (2015a) states that the Yukon boasts twice the national average of artists per capita. In 2010, the Yukon Bureau of Statistics reported that there were approximately 207 businesses whose primary activity is tied to visual and applied arts in the Yukon, with another 65 businesses contributing to the sector (Yukon Government, 2013). For more than a decade, the Yukon government has declared the visual arts and crafts and the cultural sector as important parts of the territorial economy (see Yukon Arts Centre Corporation, 2004; Yukon Government, 1999, 2001, 2004, 2013). Strategies and studies aim to support the development of both contemporary visual arts and crafts, as well as First Nation traditional arts. The strategic priorities identified for the sector by the YFNCTA are to provide a united voice for marketing, training and partnerships to advance artistic and economic aspirations (Yukon First Nations Culture and Tourism Association, 2015). The strategy advocates a training and education program that fosters artistic, technical and business excellence through the provision of business development packaging and marketing programs, training in artistic techniques, support for artists to be strong trainers and workshops to prepare young artists to understand copyright and licensing requirements in a digital economy (Yukon Government, 2013).

On the Travel Yukon website, the marketing of the arts connects the way that art is nourished by the land; it describes how First Nations culture and traditions provide a rich foundation for the arts. The Yukon is marketed as a 'melting pot' for artistic exploration, experimentation and collaborative creativity; a list of festivals, galleries and workshops for visitors is also provided (Travel Yukon, 2015a). The YFNCTA website features a section titled 'Places to See and Buy Art' that is accompanied by images and descriptions on where and how to see and buy First Nations art. Both websites also provide a direct link to the annually produced self-guided tour booklet titled *Art Adventures on Yukon Time* (Yukon Government, 2015a). This government-published booklet features artists, locations to buy art and a list of organizations supporting visual arts and crafts in the territory.

The two websites address issues of education and self-determination as a critical component of the creative industries both from a consumer standpoint – visitors seeking self-development and creative experiences – and from a production standpoint – Aboriginal communities and the Yukon territorial government actively involved in the promotion, marketing and training for the arts and crafts sector.

Conclusion

The Yukon Territory is the Canadian Arctic's most visited tourism destination. This research identifies the importance of First Nations culture to the creative industries and to the culturalization of the tourism economy (Pratt, 2008). The study aimed to identify key opportunities and challenges for strengthening the representation of the Aboriginal creative tourism product in the Yukon's tourism marketing efforts. The promotional websites highlight 14 First Nations cultures. The adoption of creative, cultural and craft strategies and the land claims settlements have provided many positive opportunities for community transformation and self-determination for First Nations.

Shaped by global forces positioning visitor motivation to experience Aboriginal people, the two tourism websites for the territory analyzed for this study reflect the increasing visitor interest in the Yukon's Aboriginal creative tourism product (Travel Yukon, 2015a; Yukon First Nations Culture and Tourism Association, 2015). The link between the creative and cultural economies and tourism, although not a new phenomenon, is receiving increasing attention in the Yukon's place marketing, especially in a context of host and guest tourism experience co-production. Results from a content and textual analysis of two primary Yukon tourism websites illustrate a number of challenges and opportunities facing the promotion and economic development of the Aboriginal creative product in the territory.

Challenges

Richards and Wilson (2006) claim that creative tourism experiences result from co-creation processes between producers and consumers in time and space. In reviewing the representation of Aboriginal tourism images in this study, it was found that the majority of the images on the two sites lacked interaction between Aboriginal hosts and guests in relation to creative and cultural activities or experiences that were suggested or implied on the tourism websites. The majority of the images presented cultural activities where the host performed while being viewed by an assumed guest (e.g. no guests were featured in the image). Providing more imagery of actual host–guest interaction would assist in strengthening the visual frequency of cross-cultural interaction and co-creation on the websites.

For many Arctic destinations, tourism development is based on the natural resources of place – scenic landscapes, biophysical features and a sense of being in a remote place. Both the Travel Yukon website and the YFNCTA website used a large number of images where there is no host/guest present. The majority of the images presented landscapes, wildlife, built heritage and services. However, it has been suggested in other research that Yukon destination planners aim to position the territory so that it is recognized not only as a natural landscape, but also as a cultural one: a landscape that has a rich heritage due to its Aboriginal peoples, its frontier-connected settlers and its diverse communities of newcomers from across the globe (de la Barre, 2012, 2013). Within that changing cultural landscape, there are advantages to having more imagery that acknowledges the larger cultural landscape and the unique contribution made by Aboriginal residents and artists.

The colonial history of the Yukon has had a profoundly negative impact on its Aboriginal people. Land claim settlements and self-government agreements have provided opportunities for the Yukon First Nations people to control and direct future development, including community and economic development linked to Aboriginal tourism. However, the significant requirements of implementing self-government have also had a detrimental impact on the development of Aboriginal tourism. Tourism experience development is hindered by the limited capacity that First Nations have to attend to this sector, especially in light of other more significant responsibilities. The transfer of responsibilities from the territorial and federal governments to Yukon First Nations requires that they engage in, for instance, social, health and education mandates and complex land use decision-making, and that they be involved in the ongoing negotiation of natural resource development. All of these responsibilities are mired by present-day manifestations of colonial legacies and relationships; not least of which include that these activities occur in the context of reconciliation and healing across individual and collective (culture and society-wide) levels and paradoxically, in the context also of ongoing resistance to Aboriginal rights leading to legal disputes that required resolution in the territorial or federal courts. As a result, it is perhaps no surprise that there is less Aboriginal tourism product than there is Aboriginal presence in the territory's tourism promotion.

Opportunities

To a significant degree, Yukon First Nations have considerable control over how their culture is represented for tourism, with the territory having its own Aboriginal tourism organization. The YFNCTA provides a feedback loop to the Yukon government and is involved in determining what and how Yukon First Nations are represented on government marketing platforms. More recently, the use of social media has played a role in ensuring that First Nation perspectives are directly relayed in promotional campaigns. At present, the YFNCTA has a website that supports the development of a creative cluster for Aboriginal tourism. It includes information on Aboriginal communities, Aboriginal businesses and cultural centers, arts and culture organizations and museum and heritage organizations. There is a need for a greater sharing of Aboriginal and Yukon government tourism links that build bridges between the Travel Yukon website and the YFNCTA website to showcase the Aboriginal creative tourism product in the territory.

The Yukon government has acknowledged that there is a need for a united voice for marketing and training, as well as a need for partnerships that advance artistic and economic aspirations linked to visual arts and crafts. Aboriginal culture and traditions are acknowledged on the tourism promotional websites as providing a rich foundation for the creative arts in the territory. Government policy and programs and increased operational funding support the promotion efforts – and Aboriginal control over representation objectives – of Aboriginal creative products through cooperative marketing campaigns.

While there is value in the 'self-guided' type of Aboriginal experiences such as attending festivals and visiting cultural centers, there is a need to increase interpreted and mediated Aboriginal visitor experiences. The latter provide a direct encounter between host and guest and may also elaborate on the heterogenous aspects of Yukon's different First Nations given the more intimate and 'active' nature of a co-created visitor experience. This type of opportunity may predispose the host and visitor encounter to the ways that traditional and lived history are intertwined in the lives of Aboriginal people and their communities. In a related way, there are opportunities to examine what policy and planning support looks like for a 'small is beautiful' approach to destination development, and specifically identify the kinds of strategies that are required at policy levels.

The Travel Yukon and YFNCTA websites both provide rich imagery and strong narrative expression of the Aboriginal creative tourism product in the territory. Activities represented on the websites support the creative and cultural industries through storytelling, the visual arts, dance, music and ceremonies/cultural practices. The documentation of these activities provides motivation for visitors to experience the Yukon's unique Aboriginal cultural identity. Detailed descriptions of annual cultural events and festivals, such as the Adäka Cultural Festival, and the promotion of First Nations cultural centers representing each of the Frist Nations in their communities, underline the diverse nature of the Yukon's First Nations population, and point also to the type of cultural experiences on offer to the visitor. Aboriginal communities are increasingly politically aware, active and informed. As a result, the Aboriginal creative tourism product is strengthening the capacity for creative production, and shaping the nature of creative consumption, all the while becoming a powerful tool for educating visitors – not to mention the territory's non-Aboriginal residents. There is cautious optimism that, alongside the ongoing work of building creative clusters, the creative and tourism sectors can contribute to fostering pride in community and provide economic benefits, even if this largely occurs on a small scale and at localized levels.

The Yukon Territory holds an increasing allure for travellers. It is a destination that is marketed as 'Larger than Life' (Travel Yukon, 2015a), and is a place where Aboriginal citizens express 'We are happy to see you' to visitors (Yukon First Nations Culture and Tourism Association, 2015). The continued success and future sustainability of the Aboriginal creative tourism product will depend on Aboriginals and other stakeholders working together to protect and share the Yukon's rich Aboriginal heritage with visitors. A continued awareness of how tourism experiences are co-created is essential.

References

Adäka Cultural Festival (2015) Festival information. See http://www.adakafestival.ca/ (accessed 14 July 2015).

Ahuvia, A. (2001) Traditional, interpretive, and reception based content analyses: Improving the ability of content analysis to address issues of pragmatic and theoretical concern. *Social Indicators Research* 54 (2), 139–172.

Albers, P. and James, W. (1988) Travel photography: A methodological approach. *Annals of Tourism Research* 15, 134–158.

Aoyama, Y. (2009) Artists, tourists, and the state: Cultural tourism and the flamenco industry in Andalusia, Spain. *International Journal of Urban and Regional Research* 33 (1), 80–104.

ATAC (2015a) *Aboriginal Tourism Association of Canada (ATAC)*, information. See http://aboriginalcanada.ca/en/ (accessed 18 July 2015).

ATAC (2015b) *Economic Impact of Aboriginal Tourism in Canada*. Ottawa, Ont.: ATAC.

Barnes, T. and Hutton, T. (2009) Situating the new economy: Contingencies of regeneration and dislocation in Vancouver's inner city. *Urban Studies* 46, 1247–1269.

Beckendorff, P., Moscardo, G. and Murphy, L. (2006) Visitor perceptions of technology use in tourist attraction experiences. In G. Papageorgiou (ed.) *Cutting Edge Research in Tourism Conference Proceedings* (pp. 1–17). Guildford: University of Surrey, School of Management.

Bramwell, B. and Sharman, A. (1999) Collaboration in local tourism policymaking. *Annals of Tourism Research* 26, 392–415.

Brouder, P. (2012) Creative outposts: Tourism's place in rural innovation. *Tourism Planning and Development* 9 (4), 383–396.

Bruner, E.M. (2005) *Culture on Tour: Ethnographies of Travel*. Chicago, IL: University of Chicago Press.

Bruner, J. (1991) The narrative construction of reality. *Critical Inquiry* 18, 1–21.

Bryman, A. (2012) *Social Research Methods*. Oxford: Oxford University Press.

Bunten, A.C. (2011) More like ourselves: Indigenous capitalism through tourism. *The American Indian Quarterly* 34 (3), 285–311.

Butler, R. and Hinch, T. (eds) (2007) *Tourism and Indigenous Peoples: Issues and Implications*. Oxford: Elsevier.

Caton, K. and Santos, C.A. (2008) Closing the hermeneutic circle? Photographic encounters with the other. *Annals of Tourism Research* 35 (1), 7–26.

Christopherson, S. (2008) Beyond the self-expressive creative worker. *Theory, Culture and Society* 35, 73–95.

Cloke, P. (2007) Creativity and tourism in rural environments. In G. Richards and J. Wilson (eds) *Tourism, Creativity and Development* (pp. 37–47). London: Routledge.

Coates, K. and Morrison, W. (1998) *True North: The Yukon and Northwest Territories*. Toronto: Oxford University Press.

Coates, K. and Morrison, W. (2005) *Land of the Midnight Sun: A History of the Yukon* (Vol. 202). Montreal: McGill-Queen's Press-MQUP.

Colton, J.W. (2005) Indigenous tourism development in northern Canada: Beyond economic incentives. *The Canadian Journal of Native Studies* 1, 185–206.

Cooke, L. (2004) Narrating Northern Places: Space, Place, and Environmentalism in Whitehorse. Yukon. PhD thesis, University of Manitoba.

Council of Yukon First Nations (2015) *First Nations' Final Agreements*, agreement information. See http://cyfn.ca/agreements/ (accessed 15 July 2015).

Creative Newfoundland and Labrador (2003) *Creative Newfoundland and Labrador: A Blueprint for Development and Investment in Culture*, strategy document. See http://www.tcr.gov.nl.ca/tcr/publications/2006/culturalplan2006.pdf (accessed 15 July 2015).

Cruikshank, J. (1998) My roots grow in jackpine roots: Culture, history and narrative practice in the Yukon. In J. Cruikshank (ed.) *The Social Life of Stories: Narrative and Knowledge in the Yukon Territory* (pp. 1–25). Lincoln, NB: University of Nebraska.

Davis, H. and McLeod, S.L. (2003) Why humans value sensational news: An evolutionary perspective. *Evolution and Human Behavior* 24 (3), 208–216.

de la Barre, S. (2005) Authenticity in cultural tourism and Aboriginal empowerment in Northern Canada. In H. Castleden, R. Danby, A. Giles and J.P. Pinard (eds) *New Northern Lights: Graduate Research on Circumpolar Studies from the University of Alberta*. Canadian Circumpolar Institute (CCI) Occasional Publication #57 (pp. 48–65). Edmonton: Canadian Circumpolar Institute Press.

de la Barre, S. (2012) Travellin' around on Yukon time in Canada's north. In S. Fullagar, K. Markwell and E. Wilson (eds) *Slow Tourism: Experience and Mobilities* (pp. 157–169). Bristol: Channel View Publications.

de la Barre, S. (2013) Wilderness and cultural tour guides, place identity, and sustainable tourism in remote areas. *Journal of Sustainable Tourism* 21 (6), 825–844.

de la Barre, S. and Brouder, P. (2013) Consuming stories: Placing food in the Arctic tourism experience. *Journal of Heritage Tourism* 8 (3), 213–223.

Echtner, C.M. and Prasad, P. (2003) The context of third world tourism marketing. *Annals of Tourism Research* 30 (3), 660–682.

Edwards, D., Griffin, T. and Hayllar, B. (2008) Urban tourism research: Developing an agenda. *Annals of Tourism Research* 35 (4), 1032–1052.

Fleming, R.C. (2009) Creative economic development, sustainability and exclusion in rural areas. *Geographical Review* 99, 61–80.

Florida, R.L. (2002) *The Rise of the Creative Class: And How It's Transforming Work, Leisure, Community and Everyday Life*. New York: Basic Books.

Garín-Muñoz, T. and Pérez-Amaral, T. (2011) Internet usage for travel and tourism: The case of Spain. *Tourism Economics* 17 (5), 1071–1085.

Garrod, B. (2003) Local participation in the planning and management of ecotourism: A revised model approach. *Journal of Ecotourism* 2 (1), 33–53.

Gershuny, J. and Miles, I. (1983) *The New Service Economy: The Transformation of Employment in Industrial Societies*. London: Pinter.

Gibson, C. (2003) Cultures at work: Why 'culture' matters in research on the 'cultural' industries. *Social and Cultural Geography* 4, 201–215.

Gibson, C. (2012a) Cultural economy: Achievements, divergences, future prospects. *Geographical Research* 50 (3), 282–290.

Gibson, C. (2012b) *Creativity in Peripheral Places: Redefining the Creative Industries*. London: Routledge.

Grekin, J. and Milne, S. (1996) Toward sustainable tourism development: The case of Pond Inlet, NWT. In R. Butler and T. Hinch (eds) *Tourism and Indigenous Peoples* (pp. 76–106). Wallingford: CABI.

Hall, C.M. and Tucker, H. (eds) (2004) *Tourism and Postcolonialism*. London: Routledge.

Hall, C.M. and Valentin, A. (2005) Content analysis. In B.W. Ritchie, P. Burns and C. Palmer (eds) *Tourism Research Methods: Integrating Theory with Practice* (pp. 191–209). Wallingford: CABI.

Hills Strategies Research (2010) *Mapping Artists and Cultural Workers in Canada's Large Cities*. Kelly Hill, Hill Strategies Research Inc. Prepared for the Cities of Vancouver, Calgary, Toronto, Ottawa and Montreal (February).

Huggins, R. and Clifton, N. (2011) Competitiveness, creativity, and place-based development. *Environment and Planning A* 43, 1341–1362.

Hume, G. (2009) *Cultural Planning for Creative Communities*. Ontario: Municipal World, Canada's Municipal Magazine.

Jamal, T.B. and Getz, D. (1995) Collaboration theory and community tourism planning. *Annals of Tourism Research* 22 (1), 186–204.

Jamal, T.B. and Hollinshead, K. (2001) Tourism and the forbidden zone: The underserved power of qualitative inquiry. *Tourism Management* 22 (1), 63–82.

Jamal, T.B. and Everett, J. (2007) Resisting rationalisation in the natural and academic life-world: Critical tourism research or hermeneutic charity? In I. Ateljevic, A. Pritchard and N. Morgan (eds) *The Critical Turn in Tourism Studies* (pp. 57–76). London: Routledge.

Johnston, M.E. and Hall, C.M. (1995) Patterns and issues in Arctic and sub-Arctic tourism. In M.E. Johnston and C.M. Hall (eds) *Polar Tourism: Tourism in the Arctic and Antarctic Regions* (pp. 27–42). Wallingford: CABI.

Kneafsey, M. (2001) Rural cultural economy: Tourism and social relations. *Annals of Tourism Research* 28 (3), 762–783.

Kong, L., Gibson, C., Khoo, L.M. and Semple, A.L. (2006) Knowledges of the creative economy: Towards a relational geography of diffusion and adaptation in Asia. *Asia Pacific Viewpoint* 47 (2), 173–194.

Lash, S. and Urry, J. (1994) *Economies of Signs and Space*. London: Sage.

Lemelin, R.H. and Blangy, S. (2009) Introduction to the special issue on Aboriginal ecotourism. *Journal of Ecotourism* 8 (2), 77–81.

Leriche, F. and Daviet, S. (2010) Cultural economy: An opportunity to boost employment and regional development? *Regional Studies* 44 (7), 807–811.

Maher, P.T., Gelter, H., Hillmer-Pegram, K., Hovgaard, G., Hull, J., Jóhannesson, G.T., Karlsdóttir, A., Rantala, O. and Pashkevich, A. (2014) Arctic tourism: Realities and possibilities. *Arctic Yearbook 2014* (pp. 290–306). Akureyri: Northern Research Foundation.

Maitland, R. and Newman, P. (2009) *World Tourism Cities*. London: Routledge.

Margulies Breitbart, M. (ed.) (2013) *Creative Economies in Post-Industrial Cities: Manufacturing a (Different) Scene*. Farnham: Ashgate.

Milne, S., Grekin, J. and Woodley, S. (1998) Tourism and the construction of place in Canada's Eastern Arctic. In G. Ringer (ed.) *Destinations: Cultural Landscapes of Tourism* (pp. 101–120). London: Routledge.

Moscardo, G. (2010) The shaping of tourist experience: The importance of stories and themes. In M. Morgan, P. Lugosi and J.R.B. Ritchie (eds) *The Tourism and Leisure Experience: Consumer and Managerial Perspectives* (pp. 43–58). Bristol: Channel View Publications.

Müller, D.K., Lundmark, L. and Lemelin, R.H. (eds) (2013) *New Issues in Polar Tourism: Communities, Environments, Politics*. Dordrecht: Springer.

Nelson, R., Duxbury, N. and Murray, C. (2013) Cultural and creative economy strategies for community transformation: Four approaches. In J.R. Parkins and M. Reed (eds) *Social Transformation in Rural Canada: New Insights into Community, Cultures, and Collective Action* (pp. 368–386). Vancouver: University of British Columbia Press.
Neuendorf, K.A. (2002) *The Content Analysis Guidebook*. Thousand Oaks, CA: Sage.
Notzke, C. (1999) Indigenous tourism development in the Arctic. *Annals of Tourism Research* 26 (1), 55–76.
Notzke, C. (2006) *'The Stranger, the Native and the Land': Perspectives on Indigenous Tourism*. Concord, ON: Captus.
OECD (2014) *Tourism and the Creative Economy*. OECD Studies on Tourism. Paris: OECD.
Padgett, M. and Hall, C.M. (2001) Tourism at the polls. In C.M. Hall and G.W. Kearsley (eds) *Tourism in New Zealand: An Introduction* (pp. 102–104). Oxford: Oxford University Press.
Petrov, A. (2007) A look beyond metropolis: Exploring creative class in the Canadian periphery. *Canadian Journal of Regional Science* 30 (3), 359–386.
Petrov, A. (2008) Talent in the cold? Creative capital and the economic future of the Canadian North. *Arctic* 61 (2), 162–178.
Pratt, A.C. (2008) Creative cities: The cultural industries and the creative class. *Geografiska Annaler B* 90, 107–117.
Prentice, R. (2004) Tourist familiarity and imagery. *Annals of Tourism Research* 31 (4), 923–945.
Pritchard, A. and Morgan, N.J. (2001) Culture, identity and tourism representation: Marketing Cymru or Wales? *Tourism Management* 22 (2), 167–179.
Rakić, T. and Chambers, D. (eds) (2011) *An Introduction to Visual Research Methods in Tourism*. London: Routledge.
Richards, G. (2010) Increasing the attractiveness of places through cultural resources. *Tourism, Culture & Communication* 10 (1), 47–58.
Richards, G. (2011) Creativity and tourism: The state of the art. *Annals of Tourism Research* 38 (4), 1225–1253.
Richards, G. and Wilson, J. (2006) Developing creativity in tourist experiences: A solution to the serial reproduction of culture? *Tourism Management* 27 (6), 1209–1223.
Scott, A.J. (1999) The cultural economy: Geography and the creative field. *Media, Culture & Society* 21, 807–817.
Scott, A.J. (2000) *The Cultural Economy of Cities*. London: Sage.
Scott, A.J. (2010) Cultural economy and the creative field of the city. *Geografiska Annaler: B* 92, 115–130.
Sennett, R. (2008) *The Craftsman*. New Haven, CT: Yale University Press.
Shields, R. (1991) *Places on the Margin*. London: Routledge.
Sirikaya, E., Sasidharan, V. and Soemez, S. (1999) Redefining ecotourism: The need for a supply-side view. *Journal of Travel Research* 38 (2), 168–172.
Smith, V. (ed.) (1989) *Hosts and Guests: The Anthropology of Tourism* (2nd edn). Philadelphia, PA: University of Pennsylvania Press.
Southcott, C. (2013) Globalization and rural change in Canada's territorial north. In J.R. Parkins and M. Reed (eds) *Social Transformation in Rural Canada: New Insights into Community, Cultures, and Collective Action* (pp. 43–66). Vancouver: University of British Columbia Press.
Stewart, E.J. and Draper, D. (2009) Reporting back research findings: A case study of community-based tourism research in northern Canada. *Journal of Ecotourism* 8 (2), 128–143.
Stewart, E.J., Draper, D. and Johnston, M.E. (2005) A review of tourism research in the Polar Regions. *Arctic* 58 (4), 383–394.

Sun, X. and Lück, M. (2015) The Internet presence of whale and dolphin watch operators in New Zealand in terms of their ecotourism attributes: A content analysis. In M. Lück, J. Velvin and B. Eisenstein (eds) *The Social Side of Tourism* (pp. 141–156). Frankfurt am Main: Peter Lang.

Teslin Tlingit Heritage Centre (2015) *Hà Kus Teyea Celebration*, information. See http://www.teslintlingitheritage.com/ (accessed 12 July 2015).

Tourism Industry Association of the Yukon (2015) *Providing a unified voice for the tourism industry in the Yukon*, information. See http://tiayukon.com/ (accessed 5 July 2015).

Travel Yukon (2015a) *Travel Yukon. Larger Than Life*, tourism promotion. See http://travelyukon.com (accessed 1 July 2015).

Travel Yukon (2015b) *Uniquely Yukon*, tourism promotion. See http://travelyukon.com (accessed 12 July 2015).

Travel Yukon (2015c) *Yukon Culture Quest*, tourism promotion. See http://travelyukon.com/Plan/Itineraries/Heritage-and-Culture/Yukon-Culture-Quest (accessed 15 July 2015).

UNESCO (2006) Towards sustainable strategies for creative tourism. *Discussion Report of the Planning Meeting for 2008 International Conference on Creative Tourism*. October 25–27, Sante Fe, New Mexico.

Urry, J. (1990) The consumption of tourism. *Sociology* 24 (1), 23–35.

Weber, R.P. (1990) *Basic Content Analysis*. London: Sage.

Williams, P.W. and O'Neil, B. (2007) Building a triangulated research foundation for indigenous tourism in BC, Canada. In R. Butler and T. Hinch (eds) *Tourism and Indigenous Peoples: Issues and Implications* (pp. 40–57). Oxford: Elsevier.

Wolfe-Keddie, J. and Reid, D.G. (1993) Tourism in the eastern Arctic: Coping with 'dangerous children'. *Journal of Applied Recreation Research* 18 (2), 143–162.

Yukon Arts Centre Corporation (2004) *Cultural Spaces and Places Cultural Tourism Infrastructure in Downtown Whitehorse*. Whitehorse: Yukon Arts Centre Corporation.

Yukon First Nations Culture and Tourism Association (2015) Information. See http://www.yfnct.ca (accessed 1 July 2015).

Yukon Government (1999) *Cultural Industries Strategy*. Whitehorse: Government of Yukon.

Yukon Government (2001) *Yukon Visual Arts Craft Strategy*. Whitehorse: Government of Yukon.

Yukon Government (2004) *Yukon's Cultural Labour Force Study*. Whitehorse: Government of Yukon.

Yukon Government (2013) *Yukon Visual Arts and Craft Strategy*. Whitehorse: Government of Yukon.

Yukon Government (2014) *Yukon Annual Statistical Review*. Whitehorse: Government of Yukon.

Yukon Government (2015a) *Art Adventures on Yukon Time*, tourist promotion. See http://yukonartguide.ca/ (accessed 16 July 2015).

Yukon Government (2015b) *Yukon Bureau of Statistics*. Whitehorse: Government of Yukon.

Yukon Government (2016) *Yukon Government and Yukon First Nations Culture and Tourism Association Sign a Partnership Agreement*, news. See http://www.gov.yk.ca/news/16-097.html#.VzoNh9ebFaV (accessed April 2016).

Zukin, S. (1995) *The Cultures of Cities*. Oxford: Blackwell.

11 Sport and Folklore Festivals of the North as Sites of Indigenous Cultural Revitalization in Russia

Vladislava Vladimirova

Introduction

It is a bright winter day in March, with the sun shining in the blue sky and the temperature around –10°C. In an open area at the outskirts of the city of Nar'an-Mar, in the Nenets Region in northwest Russia, a stage has been erected. A cloth with the inscription 'Siamiankhat Mereta, VIIth Sports Competition of the Reindeer Herders of Nenets Autonomous Region' is attached to the stage. Next to it, tarpaulin shades and long tables have been set up. On some of them, women exhibit ethnic souvenirs and reindeer boots, at others, hot drinks and products from the local bakery are being sold. Of course, like every public celebration or fair in Russia these days, barbecues are producing smoke and the spicy fragrance of *shashliki* in the crisp air. As the menu promises, there are reindeer *shashliki* on sale, which together with fish soup, are trademarks of the north and are supposed to represent ethnic culture at such events. On a stage, built at the side, a folklore program is going on. Local ethnic music and dance ensembles are performing. What is original at this particular regional event is a Nenets theater group, presenting an interpretation of an old myth about the raven. All actors, singers and dancers, including the young couple announcing the different performances, wear what are locally called 'stylized' (*stilizovannye*) folk costumes. These reproduce some basic elements and decoration of what is considered ethnic clothing, using contemporary materials and sewing technology. The performance is in Russian, with some concepts or expressive phrases in Nenets. The audience is not numerous, but people seem to be excited. And devoted, since spending a day participating in the event at –10°C even in sunny weather requires serious motivation. When the folklore performance is over, the speakers play loud popular music.

But even this music carries the ethnic themes and motifs of a Soviet genre that is still appreciated and reproduced. Sport competitions take place one after another following the program, including jumping over sleds, lasso throwing and ethnic wrestling. Strict rules for these sports have been created during the years and a referee monitors all participants (Iavtysyi, 2011). Undoubtedly, most popular of all are the reindeer races, attracting the most visitors and emotions. Men and women compete separately at a few different distances to find out who are the absolute champions of the year. The competitors are mostly Nenets reindeer herders and herders' wives from different locations working at reindeer enterprises. Motivation is high because the stakes include two brand new, powerful model Yamaha snow scooters, obtained by the governor of the region as prizes for the winners. After the reindeer races, the award ceremony takes place: best-looking ethnic costumes, prettiest reindeer harness, winners of various games. Many participants are rewarded with material and symbolic awards for taking part in the event, showing their 'traditional' culture and skills. People in the audience, mostly from the city or the villages in the region, many relatives or former neighbors, share the excitement of the participants. Besides the anthropologists, only a few outsiders make their way to this event, not so much because of a lack of interest, but mostly due to Russian border zone and infrastructure hindrances.

This chapter will look into events like the one described above, which after a short period of decline in the 1990s, are back on stage and multiplying in the Russian sub-Arctic, where 40 different groups of indigenous people live. The Russian category of indigeneity includes only ethnic groups smaller than 50,000 people who live in the area of their original historical habitation and engage in traditional economy. Despite the huge variety of diverse indigenous cultures in Russia, festivals like this show many similarities in their design, performance, representations and the emotions they involve, from the European north to the Far East.

Public and scholarly representations, however, picture such festivals in ways that can be contradictory. In an article on the Festival of the North, the village of Lovozero was presented as an arena where indigenous identity in the 1990s was reasserted through sports (Huttunen, 1995). And local media and residents present these largely in terms of values established in the Soviet past. To me, such media and scholarly accounts seem completely unrelated to the event that I have observed in recent years. To this kind of representational paradox I can add another one: anachronistic as they may look to an external analyst like me, in their ideology, symbolism, organizational forms and roles in reasserting central authority, such festivals have survived the crisis of the 1990s and are now experiencing a revival. In the process, certain changes in the design and meaning of the events have taken place, most often in the discursive presentation of certain aspects in order to enhance the image of indigenous culture revival, and to emphasize features related to tourism and its development. Nevertheless, as my study shows, continuity with Soviet practices, historical memory and epistemology are strong. This chapter begins from these paradoxes and tensions in indigenous festivals in the Russian North to explore a possible framework for more thorough study.

Indigenous Festivals in the Russian North

Festivals have received considerable attention in social research, including anthropology (Duvignaud, 1976; Falassi, 1987; Frost, 2016; Getz, 2010). Their roles as locus of protest, counterculture and transgression (Cohen, 1993; Viken & Pedersen, 2009; Waterman, 1998) or as a field for the imposition of structural power and the reproduction of the status quo have been widely discussed (Handelman, 1998). The meaning of festivals for group cohesion and the reinforcement of regional identity along with economic development has been emphasized (Gibson & Connell, 2011; Grenier & Müller, 2011). The importance of indigenous festivals for ethnic groups' revival and self-determination has also been the subject of scholarly attention (Viken, 2011; Viken & Pedersen, 2009). In this light, it is somewhat surprising that despite a growing body of anthropological research in the Arctic, indigenous festivals in Russia have not become a subject of more focused research. Existing studies relate them to the renewed interest of indigenous people in their ethnic folklore and performative traditions and classify these events under the generic label of cultural revitalization (Huttunen, 1995; King, 2011) or social movements (Gray, 2005). Such interpretation is perhaps a response to the dominant tone of the literature on analogous events in indigenous societies around the world, the nearest examples being from Fennoscandia (Forsyth & Wamsley, 2006; Krist, 2005; Phipps, 2010; Whitford & Dunn, 2014; Whitford & Ruhanen, 2013). As a result of such cross-border comparison, a range of international scholarly discussions have become relevant to the Russian case, including indigenous tourism, authenticity of ethnic performances and the existential possibility for indigenous tradition in general (Butler & Hinch, 2007; Müller & Pettersson, 2005, 2006; Olsen, 2003, 2006; Picard & Robinson, 2006; Viken, 2006; Viken & Müller, 2006). Despite a few contributions, these issues remain largely unexplored for indigenous festivals in Russia, where these topics are further complicated by a history of standardized Soviet festivals and celebrations, directed and imposed by the authorities all over the huge territory of the state (Bat'ianova, 2011; Volovik *et al.*, 1986).

Soviet festivals have been extensively studied, and show how events knit together a web of symbols and practices, both public and private, that ensure the participation of the Soviet citizenry and the legitimacy of state power. Such literature can be classified under the broad banner of cultural and political management of the state through celebrations

(Binns, 1979, 1980; Gleason *et al.*, 1989; Lane, 1981, 1984; McDowell, 1974; Retish, 2003; Rolf, 2006, 2008, 2013; Stites, 1989; Von Geldern, 1993). The focus on festivals working to carry out the goals of state power dominates sociological, historical and other research on festivals and mass public events (Berezin, 2006; Beyrau & Rolf, 2006; Lavenda, 1980; Roche, 2001: 493; Rydell, 2013).

This historical context means that claims that festivals in the Russian North reflect or demonstrate indigenous cultural revival are relatively easy to contest on the grounds of authenticity (which is the most common objection, but also – historical continuity). As I discuss below, festivals like these are seen and presented in Russia in conceptual and practical frameworks that show continuity with the Soviet period, and facilitate embedding by contemporary political elites and authorities of their hegemonic goals in the design of the events. This parallel direction of analysis contradicts the cultural revival paradigm and emphasizes the ways that elites marginalize and control indigenous people by sponsoring and directing such events. Through festivals in the north, contemporary elites reuse established symbolism and practices to reinforce and legitimate their own political and economic positions. At the same time, like their Soviet predecessors, they can point out their support of indigenous festivals as a policy reaffirming principles of indigenous recognition and multiculturalism in line with international human rights principles and the scrutiny of international observers and media.

Festivals in the north often figure in media reports and regional tourism development project descriptions as events providing resources for the economic development of indigenous and local communities through tourism. This persistent discourse poses another paradox that will be addressed below, since during my long-term fieldwork I have encountered very few tourists visiting such events.

Theoretical Considerations

In this chapter, I argue that the explanation of Soviet festivals as sites for securing popular support and building up legitimacy for the elites or for state power is not sufficient for the understanding of the variety of experiences of different participants in the festivals of the north in Russia. It overemphasizes the alleged function of such events and the assumed intentions of one group of stakeholders. As noticed by other scholars, similar analytical frameworks take for granted that certain ideas and values are leading in a society where all members perceive the same symbols in a predictable way which shapes their experiences (Handelman, 1998). Such an approach, finally, can hardly provide a satisfactory explanation of the popularity of such events after the dismantling of the USSR which otherwise brought radical changes to Russia. Nor does it suffice to see them alone as sites of rebellion and indigenous culture revival that is a rupture to the Soviet politics of festivals and ethnicity.

A step toward more inclusive analysis of festivals, this time in rural Australia, is presented by Duffy and Waitt (2011) who explore the role of such events in bringing together locale's disparate groups in ways that emphasize certain commonalities, and hence cultivate a sense of belonging to a place and a society. Still departing from a perspective of festivals as normative and representational processes, these scholars examine the importance of the cultural politics of emotion: festival events bind people together through extraordinary activity and positive emotion, but may also playfully question or challenge dominant values and symbols (Duffy & Waitt, 2011: 47). Duffy and Waitt (2011: 55) point out the potential of festivals to put meaning and discourse to negotiation through different practices by involving the sphere of emotion, affect and experience: festival events create social integration and cohesion, but also serve as sites of subversion, protest or exclusion and alienation.

To analyze these multiple dimensions, I take inspiration from recent re-articulations of the so-called Manchester School that developed under Max Gluckman, and was famous for its anthropological studies of social change through extended ethnographic fieldwork of conflict situations, or later on – events that were seen as critical junctures of change, rather than as an example of social structure, norm and pattern (Gluckman, 1956; Mitchell, 1956). Max Gluckman and his followers combined influences from structural functionalist thought and Marxism to focus on processes seen as unfolding in situations and not predetermined by norms and rules, or the laws of social development and history (Evens & Handelman, 2006). The most renowned methodological contribution of the school, the extended case study, known also as situational or event analysis, established new ethnographic praxis focusing on empirical cases rather than selecting material to support theoretical arguments. These served as the foundation of theorizing and made space for analysis of different interests and incommensurable and incoherent dimensions of human agency within lived situations as well as taking into account the constant flux of social relations and causalities between actors and contexts at local and global levels (Glaeser, 2006; Handelman, 2006; Kapferer, 2006). In combination with post-structuralist thought, Bruce Kapferer adapts ideas from the philosophy of Gilles Deleuze and Félix Guattari, to argue for the event as a creative and generative nexus of novel social possibilities. He replaces a focus on controlled agency and patterned relations for lived practice and experience as unfolding and opening up for a 'multiplicity of sensory and cognitive processes which permits all kinds of agency and effects'. The event then is the 'critical site of emergence, manifesting the singularity of a particular multiplicity within tensional space and opening toward new horizons of potential' (Kapferer, 2015: 15). It can be studied and known through its 'actualizations' or 'realizations' rather than as predetermined by historical, social, psychological or cultural processes. The event is then 'a creative crucible of new, hitherto unrealized potential' rather than a mirror of the world and a result of the past (Kapferer, 2015). It is not premised on the past and is not predetermining the future, or constituting a link from a repetitive, regulated, predictable system. Rather, a radical empirical approach to the study of the event is 'to explore the novel potentiality of a becoming that is always not yet' (Kapferer, 2015: 16).

The following sections will show some of the analytical contributions that such a methodological approach to the study of the Festival of the North in different regions of Russia can make. The application of the extended case method in the study of post-socialist transformations or to the Russian field is not novel, but well-established through the founding work of the Chicago sociologist Michael Burawoy (2009). The major break that the method can achieve from the more rigidly historicist epistemology of Soviet Festival Studies is its power to make visible the contradictions and incoherencies existing in festival designs and in participants' experiences and explore the ways that these are instrumental in negotiating existing relations and meanings and in propelling the emergence of new ones. At the same time, as Kapferer (2015: 15) stresses, a vision of the event as 'creative and generative nexus', which breaks with essentialist and deterministic thinking, does not disclaim entirely the importance of more systematic or controlled processes, of past events or of psychology. Further, festivals accommodate intentions, ideas, models and symbols that are grounded in a complex of different traditions and systems of meaning, often in multiethnic urban environments. Such multiplicity cannot easily be subjected to any strict limitation and control, prediction and channeling, either in space or in time. As empirical observation shows, it provides creative spaces for multiple (re-)interpretations and re-presentations *post factum* that can be used and reused to support new projects and political agendas.

At the risk of contradicting the methodological basis of event anthropology, I have chosen in this chapter to include empirical observations and accounts from different regions of the Russian North instead of doing a detailed study of an event. My purpose is not to offer a thorough examination of the Festival of the North in one place, but rather to show that the festivals that anthropologists of indigenous Russia have described are worth more focused research, analytic attention and theorization. Here, I explore ways to study festivals 'in their own right', and use moments and details from different festivals to show how important aspects have been obscured and under-represented by existing studies and methodologies. First, I embark on a historical analysis of the early days of the Festival of the North in one part of Russia in order to provide simultaneously a background and an example of the dominant analytical framework to Soviet mass culture. Then, I demonstrate how such festivals have preserved some continuities that make them an attractive arena for politicians and economic elites seeking popular legitimacy. The challenge posed by representing festivals as instances of indigenous cultural revitalization to the idea of historical continuities is further addressed when I discuss Bruce Kapferer's elaboration of event anthropology. I use selectively first-hand anthropological data from three regions of the Russian North: Kola Peninsula, Nenets Autonomous Region and Sakhalin Island.

Soviet Mass Festivals and the Northern Olympics

In 1934, an annual sport event was initiated in Murmansk Region, known as the Polar Olympics, with ski races as the major attraction. Later, it became famous under the name The Festival of the North. This sport competition, celebrating Soviet esthetics of a healthy lifestyle and sportsmanship, gradually incorporated more indigenous identity elements, starting with reindeer races in its third instance in 1937. As the authors of the only existing history of the Northern Olympics note, long before the local authorities showed interest in this 'ethnic sport', already in 1929 the first reindeer races took place in Murmansk attracting many participants and 'exceptional interest' from the audience (Dubnickii & Khrapovickii, 1984: 20). The event, organized by the authorities, had been motivated by a few considerations: (1) to contribute to the enlargement of the relatively new city of Murmansk, expected to subsume the ancient-old traditional Sami village (*pogost*) Loparskii; (2) to evaluate the adaptability of reindeer to different environments and to test their endurance; (3) in a minor note in the end of the press publication, the author mentions the great interest which the 2nd Regional Congress of the Soviets in February 1929 had expressed toward reindeer people and herding in Murmansk Region (V.A., 1929: 47). These, I believe, point to the initial stages of Soviet reformation of reindeer herding through collectivization (Osinovskii, 1930). However, how exactly the politicians envisioned the role of the races in achieving these aims remains undocumented.

After the official inclusion of reindeer races in the Northern Olympics and the gradual working out of stricter rules for the competition, the sport aspect of the race has obviously taken some priority. However, reading carefully through the sparse literature about the Polar Olympics, it has not attained the significance of well-established sports in Russia and internationally, such as skiing or biathlon. One history that hints at the special but secondary role of the races, states that they 'have become a real ornament of the Olympics' (Dubnickii & Khrapovickii, 1984: 26). Lovozero, with its official status as an indigenous Sami village, held the first of its annual series of sport festivals in 1934.

The scholarly literature on Soviet mass festivals helps to reconstruct the changing political and social role of the Festival of the North. The use of mass festivals and public events for ideological propaganda and expanding the legitimacy of political regimes has been explored by many scholars (Lane, 1984; Rolf, 2013; Von Geldern, 1993). Catchy phrases, such as the 'the spectacular state' (Adams, 2010) and the 'rites of rulers' (Lane, 1981) have been coined to grasp the ways these events are planned to expand political power.

A starting point for most analysts is that Soviet festive culture was centralized: the initiatives were launched and controlled from above and the regional authorities had limited space for divergence and creativity. In this light, it is easy to explain the birth of the Festival of the North, and more specifically its incorporation and featuring of reindeer herding, through the central politics of reforming indigenous peoples' economy, society and culture, which took a new 'revolutionary' stage with Stalin's policy of the 'Great Break' (*Velikii perelom*) which urged mass collectivization (*sploshnoi kolektivizacii*). After some less successful experiments of applying mass collectivization principles from the agricultural south to organize northern branches of the economy as reindeer herding and fishing, relatively less radical reforms were announced toward the mid-1930s, which coincided with the establishment of reindeer races as a traditional attraction at the Polar Olympics in Murmansk as well as in Lovozero. It is plausible to connect these events to the rapidly unfolding policies of collectivization and centralization that affected indigenous lives in a myriad of ways, some of which are now described with terms like 'catastrophe'. Reindeer races were perhaps a way to preserve an image of cultural tolerance. Reindeer races and folklore performances provided place and time for expression of what the Soviet framed as indigenous culture in ways that have been carefully controlled. As Christel Lane (1981) has emphasized in her research of Soviet celebrations and rituals, they executed careful 'cultural management'; 'identification of community as a unity through the creation of "cultural memory", displaying the leaders' authority, manifesting directives for organizing public space and time, putting standards for looks and behavior …' (Rolf, 2013: 6). As the historian Malte Rolf (2013: 7) summarizes, the variety of research on celebrations shows a common concern with their role in 'making people a part of a certain scheme, integrating them into a particular social group, and/or making them part of a particular political system'. Extending this analytical trend to festivals in the north can lead to a conclusion that they designed and taught what Soviet indigenous peoples and cultures looked like.

Following a historical approach then, the beginnings of the incorporation of indigenous cultural elements into mainstream sport and festive events such as the Northern Olympics in Murmansk Region can be connected with changes in the central policies toward northern peoples. This change has been described by Yurii Slezkin as 'immediate indigenization', referring to the Soviet politics of *korenizaciia* ('indigenization'). Korenizaciia as officially incited national policy since the 1920s, argued for the need to develop in each region an ethnic language and cultural elite (intelligentsia) that would be educated and prepared ideologically to take leadership positions in regional units of government, industry and the Communist Party (Martin, 2011: 23). The politics of korenizaciia was grounded in the official Soviet statement that national 'forms' that do not oppose the centralized unitary state will be supported, such as national territory, language, elites and culture. It was an instance of general Soviet nationalities policy and at the same time – in the territory of the north, where fast-growing natural resource exploitation and industrialization attracted a huge number of incomers from other regions – it was a way to protect indigenous people from being pushed out and deprived of their natural and traditional cultural resources (Martin, 2011: 21–24; Slezkine, 1994).

The implementation of this policy took different turns in the 1920s and 1930s, with 1935 as a pronounced point of shift in the vision that the state promoted about the status of indigenous people: from 'exoticizing' to 'modernizing' discourse, to use the vocabulary of Francine Hirsch (2005: 187–188), that is from being seen as backward they were 'promoted' to ethnic equality but in need of assistance to overcome the 'powerful pull of traditional customs (*byt*)'. In practical terms, this meant stronger attention to the economy of the north, intensification of industrialization and collectivization, as well as education and the 'production' of local 'cadres' (i.e. clerks and officials of indigenous ancestry), as well as more direct involvement of locals in administering the indigenous population. According to archival documents, reindeer races, local Olympics, together with meetings of shock-workers and conferences of the wives of polar explorers, were seen as appropriate measures to achieve the desired 'consolidation' (the leading ideological concept of the period) through incorporating 'nationals' (Slezkine, 1994: 284–285).

Without direct evidence, the coincidence of this political shift and the inauguration of the Northern Olympics in Murmansk Region, and more specifically its 'national component', the reindeer races, suggest a direct link. The typical Soviet historical account of the 1930s policies toward indigenous Sami in Murmansk Region acclaims the successful development in each of the areas emphasized above. To take one of the most outstanding examples, in the work of Kiselev and Kiseleva (1987: 84–107) *The Soviet Sami*, under the chapter 'New Every-Day Life, New Culture', the successes in indigenous education, preparation of 'national cadres', creation of Sami alphabet, folklore and cultural work have been described in some detail. Such accounts, expectedly, do not provide much detail on the practices that led to the realization of such achievements. Nevertheless, in one remark about 'ideological struggle' against 'kulaks' who manipulated ordinary people from being 'enlightened' in the newly created cultural hubs in the traditional Sami villages (*pogosti*), they mention the reindeer races in

Lovozero. The races held on January 19, became the youth Komsomol replacement of Epiphany which the Orthodox Church celebrated on the same day, and the Sami – their ancient celebration of the Sun, *vstrechi solnca* (Kiselev & Kiseleva, 1987: 100).

The 'usurpation' of earlier folk festivities by Soviet ones originated in the 1920s when politicians singled out *'stary byt'* (pre-revolutionary life routines and habits) as the 'enemy' (Rolf, 2013: 41–42). Traditional celebrations as imbued with persistent religious symbols and beliefs, they realized, could be used to reverse emotional bonds from religion to the new ideology. 'Red rituals' were designed to distract the masses from old traditions by way of emotion. As Malte Rolf (2013) argues, for the Soviet authorities, 'it seemed easier to communicate new authority by imbedding it in familiar forms', while simultaneously spreading the symbols and structures of the new order. The account of the first reindeer races in Murmansk in 1929 as part of the assimilation of the old Sami village Loparskii is revealing: the event was designed to introduce the arrangement and codify the new roles that were attributed to different participants. This function is additionally constituted through the media report on the event:

> 30 reindeer teams arrived .... The whole Kil'din Loparskii pogost, ... and 4 people from the Voronezhhskii pogost. The races attracted many people and raised remarkable interest .... No doubt it is a success. Lapp people arrived as if in a parade, with reindeer in festive harnesses and people in festive clothes – whole families ... Sled after sled raced through the city streets with sharp turns ... After the races the participants were served lunch, after which the prizes were awarded to the mass applause of all Sami, and in the evening a performance was presented. (V.A., 1929: 47)

This text, although difficult to judge for its factual accuracy, accurately presents the symbolic social order enforced by Soviet policy of the time. The urban center becomes the space of common habitation, where numerous people come to present the best they have. The event and the city as emblematic for the Soviet state, are celebrating a success. Reindeer, as the article notes, 'that run fast and calm in their native land – the tundra and the forest, often become stressed around people, shy away from humans and dogs, wildly run away from the noisy crowd' (V.A., 1929: 47). The races were also organized to evaluate the qualities of reindeer in a city environment. This account maps out another set of hierarchical relations of Soviet establishment. It can be interpreted as foundational for an ontology in which reindeer, the economic bases of indigenous Sami, belong to a more ancient and uncivilized environment, the wilderness. Taken to the city, they are out of place. Nevertheless, as the festive event and the races are meant to establish, they can be supported through scientific knowledge, hard work and strong commitment, to adapt even to this space.

Reindeer can perhaps even be seen as metonyms for indigenous people. Sami from Loparskii, like the reindeer, are not only made part of the city by merging their pogost with Murmansk, but accommodated by the state apparatus into modern life, higher culture and ultimately to civilization. The gift for their acquiescence is the promise to be given what Soviet modernity promoted as a higher standard of living and superior moral order, and a share in the fruits of civilization, as symbolically represented in the narrative of the event by the guns that were awarded to the first two male winners of the races, and the *samovar*, soap and mirror given to the female winner, and finally, the theater performance ending the event.

In the light of this interpretation, and the lack of first-hand anthropological accounts, the reindeer races can from their beginning be defined as a frame for imposition and reproduction of social roles and order. It provided an arena and a set of practices for disciplinary work that later became a routine with the institutionalization of the Polar Olympics and the Festival of the North in many residential areas in Murmansk region. The inauguration of such events in the 1930 can perhaps be seen as a time marker of the establishment of more stable regional and local administrative power (Rolf, 2013).

As I show in the next section, a pronounced analogy among the symbols and practices of contemporary festivals and those of the Soviet period is what prompts the careful fieldworker to embark first on historical analysis and to pursue continuities. In the next section, I will follow further some of the Soviet legacies that facilitate contemporary political and economic elites in borrowing Soviet symbols and strategies to impose their power. This tendency contradicts the popular representation of festivals as instances of indigenous cultural revitalization in the north. I will dwell on the contradiction in order to challenge the historical analysis as an adequate framework and introduce parallel analytical trends.

Indigenous People, Oil Companies and Regional Government

In August 2009, as part of a team fieldwork, I went to Kharkuta, the site of the only reindeer herding team on Sakhalin Island. The celebration, called The Day of the Reindeer Herder, takes place there. On the program are reindeer races, with first women and then men competing. The herding traditionally practiced in the region involves riding directly on the animal's back which was a new practice for me. While the audience was very excited and supported the sportsmen, it was not numerous. Besides the group of anthropologists and a few relatives of the participants, few people had taken the mud road to drive to the location which was some 4 km from the nearest village of Val and around 40 km from the town of Nogliki.

After the races, all participants were served warm reindeer soup or fish stew plus a few local specialties, like fish meat and fat with berries (in this case, for free). A quickly set up wooden board showed pictures of indigenous handicrafts, but to buy souvenirs turned out to be impossible. Finally, when the time for awards came, the absolute winner got an electricity generator – an equipment of utmost significance for a herding camp in the wilderness. Since only members of seven families continue to practice reindeer herding, the festival is very small and the prizes more modest than those awarded in 2013 at Siamiankhat Mereta, described above. The latter were awarded the latest models of Yamaha snow scooters procured by the governor of the region. At this event, envelopes with different amounts of cash were also given to winners in other indigenous sports, such as lasso throwing, wrestling or jumping over a row of sleds. Winners were even nominated for best indigenous costume and most beautiful reindeer harness. As the report on the first reindeer races in Murmansk in 1929 already indicates, the idea is to provide as many prizes as possible. In Soviet times, prizes came from different sources, but often from the reindeer herding state-owned enterprise (*sovkhoz*). In recent decades, there had been an often mentioned 'award crisis' in Lovozero, when reindeer herders were questioning whether there was any sound reason to participate in the races.

> Just to bring a sled team to the village is a two day trip (about 40 km). Reindeer get exhausted, but mostly the risk of them being slaughtered by poachers in the vicinity of the village is great. Why take such pains? For a metal teapot … In the old days the sovkhoz was giving big prizes, it could be even a TV set … (a reindeer herder's comment from my field notes in 2005)

While in the last years teapots have been replaced with chainsaws, it is not likely that Lovozero will reach the affluence of Nenets Autonomous Region. As rumors had it, the Yamahas of 2013 were in reality bought collectively by some of the oil companies working in the region, who seek to keep the governor in power because he supports their activities. In contrast to Kola Peninsula, Nenets Autonomous Region is an area of mineral resources of strategic importance for the Russian Federation and its economic and political elite. Since extraction often requires that herding grounds be expropriated, the co-optation of local indigenous residents, while not the first priority, is still a concern. Notwithstanding the weak legal guarantees of the rights of indigenous people against industry in Russia (Novikova, 2008, 2014), companies prefer to avoid open conflicts. Instead, different strategies are worked out that help indigenous reindeer herders 'adapt' to industrial appropriation of their lands and resources (Stammler,

2011; Stammler & Peskov, 2008; Stammler & Wilson, 2006). Support of folklore, culture or sports and festivals, are among the usual public relations activities, as shown by the Yamahas, generators, chainsaws and envelopes with cash awarded during the event.

The Day of the Reindeer Herder at Kharkuta, in its turn, has been sponsored by the American Company Exxon, drilling nearby. The active reindeer herding population on the island, as of 2009, was minute, 13 people were staying in the camp mentioned above. Their herd contained deer in the range of 150, all of which were used as transport animals. The economic rationale for keeping them was their potential use as pack animals during the hunting of wild reindeer in winter when people and reindeer move to the forest zone. In addition, they were attractive because of the material aid provided on a weekly basis by the nearby oil company.

These could be pointed to as just a few instances of continuity in the ways that present economic actors and administrators build on Soviet perceptions, hierarchy, symbolic forms and practices to draw and preserve the contours of social and economic order. As in the past, the festival provides a framework for conceptualizing what legitimate indigenous culture is and what its position is in relation to the mainstream. From what has been called in Russia 'stylized national costume', to dances and songs carefully rehearsed in the group at the local House of Culture, to indigenous sports and the exhibition of material culture (as a rule organized around the idea of the timeless 'ethnographic present'), all elements have changed little, as local residents underline. The same is true for the roles and social hierarchies set forth at the festival. The division between reindeer herders – a marginal social group – and the audience to the races can be metonymically extended from the orange plastic fence put around the reindeer teams and their owners (to stop inquisitive visitors from bothering the reindeer). In the same vain, indigenous people take the role of entertaining the public through 'exotic' culture, performances, food and handicraft. This shows parallels with the first reindeer races in Murmansk where members of Sami villages came in their festive clothes and prettiest reindeer harnesses to show themselves and their exotic 'sports' to the urban audience.

Such analysis, while legitimate in its search and striving, does not fully capture important dimensions of the experience of the event. It seems to remain blind to the positive emotions that the event produces in the participants, who now as well as in the past describe such festivals as regional landmarks. The Festival of the North in Lovozero, for example, is mostly mentioned with nostalgia when people remember its past scale, as seen in the remark of the reindeer herder quoted above. A historical analysis that focuses on power is not able to account fully for such emotions and is also ill-equipped to recognize recent festivals as sites of contestation and renegotiation of meaning, as the next section shows festivals of the north to be.

Celebrations of Indigenous Culture and Identity

Siamiankhat Mereta VII, as the number implies, is a relatively new event. It has been designed and inaugurated by initiative of the Unit of Indigenous People at the Regional Administration, together with the biggest Nenets organization 'Iasavei'. In the Kola Peninsula, two new festivals are held in addition to the traditional Festival of the North: Sami games in June and Bear games in the autumn. Although I have never attended the Bear games, the Sami games draw heavily upon their Soviet ancestors. This formal resemblance challenges the interpretation that new celebrations were designed as alternatives that resist Soviet festivals, or are in better harmony with indigenous culture and its revitalization. From the organization of the event by some of the same people, to performances and participants, the overlaps are striking. At the same time, both the new and old festivals are in various contexts identified as sites of indigenous culture and its revival, and as signs of indigenous cultural recognition and identity formation.

An analysis informed by Kapferer's conceptualization of events as 'sites of emergence' is perhaps better attuned to capture the multiplicity of meanings and practices that unfold during festivals and provides space for different innovative relations that may be actualized in the future. The play about the raven, presented at Siamiankhat Mereta, can provide a good example. As a staged performance of a myth, recorded by ethnographers and prepared in the new Nenets House of Culture in the city of Nar'an-Mar, it follows established Soviet esthetic requirements for the genre (stylized costumes, among others). Even though the Nenets theater group is newly established, its repertoire and style, as the performers stress, follow a well-established canon of standards for the presentation of folklore dance, theater and music which serve as guarantees of quality, even for non-professional performers, as in this case. The long-term experience of the group director in folklore performances is pointed to evidence of the high standards of their work. This Soviet continuity, however, is not perceived in opposition to the performers striving for authentic cultural experience. In an interview recorded before the festival, the elderly ladies performing in the group related to me at length and with genuine excitement about the great opportunity they have when participating in the newly founded first traditional Nenets theater to reintegrate their present self with their youth, living in small tundra communities with folklore still in circulation, and living closer to the natural world which such myths address and explain. The extent to which such feelings have been transmitted to the audience during this particular performance is hard to say by their diverse responses.

As I have noticed during my work with other indigenous people involved with folklore, their perceptions of and relations to what they perform can be indecisive and changing. They can claim or disclaim authenticity in quick succession, expressing pride in their folklore tradition or indicating that they do it as a ludic play without attributing to it much spiritual importance. The last stance I have encountered more often among younger people, who also tend to be more open about it. In certain cases, irony about the performance itself can be indicated by performers, by overdoing some elements of the performance, or in commentary after it.

In the case of the raven play, I felt this was a solemn and respectful performance, with few elements of a ludic play. As such, it was accepted well but calmly by the public. Musical and dance performances I have seen engage audiences in more direct way, especially during the big folklore concert that closes the Festival of the North in Lovozero. This concert I have seen many times produce a powerful effect on the audience that is later described and re-experienced through memory in diverse but strong ways. For some people, it provides a sense of indigenous revival and public recognition, while for others it perpetuates the celebration of Soviet multiculturalism – a culture that is 'socialist in content and ethnic in form', as famously put by Stalin. For external observers, the experience may be mediated by a local person, who strongly influences it one way or another. In one case, a foreign activist supporting indigenous revitalization in Lovozero interpreted a concert as the performance of a genuine folklore tradition, awakening after Soviet repression. What my long interview with the Nenets theater group vividly showed is the playful manner in which their engagement with what is locally defined as traditional culture is communicated and enacted when not on stage. Such phenomena have been captured in the phrase of another anthropologist working in Siberia, Alex King (2011), as 'playing with culture'.

One further aspect of the festival as a site of indigenous cultural revival is that such an idea easily builds on Soviet representations. Such festivals have been organized throughout the country not only to frame what indigenous culture and community is, but also to show Soviet concern about indigenous people and that they are provided with space to 'freely' exercise and develop it. In this sense, epistemologically and practically, such events are conceived by many, including indigenous people themselves, as loci of ethnic culture. Such a legacy, in some cases, helps the present authorities build the image of their support for indigenous people and cultures by funding festivals (Konstantinov, 2015; Vladimirova, 2011, 2014). The same Soviet legacy, however, helps to create attractive images of traditional culture that procure international support, which sometimes confront Russian authorities in order to strengthen the indigenous pursuit of rights.

A further contradiction to be mentioned is that indigenous groups despite often being presented as homogenous, to a large extent reproduce the basic social divisions of Russian society (Vladimirova, 2006). Thus, indigenous elites present another possible focal point for the study of festivals, because their members have been involved in the organization of

Soviet festivals and especially the design of new festivals. As products of Soviet power, one might argue, extending the type of analysis employed by historians, indigenous elites transfer to present-day festivals the whole kit of symbols and meanings of state cultural and political management. Thus, such groups within the indigenous community simultaneously embody the Soviet imposition of indigenous culture as 'socialist in content and ethnic in form', and its post-Soviet liberation and revitalization.

For political purposes, however, as well as financial support, festivals can be described and presented as the nexus of indigenous cultural revitalization. Performance under the badge of indigenous culture can become empowering if projected into the future, because of the open potentiality it has. This potentiality which might remain unrealized, can find expression in political representations, especially if they address a foreign public (international non-governmental organizations [NGOs] promoting indigenous rights, or indigenous people, like Fennoscandic Sami), who are less influenced by pre-existing Soviet contexts. To such visitors, indigenous festivals can present an experience of different dimensions, relations and associations. This openness has been exploited in the last decade by different political actors, who persistently connect festivals of the north to tourism. This relation will be explored in the next section.

Festivals of the North as Tourist Sites

Facing an ongoing crisis in the regional economy since the early 1990s, many northern areas are forced to look for alternative economic sources and innovative solutions. One such 'solution' actively advertised during the last 10 years has been tourism. With the closing of big industrial enterprises in the Kola Peninsula and the decrease of reindeer in Lovozero, the idea of tourism, and more specifically ethnotourism, as an alternative source of income has been entertained by the local authorities. Many hindrances prevent the efficient development of such economy: lack of regular and comfortable transport, lack of accommodation and restaurants and, finally, the limited variety of attractions (Vladimirova, 2011). A relatively small number of people make it to Lovozero. Nevertheless, the Festival of the North has often been pointed out as an exotic attraction that can lure tourists. The few foreigners (mostly from the Nordic countries) I have met at the festival have been pointed out with pride by local people. Most of them, however, like me, are scholars doing research, or have a personal relation with some of the participants and have been invited and hosted in Lovozero. In the case of Nordic Sami, when they come it is to show their moral and political support for Russian Sami. But even they are not numerous.

The only occasion I met a group of eight visitors from St Petersburg was at the Festival of the North in the village of Krasnoshchel'e in March

2010. They came on a one-day trip and did not stay for evening concert and prize-giving 'ceremony'. Some of them when asked if they would come back answered that they would if there was a hotel and the prices were reasonable. As it is now, lack of infrastructure is used by local entrepreneurs to demand unrealistically high prices for transportation in the tundra and for accommodation, which few people can afford or like to pay. Besides, visitors noted that they were not even able to buy souvenirs because the few items put on a table for sale were quickly sold and there was no further supply. This fact indicates local peoples' skepticism about the future of ethnotourism.

Because of all these problems it is hard to expect increasing popularity of festivals among tourists in the near future. Their persistent inclusion on the list of tourist sites can perhaps be interpreted in different way. First, it presents evidence for the work that managers and administrators are supposedly doing in order to develop tourism as an alternative economy. Second, it is said that such attention increases the popularity and value of the events. The few indigenous people who have embraced the idea of tourism or try to develop economic activity in relation to it, may benefit from the reproduction of the discourse of festivals as ethnic tourism sites. My experience, however, indicates that tourism is at best only a supplementary activity in their economy that relies most heavily on state support for indigenous organizations and, to a lesser extent at present, on funds generated by international projects.

Closing Thoughts

This chapter began with a historical exploration of the ways that Soviet festivals constituted spaces for cultural and political management, and for disciplining and educating the masses in Soviet epistemology and modern practices. A number of cultural forms and symbols have been created and established in the process. Even after the changes in the 1990s, some of these forms have persisted and have been reproduced, accommodating new meanings and ambitions. Festivals of the North got the support of new economic and political elites, who saw an opportunity to legitimize and popularize their power. At the same time, indigenous elites use and represent them as cultural revival loci and sites for ethnic tourism.

While a historical approach to festivals as institutionalized events that represent an existing social order provides tools for exploring continuities and social change, it is mostly preoccupied with a centralized state power and its capacity for control over agency and meaning (Handelman, 1998; Rolf, 2013). Such analysis misses some of the phenomenological aspects of festival activities, related to emotion and sensual experience created in the interaction between people, animals and material objects. As scholars of festivals and particularly performance studies have observed, such experiences are key attractions of festival activities (Noyes, 2011; Ryan & Wollan, 2013). Meaning, symbols and affects that are born and circulate during festivals emerge into realms of negotiation and generation outside the intentions of any group of people. Meanings and symbols become re-worked and re-contextualized under emotional upheaval in ways that are not under the control or even grasped by the understanding of particular groups or individuals.

As field research indicates, it is difficult to delineate a unitary agency behind the organization of festivals. The attitude of organizers and performers to the festival can take different expressive and meaning trajectories, and uncertainty is communicated to the audience through playfulness, or irony, as in the case of the Nenets theater group. Indigenous politicians and members of intelligentsia who I have worked with show a wide spectrum of emotionally expressed, gestured, narrated 'reflections' on their involvement with the organization of indigenous festivals, ethnic performances or politics in general, varying from pride to self-irony and disavowal.

Scholarship on festivals has attempted to develop analytical lens and theorize the affective, ludic and carnival aspects as essential to emergence, novelty and social change (Bakhtin, 1984; Chau, 2008; Duvignaud, 1976; Noyes, 2011; Turner, 1982). Event analysis and Kapferer's theorizing of the event is just one framework that connects such aspects into the wider context of social life. Contemporary theorists of event analysis put different emphasis – on its consequent processualism (Glaeser, 2006), the match between micro and macro history (Handelman, 2006), the creativity of human agency (Evens, 2006) or the opening of 'numerous pathways into various potential futures' (Kapferer, 2015: 16). I believe that its post-structuralist re-articulation can serve as guidance to a more thorough study of the dynamic of indigenous festivals in the Russian North, that often serve multiple and contradictory agendas. The idea of the event as something emerging opens up space for analysis of the processes of explicit and implicit negotiations that festivals create: negotiations of meaning, of identities, of moral values and finally of political agendas. Such negotiations never come to a certain resolute and stable configuration of positions: they rather float through different editions of festivals, constantly offering new opportunities for reconfiguration that resist stability in the contemporary Russian situation. While authoritative discourses might seek to reimpose frames, for example epistemological limits to what tradition and indigeneity can be, and thus strive for hegemonic power over knowledge and cultural management, other participants (indigenous politicians, or performers) can oppose alternative visions in a number of ways such as parody of dominant cultural forms through performance or play. Festivals can even become platforms for open protest, such as the sabotage of the reindeer races by herders in Lovozero, who after they felt that they were unjustly treated by a state agency, the Game Ward, in a notorious conflict over wild and domestic reindeer definition in 2007, refused to participate (Vladimirova, 2011).

The festival event as a 'critical site of emergence' (Kapferer, 2015: 15) provides endless possibilities for negotiation through different expressive and communicative means. The orange plastic fence dividing 'marginal' herders and the deer from the audience which I mentioned above, has been re-conceptualized, especially in the context of the protest from 2007. At the first instance that I pointed out, it was metaphorically extended to bracket as exotic and marginal the space occupied by herders and reindeer in the village. It symbolized their exclusion and constituted their non-belonging to the urban space, reserved for culture. During the act of sabotage, reindeer herders achieved a reversal of the fence: they made the audience experience that without reindeer races, the festival event becomes non-festival. The orange fence then can be imagined to symbolize the border that the group of reindeer herders erects from the outside world, to sacralize its role in the festival and for the local community more generally. Or as a retired reindeer herder once said to me in an interview: 'without deer there will be no people, nothing'.

Events like the herders' protest show how festivals present new horizons of potentiality that need to be studied in their own right rather than limited to their historical, social or psychological lines of causality. New and unpredictable relations become actualized in experience, as practice, emotion and meaning unfold. This is the 'creative and generative nexus' that Kapferer (2015: 15) stresses as anthropological 'site'. The relations that are created and realized constitute the festival ecology that Nicola Frost (2016: 571) proclaims anthropology should study by examining the many different aspects and possibilities of festivals.

Finally, as much as research on the contradictions that become negotiated and eventually resolved during festivals is necessary, a focus on the extended temporal and regional scale of festivals is needed. The Festival of the North provides an arena for interaction and communication of symbols and meanings, which are contextual and subjected to different interpretations, as the sections above indicate. Festivals of the North then appear as instances of social interaction and practice that are part of a larger context, which must be taken into account in the analysis. Despite their limitless generative potential and openness to agency, festivals constitute situations that are perceived as normatively framed, as demonstrated in their standardized structure and form. Their normativity gives social legitimacy and weight to the symbols and restatements that are communicated to the participants and allows them to become arenas for political elites looking to increase their popularity. In this sense, event analysis that attends to the normative dimensions of a situation (Kapferer, 2015: 15) as part of its liberating and creative potential is essential.

Acknowledgments

The research of which this chapter is a result has been supported by the programs CLUE 'Dynamics of Circumpolar Land Use and Ethnicity', funded by the American National Science Foundation, and INPOINT 'Socioeconomic Significance of Developmental Projects in Northwest Russia: The Insiders' Point of View', funded by the Norwegian Research Council. I am grateful to the editors of this anthology, as well as to Nathan Light, Hugh Beach and Yulian Konstantinov for comments and discussion during the process of writing.

References

Adams, L.L. (2010) *The Spectacular State: Culture and National Identity in Uzbekistan*: Durham, NC: Duke University Press.

Bakhtin, M.M. (1984) *Rabelais and His World*. Bloomington, IN: Indiana University Press.

Bat'ianova, E.P. (2011) Nacional'nye prazdniki u teleutov v XX-XXI vv. *Etnograficheskoe obozrenie* 6, 16–25.

Berezin, M. (2006) The festival state: Celebration and commemoration in fascist Italy. *Journal of Modern European History* 4 (1), 60–74.

Beyrau, D. and Rolf, M. (2006) *Dictatorship and Festivals*: München: CH Beck.

Binns, C.A.P. (1979) The changing face of power: Revolution and accommodation in the development of the Soviet ceremonial system: Part I. *Man* 14 (4), 585–606.

Binns, C.A.P. (1980) The changing face of power: Revolution and accommodation in the development of the Soviet ceremonial system: Part II. *Man* 15 (1), 170–187.

Burawoy, M. (2009) *The Extended Case Method: Four Countries, Four Decades, Four Great Transformations, and One Theoretical Tradition*. Oakland, CA: University of California Press.

Butler, R. and Hinch, T. (2007) *Tourism and Indigenous Peoples: Issues and Implications*. Amsterdam: Elsevier.

Chau, A.Y. (2008) The sensorial production of the social. *Ethnos* 73 (4), 485–504.

Cohen, A. (1993) *Masquerade Politics: Explorations in the Structure of Urban Cultural Movements*. Oakland, CA: University of California Press.

Dubnickii, M.A. and Khrapovickii, A.S. (1984) *Prazdnik Severa* [*The Festival of the North*]. Murmansk: Murmanskoe Knizhnoe Izdatel'stvo.

Duffy, M. and Waitt, G. (2011) Rural festivals and processes of belonging. In C. Gibson and J. Connell (eds) *Festival Places: Revitalising Rural Australia* (pp. 44–57). Bristol: Channel View Publications.

Duvignaud, J. (1976) Festivals: Sociological approach. *Cultures* 3 (1), 14–25.

Evens, T.M.S. (2006) Some ontological implications of situational analysis. In T.M.S. Evens and D. Handelman (eds) *The Manchester School: Practice and Ethnographic Praxis in Anthropology* (pp. 49–63). Oxford: Berghahn.

Evens, T.M.S. and Handelman, D. (2006) Introduction: The ethnographic praxis of the theory of practice. In T.M.S. Evens and D. Handelman (eds) *The Manchester School: Practice and Ethnographic Praxis in Anthropology* (pp. 1–11). Oxford: Berghahn.

Falassi, A. (1987) *Time Out of Time: Essays on the Festival*. Albuquerque, NM: University of New Mexico Press.

Forsyth, J. and Wamsley, K.B. (2006) 'Native to native … we'll recapture our spirits': The world indigenous nations games and north American indigenous games as cultural resistance. *The International Journal of the History of Sport* 23 (2), 294–314.

Frost, N. (2016) Anthropology and festivals: Festival ecologies. *Ethnos* 81 (4), 569–583.

Getz, D. (2010) The nature and scope of festival studies. *International Journal of Event Management Research* 5 (1), 1–47.
Gibson, C. and Connell, J. (eds) (2011) *Festival Places: Revitalising Rural Australia*. Bristol: Channel View Publications.
Glaeser, A. (2006) An ontology for the ethnographic analysis of social processes. Extending the extended-case method. In T.M.S. Evens and D. Handelman (eds) *The Manchester School: Practice and Ethnographic Praxis in Anthropology* (pp. 64–93). Oxford: Berghahn.
Gleason, A., Kenez, P. and Stites, R. (1989) *Bolshevik Culture: Experiment and Order in the Russian Revolution*. Bloomington, IN: Indiana University Press.
Gluckman, M. (1956) *Custom and Conflict in Africa*. Oxford: Blackwell.
Gray, P.A. (2005) *The Predicament of Chukotka's Indigenous Movement. Post-Soviet Activism in the Russian Far North*. Cambridge: Cambridge University Press.
Grenier, A.A. and Müller, D.K. (eds) (2011) *Polar Tourism: A Tool for Regional Development*. Montreal: Presses de l'Université du Québec.
Handelman, D. (1998) *Models and Mirrors: Towards an Anthropology of Public Events*. Oxford: Berghahn Books.
Handelman, D. (2006) The extended case: Interactional foundations and prospective dimensions. In T.M.S. Evens and D. Handelman (eds) *The Manchester School: Practice and Ethnographic Praxis in Anthropology* (pp. 94–117). Oxford: Berghahn.
Hirsch, F. (2005) *Empire of Nations: Ethnographic Knowledge and the Making of the Soviet Union*: Ithaca, NY: Cornell University Press.
Huttunen, A. (1995) Encountering ethnic groups and their identities in sports. In T. Korhonen (ed.) *Encountering Ethnicities: Ethnological Aspects on Ethnicity, Identity and Migration* (pp. 118–127). Helsinki: Finnish Literature Society.
Iavtysyi, P.A. (2011) *Severnoe Mnogobor'e/North Athletics/*. Nar'an-Mar: IPP.A.Kulakov.
Kapferer, B. (2006) Situations, crisis, and the anthropology of the concrete: The contribution of Max Gluckman. In T.M.S. Evens and D. Handelman (eds) *The Manchester School: Practice and Ethnographic Praxis in Anthropology* (pp. 118–158). Oxford: Berghahn.
Kapferer, B. (2015) Introduction: In the event—toward an anthropology of generic moments. In L. Meinert and B. Kapferer (eds) *In the Event: Toward an Anthropology of Generic Moments* (pp. 1–28). Oxford: Berghahn.
King, A.D. (2011) *Living with Koryak Traditions. Playing with Culture in Siberia*. Lincoln, NE: University of Nebraska Press.
Kiselev, A. and Kiseleva, T. (1987) *Sovetskie Saamy. Istoriia, ekonomika, kul'tura* [*Soviet Sami. History, Economy, Culture*]. Murmansk: Murmanskoe knizhnoe izdatel'stvo.
Konstantinov, Y. (2015) *Conversations with Power: Soviet and Postsoviet Developments in the Reindeer Husbandry Part of the Kola Peninsula*. Uppsala: Uppsala University.
Krist, S. (2005) Where going back is a step forward: The re-traditionalising of sport games in post-Soviet Buriatiia. *Sibirica* 4 (1), 104–115.
Lane, C. (1981) *The Rites of Rulers: Rituals in an Industrial Society – The Soviet Case*. Cambridge: Cambridge University Press.
Lane, C. (1984) Legitimacy and power in the Soviet Union through socialist ritual. *British Journal of Political Science* 14 (2), 207–217.
Lavenda, R.H. (1980) The festival of progress: The globalizing world-system and the transformation of the Caracas carnival. *The Journal of Popular Culture* 14 (3), 465–475.
Martin, T. (2011) *Imperia 'polozhitel' noi deiatel 'nosti'. Nacii i nacionalizm v SSSR 1923–1939* [*The Affirmative Action Empire. Nations and Nationalism in USSR 1923–1939*]. Moscow: ROSSPEN.
McDowell, J. (1974) Soviet civil ceremonies. *Journal for the Scientific Study of Religion* 13 (3), 265–279.

Mitchell, J.C. (1956) *The Kalela Dance: Aspects of Social Relationships Among Urban Africans in Northern Rhodesia*. Manchester: Manchester University Press.
Müller, D.K. and Pettersson, R. (2005) What and where is the indigenous at an indigenous festival? Observations from the winter festival in Jokkmokk, Sweden. In C. Ryan and M. Aicken (eds) *Indigenous Tourism: The Commodification and Management of Culture* (pp. 201–216). Amsterdam: Elsevier.
Müller, D.K. and Pettersson, R. (2006) Sami heritage at the winter festival in Jokkmokk, Sweden. *Scandinavian Journal of Hospitality and Tourism* 6 (1), 54–69.
Novikova, N.I. (2008) Legal anthropology of the interaction of Russia's numerically small indigenous peoples of the North with oil companies. *Anthropology & Archeology of Eurasia* 47 (2), 12–38.
Novikova, N.I. (2014) *Ohotniki i neftianiki. Issledovanie po iuridicheskoi antropologii* [*Hunters and Oil Workers. Study of Legal Anthropology*]. Moscow: Nauka.
Noyes, D. (2011) *Fire in the Plaça: Catalan Festival Politics after Franco*. Philadelphia, PA: University of Pennsylvania Press.
Olsen, K. (2003) The touristic construction of the emblematic Sami. *Acta Borealia* 20 (1), 3–20.
Olsen, K. (2006) Making differences in a changing world: The Norwegian Sami in the tourist industry. *Scandinavian Journal of Hospitality and Tourism* 6 (1), 37–53.
Osinovskii, V. (1930) Olenevodcheskii sovkhoz [Reindeer herding Sovkhoz]. *Karelo-Murmanskii Krai* 4–5, 23–25.
Phipps, P. (2010) Performances of power: Indigenous cultural festivals as globally engaged cultural strategy. *Alternatives: Global, Local, Political* 35 (3), 217–240.
Picard, D. and Robinson, M. (2006) *Festivals, Tourism and Social Change: Remaking Worlds*. Clevedon: Channel View Publications.
Retish, A.B. (2003) Creating peasant citizens: Rituals of power, rituals of citizenship in Viatka province, 1917. *Revolutionary Russia* 16 (1), 47–67.
Roche, M. (2001) Modernity, cultural events and the construction of charisma: Mass cultural events in the USSR in the interwar period. *International Journal of Cultural Policy* 7 (3), 493–520.
Rolf, M. (2006) Das sowjetische Massenfest im Stalinismus (1932–1941) [Soviet mass festivals during Stalinism (1932–1941)]. *Geschichte und Gesellschaft* 32 (1), 69–92.
Rolf, M. (2008) Constructing a Soviet time: Bolshevik festivals and their rivals during the first five-year plan: A study of the Central Black Earth Region. *Kritika: Explorations in Russian and Eurasian History* 1 (3), 447–473.
Rolf, M. (2013) *Soviet Mass Festivals, 1917–1991*. Pittsburgh, PA: University of Pittsburgh Press.
Ryan, A.W. and Wollan, G. (2013) Festivals, landscapes, and aesthetic engagement: A phenomenological approach to four Norwegian festivals. *Norsk Geografisk Tidsskrift* 67 (2), 99–112.
Rydell, R.W. (2013) *All the World's a Fair: Visions of Empire at American International Expositions, 1876–1916*. Chicago, IL: University of Chicago Press.
Slezkine, Y. (1994) *Arctic Mirrors: Russia and the Small Peoples of the North*. Ithaca, NY: Cornell University Press.
Stammler, F. (2011) Oil without conflict? The anthropology of industrialisation in Northern Russia. In A. Behrends, S.P. Reyna and G. Schlee (eds) *Crude Domination: An Anthropology of Oil* (pp. 243–269). Oxford: Berghahn Books.
Stammler, F. and Wilson, E. (2006) Dialogue for development: An exploration of relations between oil and gas companies, communities, and the state. *Sibirica* 5 (2), 1–42.
Stammler, F. and Peskov, V. (2008) Building a 'Culture of dialogue' among stakeholders in North-West Russian oil extraction. *Europe-Asia Studies* 60 (5), 831–849.
Stites, R. (1989) *Revolutionary Dreams: Utopian Vision and Experimental Life in the Russian Revolution*. Oxford: Oxford University Press.

Turner, V.W. (1982) *Celebration, Studies in Festivity and Ritual.* Washington: Smithsonian Institute Press.
V.A. (1929) *Pervye olen'i bega v Murmanske* [The first reindeer races in Murmansk]. *Karelo-Murmanskii Krai* 4–5, 49.
Viken, A. (2006) Tourism and Sami identity: An analysis of the tourism–identity nexus in a Sami community. *Scandinavian Journal of Hospitality and Tourism* 6 (1), 7–24.
Viken, A. (2011) Reinvention of ethnic identity: A local festival as a national institution on a global scene. In A.A. Grenier and D.K. Müller (eds) *Polar Tourism. A Tool for Regional Development* (pp. 179–206). Montreal: Presses de L'Université du Québec.
Viken, A. and Müller, D.K. (2006) Introduction: Tourism and the Sami. *Scandinavian Journal of Hospitality and Tourism* 6 (1), 1–6.
Viken, A. and Pedersen, P. (2009) Globalized reinvention of indigenuity: The Riddu Riđđu festival as a tool for ethnic negotiation of place. In T. Nyseth and A. Viken (eds) *Place Reinvention: Northern Perspectives* (pp. 183–202). Aldershot: Ashgate.
Vladimirova, V.K. (2006) *Just Labor: Labor Ethic in a Post-Soviet Reindeer Herding Community.* Uppsala: Uppsala Universitet.
Vladimirova, V.K. (2011) We are reindeer people, we come from reindeer: Reindeer herding in representations of the Sami in Russia. *Acta Borealia* 28 (1), 89–113.
Vladimirova, V.K. (2014) It is not our reindeer but our politicians that are wild: Contests over reindeer and categories in the Kola peninsula, Northwestern Russia. *Arctic Anthropology* 51 (1), 24–40.
Volovik, V.I., Dankov, V.V. and Erkhov, G.P. (1986) *Sovetskie tradicii, prazdniki i obriady. Opyt, problemy, rekomendacii* [*Soviet Traditions, Celebrations and Rituals. Experience, Problems, and Recommendations*]. Moscow: Profizdat.
Von Geldern, J. (1993) *Bolshevik Festivals, 1917–1920.* Oakland, CA: University of California Press.Waterman, S. (1998) Carnivals for elites? The cultural politics of arts festivals. *Progress in Human Geography* 22 (1), 54–74.
Whitford, M. and Ruhanen, L. (2013) Indigenous festivals and community development: A sociocultural analysis of an Australian indigenous festival. *Event Management* 17 (1), 49–61.
Whitford, M. and Dunn, A. (2014) Papua New Guinea's indigenous cultural festivals: Cultural tragedy or triumph? *Event Management* 18 (3), 265–283.

12 Indigenous Hospitality and Tourism: Past Trajectories and New Beginnings

Gro B. Ween and Jan Åge Riseth

Introduction: Indigenous Tourism and its Logic

These days, as indigenous peoples are increasingly written into a new economic world order, nature tourism has been described as a possibility for new economic independence in marginal areas with few other opportunities for people whose existence cannot be moved elsewhere (Butler & Hinch, 2007; Bunten, 2010; Ryan & Aitken, 2005). After an initial period of excitement over the economic opportunities offered by nature tourism, new critiques have emerged. The other side of the coin of self-determination, cultural rejuvenation (Butler & Hinch, 2007) and access to new influential contact zones (Pratt, 1992; Ween, 2002) is new colonialism, cultural and natural degradation and, as a consequence, new economic dependence (Butler & Hinch, 2007). Before we can assess the opportunities offered by indigenous tourism, the role of nature in such tourism activities must be determined. Often, indigenous tourism is simply made to equal nature tourism as these activities often take place in what we call nature.

At its best, such relationships have been profitable and offered opportunities for new understanding across worlds of difference (Ween, 2002). Sometimes, this is also a powerful motivation for engaging in tourism. Worst-case scenarios, on the other hand, involve mass tourism, alienation, loss of natural resources and, as a consequence, loss of both tangible and intangible cultural heritage (Ween, 2016). In the High North, nature tourism is by no means a new activity. Southerners have travelled north to experience nature in the company of indigenous guides for over a century. In this chapter, we compare two Norwegian Sami cases of engagement in nature tourism and hospitality, one from south Sami areas in Nord-Trøndelag, and the other from a north Sami area in Finnmark, more specifically from the Tana River on the border between Norway and Finland.

Since 2013, nature tourism has been a priority for the Ministry of Environment (Fedreheim, 2013). Wilderness is considered a national asset, something that can be marketed to foreign tourists. Natural resource management regimes have moved from emphasising the protection of nature to focus on the sustainable *uses* of nature (Ween, 2009). Both in Finnmark and Southern Sami areas, areas of great natural beauty such as national parks are singled out for new tourist use. There is no tradition of restricting nature tourism enterprise, rather the opposite, in such areas, anyone can start new initiatives. There is little recognition of the existence of systems of Sami land use or customary law. Only to a very limited extent is there a public conversation about how these natural resources should be put to use, even in areas where Sami are part of local management.

What is at stake here is not only the official insistence on treating nature as a tourism asset. As we will show, the affordances of the natural resources in question, the historical trajectories of relations between locals and tourists shape the conditions for how nature can be put to use. But there is more to tourism than extra income. Tourism may also offer other terms of engagement. The term 'contact zone' was introduced over 20 years ago (Pratt, 1992) to describe travel encounters as colonial relations. Remarkably, the term has kept its position in both humanities and social sciences and if anything, it has gained new significance (Clifford, 1997; Haraway, 2008). As a theoretical matter, contact zones highlight ideas of relationality; that people, things, events, processes and practices contribute to each other's production (see for example Law, 2009). The term describes sites of ongoing engagements between peoples, nations and locales, at a time in history when contact has already been established and has already had an effect. Such engagements often involve asymmetrical power relations but may also be sites where such power relations can be revisited, through new communication, ongoing processes of translation, of displays and performances of difference, or a becoming with (Pratt, 1992; Clifford, 1997; Haraway, 2008). Whether a contact zone simply revisits colonial relations or offers opportunities for new kinds of engagements depends, as we will show, upon the context within which these interactions are taking place, such as the historical trajectories of engagements between tourists and locals, but also the resources themselves.

We explore the differences in approaches to tourism in these two Sami areas. In Tana, tourism has age-old trajectories and is founded on salmon fishing, a nature activity based on a limited resource. In this case, on both sides of the river, tourist initiatives and regional authorities have high ambitions. Tourism to many, particularly in Finland, represents a significant source of income. In the Southern Sami case, Sami tourist initiatives are of a recent origin, and are not based on the exploitation of a limited resource, rather they are based on the communication and marketing of Sami culture, an unlimited resource. In these cases, we explore what nature is to the parties involved and how nature in these views should be engaged.

Sami and Other Nature Practices

For local inhabitants, both in Nord-Trøndelag and Finnmark, being in nature is based upon *other* ideals than the Norwegian *Friluftsliv* (Ween, 2012; Ween & Risan, 2014). To a large part of the Sami population, both in Finnmark and in Nord-Trøndelag, hunting and gathering activities continue to be significant as a way of life, for food procuring activities, pragmatically engaged practically by way of motorised vehicles (Riseth & Solbakken, 2010; Ween & Lien, 2012). The use of motorised vehicles in these activities is morally condemned in southern Norwegian and environmental authorities' approaches to nature as wilderness (Riseth & Solbakken, 2010). Official recognition of Sami ways of life has, however, opened negotiations with natural resource management authorities to secure the rights of all inhabitants of Finnmark to continue their nature practices (Ween & Lien, 2012). So far, acknowledgement of rights to natural resources has failed to take into account the contrasts between Norwegian natural resource management and Sami understandings of nature. From Kautokeino, in the Sami core areas in Finnmark, Riseth and Solbakken (2010) describe how local people were upset that the areas they used for food procurement were determined as wilderness by natural resource management institutions. In their book *The Land is Inherited*, Bye and Jåma (2011) provide a similar broad picture of the intimate relations between nature, culture and humans in the past, present and future, for the South Sami Låarten Sijte in Snåsa (Nord-Trøndelag).

The notion of wilderness as it is employed in national park contexts is problematic on several accounts. It is problematic because the use of all-terrain vehicles (ATVs) and snowmobiles, essential to Sami uses of nature, has no room in a place characterised as a wilderness worthy of protection. Second, the term 'wilderness' has strong similarities with the *terra nullius* that over a hundred years ago caused the Norwegian state to appropriate these lands as unused, unowned land. Third, it is problematic because in caring for nature, the sustainability of local subsistence practices is made invisible. People are hurt because the natural resource management authorities do not acknowledge that the areas they characterise as wilderness are worthy of official nature protection because the locals have taken good care of these areas (Joma, 2013; Riseth & Solbakken, 2010).

Tourists who come to Finnmark and Nord-Trøndelag for nature activities, on the other hand, are often carriers of the *Friluftsliv* ideal. Many do not appreciate local uses of motorised vehicles in nature practices. Hunting and fishing activities are acted out within very different ideals and, consequently, different practices. To many tourists, nature activities should ideally be simple and inexpensive, people mostly manage on their own and there are limited opportunities for engagement with locals. Given this extensive divide in understandings of nature and nature practices, how can nature tourism provide room for new communication?

Salmon Tourism in the North

In the north along the Tana River, tourism has been an issue since the 1830s. The salmon lords came by way of Alta, already a large settlement, with a significant salmon river, located in West Finnmark. When the German zoologistzoologist and phycologist E.O. Schmidt was in Alta in the summer of 1850, he describes a meeting with English anglers on route to Tana:

> Their caravan consisted of fifteen horses. Going north from Karasjok they intended to follow the Tana with hired rowers all the way to Tana fjord whereupon they would return to England on a schooner belonging to one of the two young Englishmen. (Solbakk, 2011a: 14)

More would follow, and local guides observed the tourist fishing practices and soon made aspects of angling into their own (Solbakk, 2011a,b). At first, British anglers were a welcome income supplement, but within quite a short time the anglers became a nuisance (Pedersen, 1986; Solbakk, 2011a; Ween, 2017). In the coming years, it would only get worse for local fishermen. Angling soon became popular with the Norwegian elite. Fishing rights in the Alta River were bought up entirely and it became the fishing paradise of the very rich. The Tana, however, awaited another fate. Its position as a border river complicated its management.

Angling tourism came to evolve very differently on the Finnish and on the Norwegian side of the Tana River. On the Norwegian side, local net fishing right holders remained a political stronghold. In addition, the protection of Sami reindeer herding as an industry served to ensure an economic basis. In Finland, however, Sami rights development was not as successful as in Norway, reindeer herding did not become an exclusive Sami industry, and local net fishing rights were formulated differently and did not secure local fishermen in the same way (Ween, in press). The limited number of salmon rivers in Finland has made Finnish authorities insist on the river being available to the masses (Ween, 2016). As a consequence, many reindeer herding families earn little on reindeer herding and opt to make a living on tourism, either by selling bits of their land, by renting out accommodation to tourists or by guiding fishermen.

On both sides of the river, post-war tourist operations consisted of letting southern families put up tents on private land along the river.

A few built a small cabin, sold local farm products and maybe built a sauna to cater for tourists. In the 1960s, at a time when more, predominantly urban wage-labourers from the south became able to take proper summer holidays, larger and more organised camp grounds were built on both sides of the river. Tourism then became more centralised, the contact between tourists and locals less regular. With the exception of relations with rowers, most encounters came to be short-term transactions.

Finnish Sami who live along the river argue that they have little choice other than to participate in the tourist industry. Many have already had to sell their reindeer as it is important to stay in Utsjok, on ancestral land where there are few other employment opportunities.

The more tourist-friendly staff at the local department of the Finnish Sami Parliament argue that salmon tourism is a continuation of local hunting and fishing traditions. In this sense salmon tourism offers the opportunity to continue living off the fish, and hence a continuation of their traditional rights. These days, anglers constitute 50% of the fishing population in Tana and their numbers are rising. Although an economic necessity for many, the anglers still fill local net fishermen on both sides of the river with despair. Anglers, in their sheer numbers, are very efficient fishermen. There are many of them and, unlike the local fishermen, they fish day and night. Particularly in the shallow waters of the falls halfway up the river, local fishermen loose the competition to large groups of anglers.

On the Norwegian side, very little land is sold, there are fewer tourist initiatives and less anglers. Here, three big camping grounds are situated on the entire length of the river, in Karasjok, Storfossen and Skiippagurra. On the Finnish side, there are several smaller camping outfits as well as a couple of large camp grounds. Most camp grounds are humble, with room for caravans and tents and with a few as-simple-as-it-gets cottages for rent. Tourists are generally of a similar uncomplicated kind. They often have caravans more or less permanently placed and come for holidays and weekends in the fishing season. They fish continuously and store the fish they catch in freezers available on the camp sites. They bring all their own food and their own car. Sometimes, they drive up from southern parts of Norway and Finland. More often, they come from the more populated areas of northern Scandinavia. A low prize level is important for both nations, but even more so for Finnish authorities who insist upon a low economic threshold to ensure that fishing in the Tana River is generally available to their population. In Norway, salmon fishing in other Norwegian rivers is very expensive, for example, in neighbouring Alta River, in the high season, fishing for a week at Sautso may cost 150,000 kroners with a guide and lodgings (Ødegaard, 2013). And while Alta is managed as a catch and release river, the angler tourists in Tana have no bag limit. Tourist fishermen in other words, pay very little and can fill their fridge at the same time.

This strategy appears almost sacrilegious to the classic angler. The effect on the Tana salmon population is also significant and, as a consequence, the Norwegian natural resource management institutions continuously restrict local Sami net fishing practices to compensate (Ween, 2012). There is one important reason why the tourist fisheries have not been stopped. This concerns the lack of successful renegotiations of the Tana treaty between the Norwegian and Finnish states. For over a decade there had been hope that a new treaty would secure the rights of local net fishermen and restrict the number of tourist anglers. This autumn, however, negotiations were finally successful, but the result was simply a strengthening of the status quo; Sami net fishing rights were further restricted while tourist fisheries further encouraged. Argumentation for allowing this tourist salmon bonanza to continue is also partly economic, it is about how best to maximise on uses of nature: visiting anglers provide more income per salmon than the local fishermen. To illustrate, in its time, the Proposition to Parliament (Ministry of Environment St. prp. no. 32 [2006–2007]) calculated that the profits from tourism and associated retail businesses was estimated to be as much as 1.3 billion NOK annually, while if fished by locals, the fish does not even enter into economic circulation (Ween, 2012).

On the Finnish side of the river, historical traditions and lack of other income options make work as rowers and guides for the more upscale tourists more attractive. The only tourist outfit on the Norwegian side that offers more upscale tourism, however, describes difficulties of finding people who are willing to row for tourist anglers. Some of the older fishermen might do so in their spare time, but fishing comes first, the pay is not great and as a consequence this is not a regular service. This tourist operator finds that the guides must be trained before they engage the tourists. The skills involved in rowing are not just about getting the fish for the tourists, a rower must also be able to engage and explain, share his knowledge and tell good stories. Local fishermen cherish their freedom, particularly in the fishing season, and although everyone is very polite and considerate, the southern Norwegian form of ongoing conversation feels foreign to many.

On the Finnish side of the Tana, however, there are families where generations of men have worked as guides and rowers. In conversations with these men, they often emphasise that they have an ongoing relationship with visiting anglers. It is an annually occurring working partnership. Many of the tourist anglers have also become familiar with their guide's families. The families exchange Christmas and birthday cards. Often, they get to keep their share of the fish. There is an awareness on both sides, according to the rowers, that this is not a master–servant relationship, it is not the fisherman telling the rower where to go, but rather an apprenticeship, where the tourist angler learns how to fish the Tana River. The Tana, as they say, is a special river. As compared to most other Norwegian salmon rivers, it is bigger, wider and has many streams. Most will not be able to fish the Tana River on their first try.

These days, tourism operations on the Finnish side of the river are becoming more innovative. Next to the old camping grounds and the smaller family-based outfits with a couple of smaller cottages to rent on private property, there is also a new upscale camping ground. The camp ground, *Poron Porijat*, and a few other smaller tourist operations focus on more upscale operations selling experiences rather than straightforward fishing. *Poron Porijat* (Sami: the gelder of reindeer) is, not surprisingly, run by a local reindeer owner. The place provides comfortable cabin accommodation, all maintained in a rough outdoorsy masculine aesthetics. Here, everything is geared towards comfortable fishing adventures. Beautiful wooden boats are laid ready on the beach for guides to take visiting anglers out; cabins are fitted with racks to place the rods in; in between fishing expeditions, there are barbeques and bonfire sites for grilling the catch; in the reception there is even an underwater camera providing fishermen with a close-up view of fishing conditions. Visiting fishermen not only pay more for accommodation and are assured experienced rowers, they can also learn about local Sami culture through culinary traditions, guided trips to trout fishing lakes in the mountains, have experiences with local reindeer and reindeer herders and listen to their storytelling. If this initiative pays off is hard to tell yet. What is certain is that most tourists coming to Utsjoki prefer the cheaper camp grounds with more rudimentary facilities, geared towards simple salmon fishing.

Put crudely, the River Tana primarily offers a nature tourism that exploits a limited natural resource. Over time, this tourism practice has served to restrict local access to culturally significant species. The particular tragedy of the Tana salmon must be seen in light of the insistence, primarily from Finnish authorities, of keeping fishing prizes to a minimum, to ensure the availability of salmon fishing to the masses. Today, tourists fish 50% of all the salmon in the Tana, and the size of these fisheries has dramatically reduced the traditional Sami fisheries. Despite ambitions, these fisheries presently provide only a limited opportunity to live from Sami culture and practices in new ways. Diverging understandings of what proper fisheries are and how one should engage nature also continuously produce friction between locals and visitors. This tourism, with a colonial origin, continues along similar trajectories and, outside the relation between rower and tourist fishermen, the situation offers limited opportunity to renegotiate the terms of engagement.

South Sami Hospitality and an Emerging Tourism

During the last two centuries, the South Sami has been in a much clearer minority situation than the Sami in Finnmark. It is indisputable that this region, all land in the northernmost (Namsskogan) and uppermost

(Røryvik) municipalities, as well as large parts of the other municipalities,[1] was Sami land in the early 18th century (NOU, 2008: 13). Norwegian farmer communities expanded northwards and upwards from the main valley, and in the late 18th century large parts of this Sami area was transformed to Crown land, and the Namsen Commons (Namsenallmenningen) were established without acknowledging Sami property (Hermanstrand & Kosmo, 2009). From this commons, private land allotment to non-Sami settlers started in 1770 (Støvik, 1997). During the subsequent centuries, the non-Sami expansion continued at the expense of the Sami.

At the heart of the region is the valley of the 220 km long River Namsen.[2] Snåsa is just south of this area surrounding a 40 km long lake. Like the Tana River, the Namsen River is a very good salmon river, but unlike the Tana the fishing rights today belong to the non-Sami farmers with properties along the river. In a similar way to the story of the Alta and the Tana, the Namsen River was 'discovered' by the English angler Mr Bilton who visited for the first time in 1837. Three years later, he wrote the angler guide *Two Summers in Norway*, which initiated the Salmon Lord era, when the British in practice controlled the whole fishery, lasting until the Second World War. The Namsen River was called the 'Queen of Rivers' because of its generosity and reliability. The Namsen River is still regarded as one of the top three salmon rivers in Norway (Nordeide, 2012).

At the Namsen River, salmon lords from the start stayed as guests at farms. In the late 19th century, hotels were built and some even built their own lodges (Nordeide, 2012). The Namdalen area encompassed several of the largest private land estates of Norway. Historically, these are forest estates. As profitability in forestry has decreased, several of the forest estates have increased their efforts in the outdoors and tourism sector.[3] Most tourist outfits are run with high numbers of visitors, in a rather traditional manner in line with Norwegian ideals of '*friluftsliv*' but some changes are under way. After the millennium shift, some of the estates were bought by a couple of Norway's highest profiled investors.[4] One of the actual plans is to develop an abandoned mining community, Skorovas, into a modern ski resort in the Austrian style.[5] The plan for the resort was adopted by the municipalities in the fall of 2016 and the first cottages are planned to be finished in 2018.[6]

Tourists found their way inland from the Namsen River to Snåsa during the same time period. At this point, the main transport to Snåsa was by ship along the 40 km long Lake Snåsa and several hotels and guesthouses were established. In the early 20th century, as a consequence of the hunt for predatory species, moose populations started to increase. These large animals in a beautiful, wild and available landscape made Snåsa a hunting paradise for rich men, both from the continent and the UK. Hunting tourism came to provide new income opportunities for local men as hunting guides and other helpers. The railway towards the north, Nordlandsbanen, was under construction and reached Snåsa in 1926. This increased the accessibility of the area. Possibly the most famous visitor of all, the German industrial magnate Baron Gustav Krupp, visited at least three years in a row before 1930 on his private train, staying in his newly built private moose hunting lodge (Gran, 2014). Krupp even had his own telephone line installed up to the lodge to be in contact with Berlin. The local assistants were paid well. One farmer could pay his farmhand one monthly salary for the day's work paid him by the baron.[7]

These more colonial kinds of nature tourism also included Southern Sami actors. Reindeer herding brothers, Bengt and Jonar Jåma, represented the next generation of the hunting guide enterprise over two decades, from the early 1950s to the early 1970s. The Jåma brothers served as hunting guides for visiting moose hunters from southern Norway, Germany, Austria and even a Spanish ambassador. Hunting magazines and newspapers referred to Bengt and Jonar as the world's best hunting guides. Part of their success was based on their ability to create a feast and joy both by cooking and storytelling (Åsvoll, 2010; Bye & Jåma, 2011).

In this area, the market for upscale nature tourism is still present, both with regard to hunting and angling. Both the mountain farmers and the reindeer herders have continued to tap into this market. The mountain farmers are still there, in scattered small, picturesque farms, living off small-scale agriculture and outfields harvesting, in combination with some tourism. These are roadless locations along the traditional paths and thoroughfares between rural communities such as Snåsa and Lierne, and along the national border between Norway and Sweden. For the reindeer herders, the mountain farmers have been neighbours, friends and when needed, extra work force (see Svensson & Viken, this volume). The farms are both temporary places of residence and expedition bases as well as arenas for work operations like herd roundups and slaughter (Åsvoll, 2010; Bye & Jåma, 2011). Mutual dependency between reindeer herders and other rural dwellers provide reindeer herders with a freedom, despite the asymmetrical power relations, ongoing land use conflicts and the lack of formal rights. South Sami herders argue that at least up until the 1970s, they felt as if they were 'masters of the mountains', in the sense that the Sami herders were able to maintain their close relation to their ancestral lands and their competence in utilising the lands for a living.

At the national level, the Sami position in the high mountains was, as we have argued elsewhere, reduced when Norway's second national park plan was introduced into natural resource management between the mid-1990s and into the 2000s. The national park plan caused an extensive increase in conservation areas in Norway. The plan provoked much local opposition, also in this area. Probably in response to local opposition, the Norwegian Parliament in 2003 adopted an initiative to make national parks more attractive for municipalities by legalising commercial tourism, not only in the recreational areas but also in the conservation areas of the national parks (Fedreheim, 2013). The regional processes for the establishment of two national parks in Snåsa and Lierne municipalities created a decade-long serious conflict with the South Sami (Arnesen & Riseth, 2008, 2009; Riseth, 2007, 2015, 2016; Riseth & Holte, 2008; Ween, 2009, 2010).

Despite the indigenous rights improvements proposed by the Sami Rights Commission, there are no recognised Sami land rights beyond the rights of reindeer herding in these areas (NOU, 2008: 13). There are ongoing processes with elements of reconciliation, as well as growing understandings that Sami culture is a valuable asset for rather remote communities with decreasing numbers of inhabitants. What has changed, and this is important, is that both the host municipalities for the South Sami enterprises we present below, Snåsa (from 2008) and Røyrvik (from 2013), have joined the Sami language management area (*Samisk språkområde*) making Sami language an official language in line with Norwegian in these municipalities.[8] One example of positive initiatives that this openness to Southern Sami rights is a four-year national value creation program initiated in 2009, Nature Heritage as a Value Creator.[9] The main objective was 'to contribute to the conservation areas and other valuable natural areas can be important resources in society. The program will also help to improve cooperation at all levels related to the conservation area and the local community'. Several ministries cooperated in providing funding for this program on the proviso that the participants contributed 50% of the funding themselves. Three local South Sami projects (Brennli, 2014; Reindriftsnytt, 2013) applied for funding under the common label 'South Sami culture and tradition as a tourism product'.[10] The projects involved are located in the northern part of Nord-Trøndelag County in the inland municipalities of Snåsa, Lierne and Røyrvik. The first tourism project was a one-man firm; the second, a crafts-based firm owned by three sisters; and the third represented the joint effort of a whole reindeer pasture district consisting of 10 families. All of them have reindeer herding management as their basis industry. While the three projects to some extent overlap in content with cultural outreach as a common denominator, each has their particular niches.

The one-man firm is owned by local reindeer owner Bernt Arvid Jåma. On his website,[11] he explains how he was raised in a home where food was culture, and as an adult he has been occupied with how to prepare the reindeer meat to be as good as possible. In an interview, he describes himself as 'a reindeer herder that always has had cooking as my big passion' and that he has developed cooking and hospitality to become a stable secondary source of income. He states: 'Instead of selling one kilo reindeer meat for a wholesale price I develop it into four portions of delicious food creating added-value'. His business idea is hosting groups of people at his own site or being mobile, serving them a good meal based on reindeer meat and other local nature products, creating a nice atmosphere by storytelling and other cultural elements. The project obviously offers opportunity for more sustainable tourism than the traditional hunting or fishing kind that depends upon tourists killing or catching something and bringing it home. What is sold is a cultural experience rather than the meat itself. The challenge here lies in the reaching out, finding arenas both for advertising and for the culinary experience. The three Wilks sisters have an established design and *duedtie* (Sami crafts) firm (Reindriftsnytt, 2013):

> Besides being lived in step with the reindeer our lives always have been stamped by Sami handicrafts. The inspiration for our design come from moments in our lives – it can be the first reindeer calf born in spring, or the atmosphere around the camp fire a summers night. The primordial power is transfered from tradition to exciting new design of clothes and jewelry produced of what nature gives.[12]

The sisters' firm aspires to sell Sami crafts. In this way, the firm is only superficially related to tourism, both in the sense that it is at least partly internet based, and that although the inspiration for the designs comes from nature, it is not directly related to nature engagements for tourists. Also, the brand sells both to Sami and non-Sami clientele. They also have ambitions to strengthen their firm by making it possible to offer various cultural extension services, such as lectures and courses for schools and others. Tjåhkere Sijte is the largest actor of the three. The leader of the enterprise, Algot Jåma, explains the basic idea:

> Within the traditional reindeer herding we have reached an acceptable level both for employment and utilization of herds and pasture resources. If we are going to extend our activities and utilize our raw material even better, we need to explore other directions. We have chosen to focus some of our new efforts towards nature and cultural based tourist experience. (Reindriftsnytt, 2010: 18)

Some years back, Tjåhkere Sijte bought a farm that blocked their reindeer migration route, and this has become the site they use as a platform for culturally based activities for visitors, in addition to extending the utilisation of reindeer products as skins, handicrafts and more. Within the project, they have established the firm Vaegkie Ltd as a tourism and information project. They have so far built a reindeer corral and facilities for accommodation, cultural performance and a nature information path, and more is underway (Brennli, 2014). Broadly speaking, it can be seen as the establishment of a showroom for South Sami reindeer herding culture. In an interview, Algot Jåma tells that in addition to public funding, the firm itself has invested considerably of their own funds. They are more interested in regional visitors than typical tourists, and in any case they want groups that book their visits in advance to be able to handle it as an activity additional to reindeer herding. The firm has also recently employed a manager with a South Sami background and outreach competence.

What is interesting to us in the comparison with our Northern Sami example is that despite differences, these three actors have some important features in common. Hospitality and tourism activities are here only supplementary to reindeer herding as a livelihood. All activities are also motivated by an ambition to pass Sami culture on to others. These three tourist outfits are also more focused towards a regional and local tourist market than long-travelling tourists. They aim to attract groups through culturally based activities, such as seminars and courses, good food, storytelling, traditional knowledge, handicrafts and accommodation in traditional buildings. All initiatives are located close to one or two national parks and thereby they can serve as additional destinations for park visitors. Belonging to three different neighbouring reindeer pastures districts (Tjåehkere Sijte, Låarte, Skæhkere) and having all together three national parks on their lands (Børgefjell, Lierne, Blåfjella/Skjækerfjella) and also having one natural park centre (Lierne nasjonalparksenter) for two of the parks, the possibilities for attracting visitors are good. Further, there are two Sami cultural centres and museums (Snåsa and Hattfjelldal) that can attract potential visitors into these destinations.

All of the Southern Sami groups have developed their activities over many years, independent of governmental programs. The opportunity to receive extra funding has been welcomed, but has not influenced their priorities. Another observation is also that they are not rushing, rather they seem to build up their activities and enterprises step by step. The aim of these activities is two-fold. Tjåehkere Sijte perceives hospitality and tourism activities both as income and employment possibilities of relevance to their youth and to smaller herders, i.e. extending their current economic basis, and as an opportunity to invest in increased understanding in the majority society. Strikingly, these actors are as occupied with the opportunity to present their life world and culture to outsiders, locals and from far away as much as they are welcoming of the extra income. For them it is clear that increasing understandings of Sami life worlds especially to non-Sami locals and regional inhabitants may contribute to the improvement of their position in society.

A Comparative Approach to Indigenous Tourism

These two cases illustrate that we cannot take indigenous nature tourism for granted as one category of activities. Nature tourism may provide new economic independence, new contact with the majority society and in the long term new emancipation, but not necessarily so. In Finnmark, tourism has obvious colonial roots. The Tana River salmon fisheries example is by no means the sustainable nature tourism that the Norwegian environmental authorities advocate. Despite the fact that salmon in the Tana is on the verge of extinction, masses of tourists fish there every year without the kinds of restrictions that are in place in other rivers, such as bag limits or catch and release (for comparative cases from other rivers with bag limits or practicing catch and release, see Nordeide [2012] and Ødegaard [2013]). In the case of salmon fishing, very little is done to ensure that Sami salmon fisheries can be continued (Ween, 2016). Sami academics and politicians are concerned that when natural resources become available to all, they become less available to locals (Ween, 2012; Ween & Lien, 2012), so also for locals in Sami communities (Myrvoll *et al.*, 2013). Can the loss of access to a core cultural and natural resource be balanced by tourist income? Our comparison illustrates the need for nuancing this understanding, which is both prevalent on the Finnish side of the Tana River, and also increasingly so within Norwegian nature management. As is the case with the Tana River salmon tourism, tourism became an option when traditional life styles became less available. Nature tourism does not necessarily provide a sustainable economy, even when it takes place without the use of motorised vehicles. In Tana, hunting and fishing tourism are enacted according to the kind of Scandinavian ideal of a simple outdoorsy life (*friluftsliv*). For this kind of being in nature (trekking, fishing, hunting or skiing) very little is necessary next to basic accommodation. Those interested buy hunting and fishing licences and organise their own transport and accommodation (which is inexpensive and simple). This kind of tourism also involves little interest in local people and the limited contact that takes place between tourists and locals does not contribute to a greater understanding of Sami rights, working against the spirit of the southern Norwegian *friluftsliv* – outdoors activities where nature is a space apart (Ween & Abram, 2012).

Enacted within a wilderness landscape, then, tourist activities naturally provide limited opportunities for engagement with individual tourists on Sami terms (for similar descriptions see Pratt [1992] and Ween [2002]). These areas also, as we have seen, encompass tourist activities with a limited opportunity for innovation. Tourists to the Tana River know exactly what they are after and are mostly not willing to pay more. Nature tourism of this kind instead increases the pressure on limited natural resources, bringing not only the risk of resource depletion but also interruption to cultural practices (Nustad, 2012; Ween, 2012; Ween & Lien, 2012), as well as disturbing local customary understandings of acceptable use.

Innovation in tourism activities is, however, apparently easier to accomplish in Southern Sami areas, where new tourist activities are not building on well-established tourist expectations. Public funding for new innovative activities also opens up new opportunities for creative grass-roots initiatives. Here, as well as in the sole example from the Finnish side of Tana, attempts are made to carve out a new way of living off natural resources, more in line with traditional cultural relations and ways of life. As a niche market, such activities may not provide significant incomes, but as the Southern Sami example illustrates, this might not be the only ambition.

In the Southern Sami case, all activities originate with a consideration of what it is that local reindeer owners would like to sell and what they think would be of interest to visitors and tourists. As is the intention with the example of the more upscale Finnish Sami tourism outfit from Tana, Southern Sami tourism ambitions are more sustainable in the sense that they sell cultural experiences rather than natural resources. The projects are smaller scaled and based on the premise of close-up encounters involving the sharing of Sami culture (for similar examples, see Butler & Hinch, 2007). Tourism in these cases offers the potential to provide opportunities to put culture on display, both for external and internal purposes. Guests and trading relations with people from elsewhere have always been a Sami tradition. These have brought stories from elsewhere about people, knowledge, skills and cultures (Butler & Hinch, 2007; Pratt, 1992). Similar situations can occur when indigenous peoples themselves are able to set the terms for interaction. Only then can new meetings occur that transgress colonial tourist scripts, providing indigenous peoples with the opportunity to reach out to the non-indigenous, to articulate and hopefully create understandings of their life projects and cultural understandings (Butler & Hinch, 2007).

A comparison between tourism in Tana and Snåsa in other words, reminds us that indigenous tourism includes concomitant influences of decolonisation and recolonisation. Mass tourism rarely enables indigenous emancipation (Bruner, 2001; Pratt, 1989). Operations based on a limited natural resource provide less opportunity for cultural exchange and threaten indigenous access to key resources, and are hence not sustainable even if such initiatives can become economically viable. Indigenous tourist activities making use of cultural knowledge as a commercial product may, however, provide economic opportunities by using experience and cultural exchange as a renewable resource. The decolonising potential of such exchanges, however, depends on how culture and experience become an economic merchandise, the existence of colonial tourist trajectories and the ability of indigenous tourist agents and entrepreneurs to successfully produce and control their own terms of engagement.

Notes

(1) Lierne, Grong, Snåsa and Overhalla.
(2) South Sami: Nååmesje which gives the name Namdalen (dalen=valley).
(3) http://www.vg.no/spesial/2012/landbaroner/landbaronene.html.
(4) http://www.adressa.no/nyheter/okonomi/article1710333.ece.
(5) http://skorovas.no/.
(6) http://www.namdalsavisa.no/artikkel/2016/10/06/Seks-%C3%A5rs-ventetid-er-over-%E2%80%93-n%C3%A5-starter-bygginga-av-fjellandsbyen-Skorovas-13609316.ece.
(7) Thor Riseth, personal information.
(8) https://www.regjeringen.no/nb/tema/urfolk-og-minoriteter/samiske-sprak/samelovens-sprakregler-og-forvaltningsom/id633281/#2.
(9) http://www.miljodirektoratet.no/no/Tema/Verneomrader/Naturarven-som-verdiskaper/Om-programmet/.
(10) http://www.ntfk.no/SiteCollectionDocuments/Naturarven%20som%20verdiskaper.pdf.
(11) http://www.reinoggodt.no/.
(12) http://wilksdesign.no/.

References

Arnesen, T. and Riseth, J.Å. (2008) Konfliktfyllda naturskyddsprocesser i Norge. In C. Sandström, S. Hovik and E.I. Falleth (eds) *Omstridd natur. Trender och utmaningar i nordisk naturförvaltning* (pp. 83–104). Umeå: Borea.

Arnesen, T. and Riseth, J.Å. (2009) Protection and equity? Local and indigenous encounters with the grand scheme of area protection in Norway. In C.M. Hall, D. K. Müller and J. Saarinen (eds) *Nordic Tourism: Issues and Cases* (pp. 77–80). Bristol: Channel View Publications.

Åsvoll, O. (2010) *I fjelli. Minner og meninger fra et langt liv i Snåsafjella.* Steinkjer: Embla.

Brennli, E. (2014) Vagekie på Røyrvik vokser til. *Ságat* 228 (28.11.2014), 14–15.

Bruner, E.M. (2001) The Maasai and the Lion King: Authenticity, nationalism, and globalization in African tourism. *American Ethnologist* 28 (4), 881–908.

Butler, R. and Hinch, T. (eds) (2007) *Tourism and Indigenous Peoples: Issues and Implications.* Oxford: Elsevier.

Bunten, A.C. (2010) More like ourselves: Indigenous capitalism through tourism. *The American Indian Quarterly* 34 (3), 285–311.

Bye, A. and Jåma, B.Å. (2011) *Mijjieh Låartesne. Landet går i arv. Rein, folk og natur i Luru.* Levanger: Låarte Reinbeitedistrikt, Nord-Trøndelag.

Clifford, J. (1997) *Routes: Travel and Translation in the Late 20th Century.* Harvard, MA: Harvard University Press.

Fedreheim, G.E. (2013) *Value Creation on Norway's Green Gold: An Analysis of Policy Formulation and Implementation in the Fields of Nature Conservation.* Bodø: University of Nordland.

Gran, E. (ed.) (2014) *Snåsabilder.* Snåsa: Snåsa BygdaMuseum og Snåsa Historielag.

Haraway, D. (2008) *When Species Meet.* Minneapolis, MI: University of Minnesota Press.

Hermanstrand, H. and Kosmo, A. (2009) *Røyrvik: Samene i Østre Namdal.* Røyrvik: Røyrvik kommune.

Joma, N.P. (2013) Speech at the 50 years anniversary of Børgefjell National Park.

Law, J. (2009) *Collateral Realities.* See http://www.heterogeneities.net/publications/Law2009CollateralRealities.pdf (accessed April 2016).

Myrvoll, M., Riseth, J.Å., Thuestad, A., Evjen, B., Dalsbø, E.T., Samuelsen, T. and Hansen, H. (2013) *Felt 4 Káráŝjohka/Karasjok: Sakkyndig utredning for Finnmarkskommisjonen.* NIKU Oppdragsrapport10/2013. Tromsø: Norsk Institutt for Kulturminneforskning.
Nordeide, A. (2012) *Møte mellom menneske og laks: Om laksefiskepraksiser Namsenvassdraget.* Master's thesis, Department of Social Anthropology, University of Oslo.
NOU (2008) *Den nye sameretten.* Oslo: Justisdepartementet.
Nustad, K.G. (2011) Performing natures and land in the iSimangaliso Wetland Park, South Africa. *Ethnos* 76 (1), 88–108.
Ødegaard, M. (2013) *'I samme båt' – stakere, fluefiskere og maskulinitet i laksefisket i Alta.* Master's thesis, Department of Social Anthropology, University of Oslo.
Pedersen, S. (1986) Laksen, allmuen og staten. Fiskerett og forvaltning i Tanavassdraget. *Diedut* 1986 (2). Kautokeino: Sami Instituhtta.
Pratt, M.L. (1992) *Imperial Eyes: Travel Writing and Transculturation.* London: Routledge.
Reindriftsnytt (2010) Destinasjon Dærga-spennende visjoner i nordtrøndersk reindrift. *Reindriftsnytt* 2010 (1), 17–22.
Reindriftsnytt (2013) Naturarven som verdiskaper: En smak av det sørsamiske Nord-Trøndelag. *Reindriftsnytt* 2013 (1), 49–51.
Riseth, J.Å. (2007) An indigenous perspective on national parks and Sami reindeer management in Norway. *Geographical Research* 45 (2), 177–185.
Riseth, J.Å. (2015) Nasjonalparker og samisk reindrift. In T. Skjeggedal and K. Overvåg (eds) *Fjellbygd og feriefjell – 50 år etter* (pp. 155–179). Bergen: Fagbokforlaget.
Riseth, J.Å. (2016) Sami reindeer herding and Norwegian national parks: A space for Sami values? In L. Elenius, C. Allard and C. Sandstrom (eds) *Sami Customary Rights in Modern Landscapes: Global Influences on Nature Conservation and Indigenous Rights* (pp. 146–166). London: Routledge.
Riseth, J.Å. and Holte, A. (2008) Rennäring och nationalparker i Norge. In C. Sandström, S. Hovik and E.I. Falleth (eds) *Omstridd natur: Trender och utmaningar i nordisk naturförvaltning* (pp. 269–290). Umeå: Borea.
Riseth, J.Å. and Solbakken, J.I. (2010) Naturbruk i Kautokeino i et kulturhistorisk lys. *Utmark* 2010 (1), n.p.
Ryan, C. and Aitken, M. (eds) (2005) *Indigenous Tourism: The Commodification and Management of Culture.* Oxford: Elsevier.
Solbakk, A. (2011a) *The Salmon Lords take over Deatnu/Tana River: English Angling 1850–1900.* Káráŝjohka: CálliidLágádus.
Solbakk, A. (2011b) *Deatnu-Tana den beste lakseelva: utviklingen av laksefisket med vekt på stangfisket – oaggun – I Tanavassdraget: Praksis, bruk og forvaltning etter 1750. Tana: Deanu luossabivdosearvi.* Tana: Deanu luossabivdosearvi.
Støvik, P. (1997) *Namseallmenningen gjennom 1000 år.* Steinkjer: Namskogan Fjellstyre.
St.prp. nr. 32 (2006–2007) *Om vern av villaksen og ferdigstilling av nasjonale laksevassdrag og laksefjorder.* Oslo: Miljøverndepartementet.
Ween, G.B. (2002) *Bran Nue Dae: Indigenous Rights and Political Activism, in Kimberley, Western Australia: A Study of Indigenous Activism.* D. Phil. Dissertation, Oxford University, UK.
Ween, G.B. (2009) Blåfjella-Skjækerfjella nasjonalpark: Naturforvaltning som produksjon av natur/sted. *Norsk Antropologisk Tidsskrift* 20 (1–2), 95–109.
Ween, G.B. (2010) Sjølaksefiskets betydning for sjøsamisk kultur. I *Sjølaksefisket i Finnmark i et historisk perspektiv. Utredning for Finnmarkskommisjonen.* Kautokeino: Sami Allaskuvla and SEG.
Ween, G.B. (2012) Performing indigeneity in human–salmon relations: Salmon, Sami and natural resource management. In R. Ellefsen, G. Larsen and R. Sollund (eds) *Eco-global Crimes: Contemporary and Future Challenges* (pp. 295–312). Farnham: Ashgate.

Ween, G.B. (2017) Protection of Sami intangible cultural heritage and intellectual property rights, and its relation to identity politics in a postcolonial Norway. In C. Antons and W. Logan (eds) *Intellectual Property: Cultural Property and Intangible Cultural Heritage*. London: Routledge.

Ween, G.B. (in press) *Laksehistorier. Fiskere, forskere og forvaltere i Tana*. Karasjok: Callidlagagud forlag.

Ween, G.B. and Abram, S. (2012) The Norwegian Trekking Association: Trekking as constituting the nation. *Landscape Research* 37 (2), 155–171.

Ween, G.B. and Lien, M.E. (2012) Decolonisation in the Arctic? Nature practices and rights in sub-Arctic Norway. *Journal of Rural and Community Development* 7 (1), 93–109.

Ween, G.B. and Risan, L.C. (2014) Exploring heritage lives: Indigenous peoples in world heritage sites. *Primitive tider* 2014 (2), 61–72.

Part 4

Tourism Negotiating Sami Traditions

13 What Does the *Sieidi* Do? Tourism as a Part of a Continued Tradition?

Kjell Olsen

A paper presentation I attended at the Nordic Symposium in Tourism and Hospitality Research some years ago, inspires this chapter. The presentation was about the Winter Festival in Jokkmokk, Sweden; in terms of visitors, probably the largest Sami tourist attraction in northern Scandinavia. As I recall, several researchers, some of whom edited this book, had individually singled out what they regarded as the Sami cultural features at this market and what they regarded as Swedish. To me, the rather puzzling result they presented was that, among other things, they found that reindeer meat sold at a stall was a Sami cultural artifact, while the hamburgers sold on the opposite side were Swedish! Rather, the hamburgers being Swedish was a logical consequence of the dichotomy between Sami and Swedish that guided the research rather than something the presenter reflected upon. As usual, culture belonged to the other and not the majority population. The categories they had probably unconsciously applied were the opposition between tradition and modernity, a dichotomy commonly in use when the indigenous becomes a tourist product (Olsen, 2003, 2008a, 2008b).

To ascribe the indigenous to the past or a timeless present (Fabian, 1991) and embody peoples in that category with certain features of the Western imagination are not only done in highly modern markets in Sápmi, the Sami area that covers northern Scandinavia. As Muehlmann (2008: 42) describes in the article '"Spread your ass cheeks": And other things that should not be said in indigenous languages', some scholars (i.e. Broady, 2000) claim that swear words in general are not found in indigenous vocabulary because anger is considered 'childish' and therefore suppressed in such cultures. As the title of Muehlmann's (2008) article indicates, such claims are not substantiated by empirical research. More important is her analysis of how, in the neoliberal climate of Latin American politics, language competence has been elevated by policymakers to a primary criterion for recognition of indigenousness (Muehlmann, 2008: 34; Olsen, 2016).

What these two examples demonstrate is the long tradition in a Western colonial culture of ascribing other populations to certain categories and delimiting them by categorizations based on cultural features created by the Westerners' own ontology of cultural evolution and hierarchies. As Chow (2006: 53) describes this tradition, the indigenous becomes 'a sign of his own foreclosure from a system that will simply go on "imagining" him – that is, keeping him afloat utopically, as it were, in a nonplace?' Furthermore, these examples also reveal how such categorizations are going on in different fields, in these cases in the governmentality of neoliberal politics and the politics of research. This renders that Kirshenblatt-Gimblett's (1998: 176) punchline 'Where do old ideas go to die? Tourism, a museum of the consciousness industry' must be seen as an attempt to uphold another important tradition of categorization in a Western tradition. This is the aim of establishing hierarchies between categories of high and low, tourism definitively belonging to the latter seen from the ivory towers of scholarly activities. However, I claim that, as my colleagues eloquently demonstrated in Jokkmokk, these boundaries are also blurred.

In this chapter, I aim to analyze such blurred boundaries and intertwining between different discourses on hierarchies and dichotomies. The pivotal point in this analysis is a sacrificial stone, Áhkku, which means grandmother or old woman (Hætta, 1994),[1] situated on the outskirts of Alta, the largest town in the county of Finnmark, in the Norwegian part of Sápmi. Áhkku is, in Northern Sami dialect, a *sieidi*, a sacrificial place usually connected to pre-Christian Sami religion.[2] What my analysis will reveal is that Áhkku seems to have her own agency by simultaneously engaging in discourses that elsewhere are kept separate. Therefore, Áhkku is a starting point for analyzing Western discourse on the indigenous in tourism as well as how an attempt is made to keep supposedly different institutions such as scholarly activity, politics and tourism separate by binaries and hierarchy. These are categorizations that Áhkku seems to deny by the performances and relationships where the rock plays a part.

First, I briefly discuss how Sami *sieidis* have been historically and contemporarily described by scholars and point out some of the contemporary uses of *sieidis* and the discourses that frame them. The main point is that *sieidis* are contested spaces where different actors aim to impose their own categorizations of how *sieidis* are to be understood. The second part describes my own findings at the site of Áhkku. These findings cover a period of 15 years during which I have visited this *sieidi*, and I describe the changes and continuities of findings. What these findings demonstrate is what I will label a relationship to conventional practices, a practical engagement with Áhkku found among locals as well as tourists (Edensor, 2000, 2007; Ingold & Kurttila, 2000). The last part discusses the discrepant discourses related to *sieidis* and, in particular, how tourism, to lend a phrase from Bauman (1996), has moved from the periphery into the center of modernity. Anderson (2014: 15–16) suggests that 'offerings' and 'placing' of items on the land is a characteristic of what he labels 'models of reciprocity' between people, animals and land, found in the circumpolar cultures. The questions then are: what kind of relationships does Áhkku establish and in what way do these relationships blur the binaries found in the discourse on indigeneity?

Sieidis: Past and Present

Sacrificial places – *sieidis* – are found all over Sápmi. Archeologists broadly divide Sami sacred sites into three classes. These are structures, terrain formations and natural objects. Structures include carved stubs, erected stones, wooden poles and stone circles. Terrain formations consist of mountains, fell tops, rock formations, islands, lakes and headlands. Cleaved boulders, stones, springs and small caves or clefts are natural objects. The *sieidi* offering sites usually consist of a rock or a rock formation like a distinctive stone, a cliff or a cleaved boulder (Äikäs, 2011: 16). Historically, wooden idols were also used as *sieidis* but, at least in the Swedish part of Sápmi, they vanished during the 17th century because 'Christianity gradually impinged on Sami life and indigenous ritual space' (Bergman *et al.*, 2008: 22). Common locations for the sacrificial sites are holy mountains, shores of lakes, islands and headlands or in areas used for hunting, gathering and fishing as well as in areas important for reindeer pastures.

Äikäs and Salmi (2013: 69) state that archeological findings demonstrate that *sieidis* have been used all over Sápmi, but the findings also reveal huge temporal and spatial variations.[3] The different groups' varying spatial orientation and changing inclusion in trade systems are evident as well as religious and societal differentiation (Äikäs & Salmi, 2013: 68–69, 76; Hansen & Olsen, 2004: 75, 123–125, 138). A major change is the increase in the discovery of bone from reindeer set in the context of the transition to a pastoral adaptation that, for some Sami groups, happened from 1400 to 1500 AD. The transition to a full pastoral adaptation for these groups ended in the 1600s but continued in combination with hunting and trapping of varying intensity in different areas (Äikäs & Salmi, 2013: 69–70).

The Sami's first contact with Christianity probably dates back to the 11th century because of the fur trade with Novgorod and thereby contact with Russian orthodoxy. Such a contact continued through the Medieval Ages, while along the Atlantic coastline and in Swedish areas there was also an influence from the Catholic Church (Hansen & Olsen, 2004: 151, 200; Kylli, 2012: 35). As Äikäs and Salmi (2013: 72) state, 'the new religion did not replace the old ways. Instead, people turned both to old gods and to the new one. In addition, the old offering practices moved into a new context'.

Turning to the written sources on Sami religion, most of them date from between the 17th and the early 20th century. Documenting the *sieidis*, they refer to offerings of live animals (deer, sheep, cattle, horses and dogs) and slaughtered animals. They also describe food, meat, fish, blood, fat, cod liver oil, butter, milk, flatbread, liquor and herbs, tobacco, komagband,[4] bone spoons, reindeer tendons, reindeer skins and coins (Äikäs & Salmi, 2013; Myrvoll, 2008: 10). Maybe more important than the actual findings are how the practices have been described. Mathisen (2010: 13) points out that the older written sources that describe Sami religion and thereby document beliefs concerning the *sieidis* are usually written by Lutheran missionaries and thereby reflect their aim of converting the population to their brand of Christianity. As a result, most such sources have a certain perspective; 'The viewpoint was often that of an outsider who saw the Sami as "the other" and their religion as pagan traditions that antedated the arrival of Christianity in the north' (Äikäs & Salmi, 2013: 65). As Mathisen (2010: 14) puts it: 'The missionaries' reports constructed a Sami religion that was antithetical to Christian religion. By demonizing Sami beliefs and religious practices, missionaries also constructed legitimate grounds for their own proselytizing activities'.

Their descriptions possibly exaggerated the *sieidis'* importance for Sami beliefs and the same applies to their antithetical position to Christianity (Äikäs & Salmi, 2013: 72). The impression of heathen and demonic places is not prominent in the later ethnographic material based on local Sami's narrations. Mathisen (2010: 13–14) reveals that in these sources there is usually a simple and uncomplicated structure and very similar plots. Mostly, the actors' use of *sieidis* is described as for their own goals and luck in fishing, hunting or reindeer herding, or they cause unhappiness when traditions and prescribed practices are violated. Happiness or misfortune is mainly dependent on the actors themselves. In particular, that sacrifices must lie undisturbed is something that still seems to be a rule (Myrvoll, 2008: 30).

Nevertheless, to regard the relationship only as a matter between *sieidis*, the land and those who sacrifice is too simple. As already described, *sieidis* became entangled in the relationship between colonizing forces and the population they aimed to Christianize. Later on, they also became intertwined in the local discourses whereby local Laestadian congregations, a lay Christian movement in northern Scandinavia that recruited among the Sami as well as the Finnish and Norwegian population, saw them as one of several signs of a heathen tradition that the Sami should leave to the past. Yet, *sieidis* were not only a part of the religious discourses and the discourses on colonialization and the need to develop the pre-modern.

Some *sieidis* also found their place in the rivalry between nation states in the north. Mathisen (2010) describes how the Finnish folklorist Samuli Paulaharju in 1920 wanted to relocate a stone he had documented in Skilggašjávri on the Norwegian side of Sápmi to the National Museum of Finland in Helsinki. Due to the historical value of the stone, the Norwegian government turned down the request. Mathisen discusses how this incident can be understood inside the frame of contemporary Finnish ambitions for extending the national influence and the political purpose of giving a helping hand to one of their 'smaller brothers' in the Finno-Urgric language group. On the Norwegian side, this incident must be understood against the background of nationalistic politics as well as the politics of heritage preservation (Mathisen, 2010: 20–21). However, Mathisen (2010: 19ff) describes how the *sieidi* that is still listed as heritage, according to the local narratives, is just a fraud (see also Hætta, 2003). The local narratives say that the local guides that showed Samuli Paulaharju the stone put it up themselves. Paulaharju had been so persistent in his claim that a *sieidi* was to be found in Skilggašjávri that it was set up just to please him and probably so as not to spend any more time on the matter. According to the local narratives, contrived attractions are obviously not something new in Sápmi and not always connected to tourism but visitors in general.

As Regina Bendix states, the scholarly interest of the past does intertwine with other societal processes:

> Cultural scholarship and inquiry, furthermore, fuelled societal interest in cultural fragments and cultural wholes, becoming one force in the 'artifactualization' of facets of culture [...]. In doing so, scholarship prepared the way for the vibrant market and politics in commodified cultural authenticities, which, in turn, are becoming the new disciplinary subject. (Bendix, 1997: 5–6)

In one of these fields of politics in the authentication of objects in Sápmi, the Norwegian Directorate of Heritage has an important role. In a work on Sami sacred places, Elin R. Myrvoll (2008: 46, my translation) points out the irony that according to the Norwegian law on Heritage, sacrifice to *sieidis* is forbidden: '§ 3 of the Cultural Heritage Act prohibits changes to automatically protected cultural heritage. Thus, it is illegal to pour fish fat that results in a change to the sites. The same applies to gifts such as fish lures, jewelry, photographs and letters which are new types of gifts found'. Important for Myrvoll's (2008: 45–46) discussion is a dilemma in the relationship between the scholars' and heritage authorities' need for preservation and research and the potential local desire for knowledge of one's own history. Many of the listed objects are not known to the public and are even registered in a database withheld from the public. In areas with a history of more than 100 years of colonial policies of integrating peoples into the dominant cultures of the nation states, this causes a dilemma for the authorities. As Myrvoll (2008: 45–46) discusses, it is a quandary that the public is unaware of this part of their local history that could be used to strengthen and revitalize a Sami identity.

Not only have the heritage authorities had an interest in secrecy. Fonneland, in her study of neo-shamans that relate to a Sami tradition, demonstrates how secrecy of the *sieidis* can also be important for religious purposes:

> The fact that sacred places in Northern Norway to a large extent have been out of reach for tourists, and are not located along well-beaten tourist tracks, enhances the authentic and magic aspect of the places. This is not a place for the masses but for the selected few. (Fonneland, 2010: 119, my translation)[5]

As Fonneland (2010: 115–117) describes this neo-shamanistic use of *sieidis*, it is an occasional use, either by individuals or when giving courses in shamanism. What is important for this religious purpose is the continuity with former use and an idea of continuity with older traditions that increases the energy and power of the place. As Fonneland (2010: 116–117, my translation) puts it when describing the act of sacrifice of coins, bones or pebbles: 'The holy places are thus something that the actor must get to happen. They require actions for the force to be triggered. These are actions that simultaneously affirm the individual informant's identity and a neo-shamanistic belonging'.[6]

Who should know about *sieidis* and who is entitled to make sacrifices are contested matters in Sápmi. An event that illustrates this occurred at a tourism conference in the Norwegian part of Sápmi. An owner of a company selling nature-based experiences told the audience that he and his guests always sacrificed to *sieidis*. It was a part of the product of guiding people in Finnmark. This immediately stirred a debate on whether he, as a Norwegian, could do this or if anyone had the right to make *sieidis* into tourist products. Originating from an area with an ethnic composition of population that made it likely that the speaker could enroll for the electoral list to the Sami Parliament makes this into a good example of what Gaski says always lurks in the background in Sami political debates:[7]

> the always present issue about identity and authenticity; the often subtle recriminations about who is and who is not – who is the right kind of – too much – or not enough Sami. This never ending internal debate in Sami society, with its roots in the social sciences, is a natural consequence of the earlier assimilation politics and the ongoing process where increasing parts of the Sami population are striving to regain and create a Sami identity for themselves. (Gaski, 2008: 13–14)

Nevertheless, the use of Sami culture in tourism in Sápmi is usually informed by debates on this matter in the Finnish part of the area. In particular, the Lapp baptism, an invented tradition in the tourism industry in Finland not found in other parts of Sápmi, is commonly referred to as an example of the horrors of the tourism industry in political as well as scholarly debates (see e.g. Comaroff & Comaroff, 2009). In the Lapp baptism, tourists are jokingly baptized by an apparent Sami often playing the role of shaman, usually when the Arctic Circle is crossed but also on other occasions. Ruotsala (2008) describes such a Lapp baptism product sold at the location of the *sieidi* of Taatsi, situated 60 kilometers from the winter tourism hub Levi. In Ruotsala's opinion:

> Lapp baptism at the *sieidi* of Taatsi is an example of how local people's sacred place, their magic or historical landscape, has been transformed to a product, a play for tourists without any ethnographic origin – or rather, to a stage of a play. In that sense it enters the area of trans-cultural and trans-local politics of ownership, monetary value and representation. The *sieidi* of Taatsi presents an ideal stage for Lapp baptism in the tourist industry, because the place is 'authentic'. In a sense Lapp baptism is also authentic: it is an authentic part of the history of tourism in Lapland, but it does not belong to the culture or history of Lapland or Sápmi in any other form. (Ruotsala, 2008: 52)

As Okkonen (2007: 38) also points out, some of these sites receive many tourists in search of experiences and 'these people attach their own meanings to the place and make their own interpretations of what they have seen quite independently of the experts'. It might be added that these are interpretations that can also be independent from local traditions that might be discrepant from the view of the experts.

Before commenting on this issue and other discourses outlined in short previously, I will present some archeological facts concerning the practices of *sieidis*. Äikäs and Salmis' (2013) research is based on excavations at offering places in Kittilä Taatsi, in the Finnish part of Sápmi. As they put it 'We expected to find ancient bone material as traces of old offering practices, but, somewhat to our surprise, we also came across an abundant amount of modern objects' (Äikäs & Salmis, 2013: 65). The most common of what they label modern post-1950s finds were coins. Such offerings are a Sami tradition, and the amount of coins found at locations far from what they label tourist routes and roads indicates that it continues as a local tradition. Hence, the amount of modern non-euro coins, from five countries other than Finland, indicates visitors from abroad.[8] Other items that are both framed by tradition and indicate tourism were bottles, which indicate alcohol offerings or drinking, and fishing gear. In addition, they found cold smoked reindeer meat and elk skulls. Furthermore, they show that the tradition of offering personal objects continues. Items that they found were rings, earrings, an eyeglass lens and a yo-yo (Äikäs & Salmis, 2013: 73–75), objects that Äikäs and Salmi (2013: 74) claim indicate a closer relationship with a *sieidi* than coins do because of their personal use. Nevertheless, the excavation cannot give evidence on who has put the items on the *sieidis*, and as Äikäs and Salmi (2013: 74–75) write 'it can sometimes be difficult to separate touristic behavior from the continuation of local traditions – they might go hand in hand. There is a continuation of traditions that have also been borrowed by tourists'.

Turning to the discourses that have integrated the *sieidis* into their attempts of ontological ordering, several binaries are found. There is the dichotomy between the savage and the colonizer symbolized by religious differences that ascribe people into separate hierarchical positions. This symbolism persists but is now framed by Laestadian lay Christianity as an opposition between the faithful and the condemned. The clear-cut ethnic categorizations promoted by the ethnographers of the 19th century are still present, even if written sources document that some stones, historically, also received offerings from settlers (Äikäs & Salmi, 2013: 75) and that locals without regard for ethnic belonging seem to use them today. The ambivalent, but still hierarchical, relationship between scholars and the locals is still found. In the early part of the 20th century, the locals at Siebe could, maybe, trick Samuli Paulaharju because of their knowledge. Today, scholars and heritage institutions have the knowledge and have to balance the interests of scholarship and local identity work. Last but not least is the question of who has the right to use the *sieidis*? Is it the locals or the tourists or tourist entrepreneurs and, if so, do they have to be Sami? Or should they be used at all? All these questions and several more can be deduced, but what does the *sieidi* say?

Áhkku

Many *sieidis* are not easily accessible, but Áhkku is located on a headland quite close to the center of Alta.[9] It is just a five minute drive from the town center and then a 15 minute walk at a moderate pace along a marked trail. On the opposite side of the fjord, on a headland in Kåfjord, another *sieidi* is located. The walk toward Áhkku starts from a parking lot on the outskirts of a housing estate at Amtmannsnes. It is a nice walk in good weather. You have a view of the fjord, can watch the moderate activities at the local airport and sometimes, if lucky, watch small whales in their activities. On the point where Áhkku is situated, there is a place for a campfire, and it is possible to sit down for a picnic (Figure 13.1). In the autumn, blueberries are usually abundant on the mountain slope. The conditions for fishing from the shore are not good. Along the seashore that the trail follows, one can, at the right time of the year, see the fishing nets of salmon- and trout-fishers aiming to catch fish in the traditional way.

My first encounter with Áhkku was early in June 2000. My colleague Stein R. Mathisen and I had prepared a symposium for a Nordic project (Antonnen *et al.*, 2000; Sikkala *et al.*, 2004). As part of the evening program,

Figure 13.1 Áhkku (Photo: K. Olsen)

we planned an excursion to a Sami *sieidi*. The cunning plan was to put some 'offerings' on Áhkku for the surprise of the distinguished scholars who were visiting us. This was shown not to be necessary. As in the case of the excavations of Äikäs and Salmi (2013: 65), somewhat to our surprise, we found quite a few items on the rock. There were coins, stones and a letter. The letter that we found about one meter from the *sieidi* was probably, because of the style of the handwriting, written by two young girls with the hope of getting in touch with a certain boy.

Some years later, the local authorities placed a small signpost in a tree close to the *sieidi*. In addition to general information about *sieidis* and pre-Christian Sami religion, the poster gives some brief information about Áhkku. It tells how the Coastal Sami used to offer fish and fish fat at the stone. The text in Norwegian describes Áhkku as a freestanding rock formation with a height of five to six meters and as situated at the north-west side of the Komsa Mountain, a popular recreational area. Finally, it is written: 'During visits to Seidekjerringa one can see that on a small shelf at the back of the sacrificial stone are lights, flowers, coins, small rocks. A letter has also been observed on the shelf'.[10] In this way, Mathisen and my observation dating back to 2000 has become a part of the information provided for visitors. Even more, this information poster has also become a part of the sacrifices. In June 2014, somebody had placed the poster on Áhkku with a stone on top, and, following the local tradition of not removing what has been sacrificed, it will probably remain there as long as the wind allows it.[11] In that way, it will add to the heaps of different items that people have placed on her. From the relatively modest amount of items that we saw on the rock in June 2000, today it has become a rather large collection of sacrifices (Figure 13.2). Some of them have been there for years.

Since our first visit, I have visited Áhkku at least once a year to see how and if it is possible to observe any changes in use. I have seldom met people at the site. Sometimes, I have encountered people on the trail, the only route to visit the stone except by boat or, for fit locals who can, by approaching it from the opposite direction or climbing down from the mountain behind. Those few I have met have been locals, often families with small children, and more seldom tourists. As for the archeological excavation described by Äikäs and Salmi (2013) on the Finnish side of Sápmi, neither do I have any information from those who sacrifice to Áhkku on the Atlantic coast of Sápmi. Nevertheless, from what I have seen of the items, they are very much in line with the post-1950s findings of Äikäs and Salmi (2013: 73–75).

Concerning the findings, the first years I mainly observed coins and small stones put on the *sieidi*. Increasingly, also because of the

Figure 13.2 Sacrifices to Áhkku (Photo: K. Olsen)

accumulation of objects, Áhkku seems to be filled with offerings. The change from a more discrete practice to that of now, whereby the *sieidi* is close to being overshadowed by its own success, seems to have started in about 2013. A picture from 2012 shows a rather modest collection of items on the shelf where most sacrifices are made. In the 2012 picture, shells, a broken eyeglass, jewelry, a bottle cork and a hair scrunchie are found together with the more common coins and stones. The June 2014 picture shows an abundance of items. Many of the 2012 items are still present, but there were also three bottles, a ball, flowers, more jewelry, packing material for sweets and cigarettes, a lighter, crayons, moss, an information poster, a receipt from the nationwide company Biltema and several other items. From the previous year, a bicycle screen seemed to have disappeared. Other small shelves on the stone had also filled up. Coins, Scandinavian snuff but also many personal items were in different places at Áhkku. The more personal items seem to have increased between 2012 and 2014. A plastic tube, seen the previous year on the main shelf, had been moved and now shared a shelf with stones, sweets, a rope, a wooden stick, a plastic cup with some toys in it and a sea urchin turned into a flower pot. Small toys were in high numbers, and there were many crayons and flowers spread around on the stone; two pieces of graffiti sprayed on, which I saw for the first time in 2011, are still there. Many of the personal objects gave me the impression that a kindergarten must have had an excursion to the site. However, as Äikäs and Salmi (2013: 74–75) write, it is difficult to separate tourists from locals and whether local traditions have just been borrowed by visitors. Anyhow, the amount of objects found in June 2014 compared with the moderate increase in August 2013 indicates that there are local traditions that have amplified in the last couple of years. Whether this is a tradition maintained by people who can trace their ancestry to Sami or other populations with a tradition in the area is even more difficult to say. The locals today are a result of Alta's growth into the hub of Finnmark, with people coming from many places in Norway and abroad. The majority of those offering to Áhkku might just be newcomers like myself. Nevertheless, there are also traces that point to tourists; there were foreign coins, and at least one of the bottles of alcohol was foreign (Figure 13.3).

Still, the practices in relation to Áhkuu can be seen in a Sami tradition. The offering of coins is part of how the tradition has been described. This also applies to offerings of alcohol and, as observed on Áhkku in 2011, sacrifices of smoked fish. The same continuation is the offering of personal objects; a continuation first observed with the letter from the two girls back in 2000, and today what can be labeled personal items seem to dominate the practice. Whether such items demonstrate a closer relationship with a *sieidi* than the coins do is hard to say. Nevertheless, they document a rapid change in this current tradition in relation to Áhkku.

Figure 13.3 Tourist offerings? Foreign coins and bottles of alcohol (Photo: K. Olsen)

The practices in relation to this particular *sieidi* occur irrespective of many of the discourses outlined by scholars. Most of the pictures in this chapter demonstrate a violation of § 3 of the Norwegian Cultural Heritage Act that prohibits changes of automatically protected cultural heritage sites (Myrvoll, 2008: 46). For example, graffiti on the Norwegian religious symbols, the stave churches or burial mounds would probably have caused investigations by the police. The heritage authorities' dilemma of secrecy and thereby protection of information for contemporary identity work which are both important in the scholarly debate (Myrvoll, 2008: 45–46) is not the matter in the case of Áhkku because of her location. Nevertheless, whether the practices at the site can be understood as a way of reclaiming a past is hard to say, but there is definitively a relationship between Áhkku and the local population that some tourists also enter.

As Áhkku does not have the remote location of many other stones preferred by neo-shamans who attach their authenticity and ideas of magic to places for the 'select few' (Fonneland, 2010: 119), it has definitively created a relationship with the 'masses'. Nevertheless, the practices at the site make the place happen, but whether these performances affirm the visitors' identity and belonging (Fonneland, 2010: 116–117), or rather what kind of identity and belonging, is impossible to know. Neither does it seem that Áhkku discriminates between locals and tourists. The site is open and even, to me, the most unlikely objects put on the shelf do seem to stay there as long as the item can withstand the sometimes harsh natural forces. Neither does it seem that the issue of ethnic identity, always present elsewhere, is at stake. The question of who is and who is not 'real Sami' (Gaski, 2008: 13–14) and therefore entitled to use the rock has never, to my knowledge, been debated. Not even the dichotomy between the pagan and the faithful Christian seems to be at play in a local context. The heap of non-spectacular and familiar items just adds up, whether they are in terms of a scholarly described Sami tradition or not. What all these physical items do, I will argue, is to dissolve the hierarchical binaries that elsewhere dominate the discourses on the *sieidis*. Instead, they indicate a way of creating a relationship with the site and thereby with the land.

Models of Relationships

As Chow (2006) claims, the indigenous has become a sign constituted by its temporal and spatial exclusion from the discourse that invented it (Fabian, 1991: 197–198), i.e. the idea of modernity. This is an idea that underpins the discourses that single out the indigenous by emblematic signs as in Jokkmokk or elsewhere, reinforces debates on identity politics, the religious fields of neo-shamanism as well as lay Christianity in Sápmi and is a pivotal force in institutional heritage policies and tourism (Fonneland, 2010; Gaski, 2008; Mathisen, 2010; Muehlmann, 2008; Myrvoll, 2008; Olsen, 2003, 2008a, 2008b).

What about Áhkku? A five to six meter high rock formation is not actually 'a sign kept afloat utopically as it were a nonplace' as Chow (2006: 53) describes the indigenous as a sign. The concrete and solid appearance of this *sieidi* necessitates an attention to materiality. The same applies to all those items that, because of concrete acts whose meaning we do not have access to outside our own imagination, result in Áhkku being covered with objects. Nevertheless, these acts and the items involved are part of performances that, as Fonneland (2010: 116–117) puts it, make the place happen. Edensor (2000, 2007) has suggested a perspective on tourism as performance. According to him, tourism is a performance guided by rigid conventions that guide habits and routines, which shapes experiences. Nevertheless, tourist performance simultaneously is 'a time in which a heightened reflexivity is sought in the confrontation with sensual and cultural difference' (Edensor, 2007: 200).

When performing as a tourist, one interacts with a discourse of tourism framed by material structures and signs (Edensor, 2007: 203ff). Tourism then becomes a performance in what Edensor (2007: 202–203) labels *touristscapes* that are shaped by 'an array of techniques and technologies which are mobilized' to give directionality to the performance. Building on Ingold and Kurttila's (2000) idea of a *taskscape* that is consistently reproduced by habitual acts by its dwellers, Edensor (2007: 206ff) describes mainly such spaces, or rather serial spaces, which are highly regulated and deliberately constructed for the purpose of a certain kind of tourism. At such places a common sense of understanding alternates with a heightened reflexivity, unpredictable sensations and 'openings for thinking otherwise' (Edensor, 2007: 212).

To regard tourism as a role that is historically situated and learned, a role that people on tour enter and leave in relation to the proper circumstances (Löfgren, 1994: 122; Olsen, 2002: 169) also opens an analysis of less predictable encounters. These are encounters with spaces organized differently by others' acts inside the frames of other relationships to the landscape than tourism and often necessitate that visitors think otherwise. These spaces might engage people to perform in other ways than habitual and practical enactions that reflect common sense understandings of how to be a tourist. In many ways, this opens one up for what Cohen (2007) describes as experiences caused by 'the flow of life' that are also an important part of tourism. In Cohen's (2007: 77) perspective, 'exactly the unplanned or unexpected sights and events on the tourist's trip, will be the most authentic, and will become the most memorable experiences of the trip'.[12] Whether these sights and events will be the most memorable or not is not easy to say and not an issue in this chapter. The issue here is to describe such a relatively non-interfered with and unplanned site as Áhkku or other such landscape formations, which are multitudinous in Sápmi as taskscapes that only occasionally present the highly regulated man-made sites of a touristscape. Rather, people on individual tours in Sápmi have to relate to a circumpolar landscape shaped by other purposes. The landscape and traces of other actors' tasks performed in relation to it, to a large degree, open up the possibility of a heightened reflexivity and interaction with the landscape mobilized by arrays of techniques and technologies not common in tourism.

Furthermore, tourists and probably many locals encounter Ákkhu at a certain time. Wang's (2005: 50) attempt to 'liberate' the concept of authenticity as applied in tourism research from its commonly object-related use is to regard existential authenticity as 'a potential existential state of Being which is about to be activated by tourist activities'. As Shepherd (2015: 68, my emphasis) points out, Wang's perspective turns the attention from the tourist site, because 'for many such tourists the site is less important than the sociability of experiences at the site. [...] Such experiences are as much a question of *when* as they are *where*'. Turning away from Shepherd's (2015), in the realm of tourism research, unusually well-informed work on existential authenticity, one might question if the *when* can be divided from the *where* if one wants to give an account of tourist experiences. Rather, in my opinion the *where* interacts with the *when* in creating relations to people and, I will argue, to the landscape. Landscapes are not an empty stage, neither are they visited on a random when!

Nevertheless, tourists sometimes, and some tourists in some places more often than others, also relate to taskscapes with a heritage shaped and consistently reproduced by habitual acts by its dwellers (Ingold & Kurttila, 2000: 193). As Ingold and Kurttila (2000: 192) assert, local traditional knowledge is a process that is not 'inside people's heads' but 'is none other than that of people's practical engagement with the environment'. As they describe local traditional knowledge, it is skills, skills in relation to a certain environment learned 'through a mixture of imitation and improvisation in the settings of practice. What happens, in effect, is that people develop their own ways of doing things, but in environmental contexts structured by the presence and activities of predecessors' (Ingold & Kurttila, 2000: 193). I will add, in an environmental context structured by the physical formations as well. Furthermore, some occasions on certain sites enable tourists to relate to a heritage that is present due to former interaction between people and an architecture of the landscape that provides potentials or certain affordances that allow relationships to be built (Anderson, 2014: 21). These might be relationships to the landscape as well as the heritage manifest in the landscape, made up of previous small acts. Nevertheless, the meaning attached to such acts might be highly individual.

The circumpolar area has been described by Anderson (2014: 11) as having an 'architecture' that shapes 'co-operation and co-domestication between, humans, animals and landscapes'. As I understand Anderson (2014), the physical formation of the landscape interacts with humans and animals for certain tasks. Like a headland in a lake interacts with humans and reindeers in the process of dividing herds, if the headland is located in a suitable place according to the changing rhythm of the human–animal relationship that constitutes pastoralism. Here, I will follow up on the aim to overcome the dichotomy between nature and man-made spaces and investigate by what kind of architecture or structure Áhkku provides potentials and affordances for creating relationships with people who visit the site usually at certain 'whens'. The question then becomes: what potentials or affordances for building relationships does Áhkku as an 'architecture' allow and what kind of imitation and improvisations do the presence and activities of predecessors imprint in its environmental context (Anderson, 2014: 21; Ingold & Kurttila, 2000: 193)?

Áhkku and the site of its location is not a 'taskscape' that is unmarked. The technologies characteristic of modern leisure and tourism have made their imprint by marking it on maps available in the tourist information office as well as the site being one of the seven possibilities of *Visit a heritage site* suggestions from The World Heritage Rock Art Center – Alta Museum.[13] On this list of sites in the municipality of Alta, Áhkku competes with two World Heritage Sites. Available parking spaces and a marked path add to the emerging incorporation into a touristscape or rather how the scape for conducting the Norwegian *friluftsliv* tradition (see Gurholt, 2008), which translates roughly in English terms to 'outdoor life' or 'outdoor recreation', merges with tourism. The previously mentioned information poster is in Norwegian only and adds to the impression of a site just scarcely regulated for tourism. The same applies to the seating at the site, conveniently made out of a log of driftwood.[14] This is something that can be seen many places along the seashore where it is suitable to rest. The place for making a fire, a ring of stones, is no different from such structures found in most places where people sit down in the Sápmi countryside. Therefore, the most obvious relation is to local people's practices in nature, and the facilitation is in terms of most other places close to the town center where the municipal authorities organize physical activities for better health among its inhabitants. In this way, the site is a familiar taskscape for local people. It has all the characteristics of a place used for a short rest, previously when hunting, fishing or gathering for a livelihood, today for the leisure activity of *friluftsliv*, and resembles many other such facilities close to settlements of a certain size in Sápmi. In this way, there is an interaction between the local tradition, in the past and present, of using nature, the national and municipal authority's contemporary policies of providing for people's use of nature and the site itself – an architectural structure – that is suitable for such activities and tourism as well, which makes the site.

What makes this site different from other places is Áhkku. Today, the site is not interesting because of the view, the possibility for fishing, hunting or harvesting or as a place to moor a boat. It is only Ákkhu and the acts that have been done previously that are of interest. What the rock does is practically engage people with an environment in a performance of what Ingold and Kurttila label (2000: 193) 'environmental contexts structured by the presence and activities of predecessors'. This is a practice where people, in their own way of doing things, developed through a mixture of imitation and improvisation that engages people through their presence at a site. As Äikäs and Salmi (2013) demonstrate for the sites in Kittilä Taatsi and as supported by my findings, the practice of putting objects on the *sieidis* seems to be in terms of Ingold and Kurttila's (2000: 192) understanding of a tradition as being continuous without taking any fixed forms through peoples' practical engagement with the environment. Such sacrifices as coins, bottles indicating alcohol offerings, fishing gear, stones, foodstuff and a long list of personal objects indicate performances structured by previous activities regardless of who is involved in the performance. However, the idea of a timeless Sami religious tradition does not fit in with Äikäs and Salmi's (2013) or my findings. As they put it, 'Religious change has been interconnected with changes in social structure, landscape use, trade relationships and subsistence strategies that have taken place at the same time' (Äikäs & Salmi, 2013: 76). These changes continue into the present and shape the relationship between people and *sieidis* today:

> Nowadays tourism, livelihood and ritual are tied together. Offerings connected to old means of livelihood have been continued by locals as well as by tourists. In addition, tourism as a livelihood uses sacred places and imitations of old rituals as tourist attractions. From visiting authentic sites, tourists have moved to invented sieidi stones, sieidi climbing walls, bars and hotels. (Äikäs & Salmi, 2013: 76)

This is a matter of practices, as in Ingold and Kurttila's (2000: 192) perspective on traditions, and not a matter of persistent content of ideas. As Okkonen (2007: 28) also points out, tourists make their own interpretations independently of the experts, and the same probably applies to the multitude of locals putting an offering on Áhkku. As stated before, it is not easy to say what meanings the act of putting items on Áhkku imbues, but it is easy to see that locals as well as tourists do so and continue to do such acts, thereby creating a meaningful site in relation to a rock with an architectural structure that invites certain actions.

Anderson (2014: 15–16) claims that 'offerings' and 'placings' of food or other items to visible or non-visible entities on the land is a characteristic of models of reciprocity found in the circumpolar cultures. He claims that there is a relational character to the *sieidis*: 'Common to many Northern places is a sort of reduction or super-imposition of small local acts onto global problems. Thus ritual gifting signifies a global relationship of balance and respect' (Anderson, 2014: 16). What I will argue is that Áhkku herself is crucial in creating such a possibility and relationship of balance and respect by having an architecture currently set in a relationship of scapes that enables performances by locals as well as tourists. It has become a site such as what Edensor (2007: 206 ff) describes as a scape in a series of scapes that opens up for a heightened reflexivity immanent in tourism, which makes it possible to perform acts of relationship to the circumpolar landscape. This is performance not only shaped by the tourist role but structured by contemporaries' and the predecessors' activities (Ingold & Kurttila, 2000: 193) as well as the architectural structure that shapes a relationship between humans and the landscape (Anderson, 2014: 11).

That such relationships, framed by roles structured otherwise, also formed between the landscape and locals performing the local brand of *friluftsliv* or practices where tourism has become a part of everyday life because of dedifferentiation (Bauman, 1996; Edensor, 2007: 200) add to the activities that keep the relationship going. Furthermore, such a perspective enables an analysis that not only regards the circumpolar landscape in Sápmi as something that people act upon (Äikäs & Salmi, 2013: 76) but also as a landscape that in itself promotes actions, a landscape that one acts with and creates a relationship to. This also brings attention to the reduction through many small local acts that Anderson (2014: 16) claims is common to many Northern places. As Carú and Cova (2003: 281) state in relation to what they label a contemporary experience-oriented society in general, there is 'a romantic trend which seeks to over-saturate all events to construct a long series of strong emotions and unforgettable and extraordinary experiences'. Attention to such small acts performed in a relationship with Áhkku gives a rather different understanding of the mundane practices of traditions (Ingold & Kurtilla, 2000: 192) both in everyday life as well as in the encounters with a series of scapes in tourism where *some sites* present the possibility of thinking otherwise (Edensor, 2007: 212). Attention to these encounters opens up relationships not usually found in the discourse on the indigenous and often blurs the dichotomies that usually guide categorization in most fields. Such performances rather seem to overcome what Chow (2006: 67, italics in original) labels 'The flip side of amazement [...] the displacement and postponement of the other to an exceptional, indeed utopian, realm whose epistemic function lies in the idea that it cannot speak or materialize *in the present*, that it is currently absent'. The indigenous, the local and the visitor dissolve as distinct categories in the small rather mundane acts in relation to the circumpolar landscape that in itself has potentials and affordances for relationships with humans.

Even destabilizing such discourses and their binaries, attention to nature as an agent does not necessarily solve the challenge of power in a contemporary time. As Ruotsala (2008: 53) in accordance with Rodman (2003: 207) claims, 'researchers should empower place by returning control over meanings of place to the rightful producers, and empower their own analysis of place by attending to the multiplicity of local voices found about place'. In my opinion, such an empowerment today predominantly happens through the application of ideas that locate the indigenous in the past, or a timeless present and embody them with certain features of the Western imagination in the landscape of Sápmi as well as in the northwest of Mexico (Chow, 2006; Muehlmann, 2008). This often gives indigenous people a certain kind of empowerment and, sometimes, returns, rightfully or not, control of meanings of place to some producers. However, all too often what Chow (2006: 67) labels the flipside of amazement becomes important for the relationship between those people who gain a certain control of meaning and place. Paying more attention to Áhkku with her ontology probably shaped in a pre-Herderian and more Protean intellectual climate, as an actor in shaping such relationships, might also be a way to destabilize such boundaries in the realm of everyday politics. Maybe such attention might enable analyses of power relations that can overcome the hierarchical binary categorization that dominates contemporary thinking on the indigenous in most fields.

Notes

(1) The Norwegian name is *Seidekjerringa*. The Norwegian word *kjerring* translates as a hag or a woman. In the local dialect, woman is probably a better translation than the derogative hag which would be a more precise description in some places in the southern part of Norway. Both Seidekjerringa and Áhkku are in use in a local context.

(2) *Siejdde/värromuorra* in the Lule dialect and *sjielegierkie/värrogierkie* in Southern Sami.

(3) Hansen and Olsen (2004: 123) state that findings at a site in Jokkmokk in the Swedish part of Sápmi indicate that such sacrifices date back to the 3rd or 4th century AD.

(4) Colorful woven fabrics used to attach traditional footwear.

(5) In Norwegian: 'At dei heilage stadane i Nord-Noreg i stor grad har fått stått urørte for turistar, og ikkje ligg langs kjende og opptrakka turistløyper, forsterkar det autentiske og magiske aspektet ved staden. Dette er ikkje stader for massane, men for dei utvalde få' (Fonneland, 2010: 119).

(6) In original: 'Dei heilage stadane er såleis nok utøvarane må få til å henda. Dei krev handling for at krafta skal verta utløyst. Dette er handlingar som samtidig stadfestar den enkelte informant sin identitet og tilhøyrsle til eit nysjamanistisk miljø' (Fonneland, 2010: 116–117).

(7) In Norway, you can enter the electoral roll for the Sami Parliament if at least one of your great grandparents spoke Sami and you fulfill the personal criteria of feeling Sami.

(8) The drift of euros makes it difficult to separate the deposits of Finns from foreigners (Äikäs & Salmi, 2013: 73).

(9) The municipality of Alta has about 20,000 inhabitants, close to 27% of the population in Finnmark.

(10) In Norwegian: 'Ved besøk til Seidekjerringa kan man se at det på en liten hylle på baksiden av offersteinen ligger lys, blomster, mynter, små steiner. Det har også vært observert et brev på hylla'.

(11) In May 2015, the signpost was put on a shelf in the mountain close behind Àhkku. Due to the lack of space on the rock itself, shelves in the cliffside close by are used for sacrifices.

(12) As Cohen (2007: 76) puts it, it is one of the many ideas revolving around the concept of authenticity: 'Authenticity as "flow of life", not interfered with by the "framing" of sights, sites, objects and events for touristic purposes, by various overt or covert markers. According to this usage, the quintessential authentic sight is one, which is utterly unmarked'.

(13) The rock carving panels in Hjemmeluft, the Lille Raipas Mountain and the Struve Meridian Arc are UNESCO World Heritage Sites. The four other suggestions are the storehouses in Lille Lerresfjord, the evacuation cave in Lille Lerresfjord, the Alten Copper Works in Kåfjord and the Northern Lights observatory on Haldde, which are all, except the rock carvings in Hjemmeluft, located at a greater distance from the town center and the tourist tracks than Áhkku (Alta Museum, n.d.)

(14) The timber log had disappeared in May 2015, but will probably be replaced by a new driftwood log.

References

Äikäs, T. (2011) From fell tops to standing stones: Sacred landscapes in Northern Finland. *Archaeologia Baltica* 15, 16–21.

Äikäs, T. and Salmi, A.-K. (2013) 'The Sieidi is a better altar. The Noaidi drum's a purer church-bell': Long-term changes and syncretism at Sami offering sites. *World Archaeology* 45 (1), 64–85.

Alta Museum (n.d.) *Visit a Cultural Heritage Site*. See http://www.alta.museum.no/sider/tekst.asp?side=139&valgtmenypunkt=145 (accessed 5 March 2015).

Anderson, D.G. (2014) Cultures of reciprocity and cultures of control in the circumpolar North. *Journal of Northern Studies* 8 (2), 11–27.

Antonnen, P.J., Siikala, A.-L., Mathisen, S.R. and Magnusson, L. (eds) (2000) *Folklore, Heritage Politics and Ethnic Diversity. Festschrift for Barbro Klein*. Botkyrka: Multicultural Centre.

Bauman, Z. (1996) From pilgrim to tourist – or a short history of identity. In S. Hall and P. Du Gay (eds) *Questions of Cultural Identity* (pp. 19–36). London: Sage Publications.

Bendix, R. (1997) *In Search of Authenticity: The Formation of Folklore Studies*. Madison, WI: The University of Wisconsin Press.

Bergman, I., Östlund, L., Zackrisson, O. and Liedgren, L. (2008) Värro Muorra: The landscape significance of Sami sacred wooden objects and sacrificial altars. *Ethnohistory* 55 (1), 1–28.
Broady, H. (2000) *The Other Side of Eden: Hunters, Farmers, and the Shaping of the World.* Vancouver: Douglas and McIntyre.
Carú, A. and Cova, B. (2003) Revisiting consumption experience: A more humble but complete view of the concept. *Marketing Theory* 3 (2), 267–286.
Chow, R. (2006) *The Age of the World Target: Self-Referentiality in War, Theory and Comparative Work.* Durham, CA: Duke University Press.
Cohen, E. (2007) 'Authenticity' in tourism studies: Aprés la Lutte. *Tourism Recreation Research* 32 (2), 75–82.
Comaroff, J.L. and Comaroff, J. (2009) *Ethnicity, Inc.* Chicago, IL: University of Chicago Press.
Edensor, T. (2000) Staging tourism: Tourists as performers. *Annals of Tourism Research* 27 (2), 322–344.
Edensor, T. (2007) Mundane mobilities, performances and spaces of tourism. *Social & Cultural Geography* 8 (2), 199–215.
Fabian, J. (1991) *Time and the Work of Anthropology. Critical Essays 1971–1991.* Chur: Harwood Academic Publishers.
Fonneland, T.A. (2010) *Samisk nysjamanisme: i dialog med (for)tid og stad.* Bergen: University of Bergen.
Gaski, L. (2008) Contesting the Sami polity: Discursive representations in the Sami electoral campaign in 2005. *Acta Borealia* 25 (1), 1–21.
Gurholt, K.P. (2008) Norwegian friluftsliv and ideals of becoming an 'educated man'. *Journal of Adventure Education and Outdoor Learning* 8 (1), 55–70.
Hansen, L.I. and Olsen, B. (2004) *Samenes historie fram til 1750.* Oslo: Cappelen Akademisk Forlag.
Hætta, O.M. (1994) *Samenes gamle religion og folketro.* Alta: Alta Museums.
Hætta, O.M. (2003) Siebe – etterkrigstid og marginalitet i grenseland. In M.A. Hauan (ed.) *Karlsøy og verden utenfor: Kulturhistoriske perspektiver på nordnorske steder* (pp. 149–161). Tromsø: Tromsø Museum.
Ingold, T. and Kurttila, T. (2000) Perceiving the environment in Finnish Lapland. *Body & Society* 6 (3–4), 183–196.
Kirshenblatt-Gimblett, B. (1998) *Destination Culture: Tourism, Museums and Heritage.* Berkeley, CA: University of California Press.
Kylli, R. (2012) *Saamelaisten kaksi kääntymystä: Uskonnon muuttuminen Utsjoen ja Enontekiön lapinmailla 1602–1905.* Helsinki: Suomalaisen Kirjallisuuden Seura.
Löfgren, O. (1994) Learning to be a tourist. *Ethnologia Scandinavica* 24, 102–125.
Mathisen, S.R. (2010) Narrated Sami sieidis: Heritage and ownership in ambiguous border zones. *Ethnologia Europaea* 39 (2), 11–25.
Muehlmann, S. (2008) 'Spread your ass cheeks': And other things that should not be said in indigenous languages. *American Ethnologist* 35 (1), 34–48.
Myrvoll, E.R. (2008) *Samiske helligsteder: Tradisjon-registrering-forvaltning.* Tromsø: NIKU.
Okkonen, J. (2007) Archaeological investigations at the Sami sacrificial site of Ukonsaari in Lake Inari. *Fennoscandia Archaeologica* 24, 29–38.
Olsen, K. (2002) Authenticity as a concept in tourism research: The social organisation of the experience of authenticity. *Tourist Studies* 2 (2), 159–182.
Olsen, K. (2003) The touristic construction of the 'emblematic' Sami. *Acta Borealia* 20 (1), 3–20.
Olsen, K. (2008a) The Maori of tourist brochures. *Journal of Tourism and Cultural Change* 6 (3), 161–184.
Olsen, K. (2008b) Urfolksturisme – en reise i tid eller rom? *Norsk Antropologisk Tidsskrift* 19 (1), 20–37.

Olsen, K. (2016) Hva er det samiske? Og hva betyr svaret for samisk institusjonsoppbygging og næringsutvikling? In E. Angell, S. Eikeland and P. Selle (eds) *Nordområdene i endring. Urfolkspolitikk og utvikling* (pp. 62–82). Oslo: Cappelen Akademisk Forlag.

Rodman, M.C. (2003) Empowering place: Multilocality and multivocality. In S.M. Low and D. Lawrence-Zúñiga (eds) *The Anthropology of Space and Place. Locating Culture* (pp. 204–223). Oxford: Blackwell Publishing.

Ruotsala, H. (2008) Does sense of place still exist? *Journal of Ethnology and Folkloristics* 1 (2), 43–54.

Shepherd, R.J. (2015) Why Heidegger did not travel: Existential angst, authenticity, and tourist experiences. *Annals of Tourism Research* 52, 60–71.

Sikkala, A.-L., Klein, B. and Mathisen, S.R. (eds) (2004) *Creating Diversities. Folklore, Religion and the Politics of Heritage*. Helsinki: SKS.

Wang, N. (2000) *Tourism and Modernity*. Amsterdam: Pergamon.

14 Sami Tourism in Northern Norway: Indigenous Spirituality and Processes of Cultural Branding

Trude Fonneland

Introduction

Tourism is a well-known way of life for an increasing portion of the world's indigenous communities, and indigenous cultures and people are seen as important in promoting tourist destinations (Butler & Hinch, 1996, 2007). So is the case in Norway where the interest in indigenous tourism products has increased over the past decades. Contemporary Sami culture is one of four main areas within the brand strategy of *Northern Norwegian Tourism* (Nord Norsk reiseliv). For this reason, a stronger focus on developing Sami experience products based on Sami cultures and histories has been recognized as a necessity.

This chapter focuses on precisely such a development project, namely Sami Tourism in Northern Norway (Samisk reiseliv i Nord-Norge; henceforth STiNN). STiNN is an initiative to organize and promote sustainable and ethical Sami tourism products. The goal of the project is to create better coordination between the different tourism entrepreneurships, to develop a clearer Sami product profile, to ensure the quality of the various offerings and, finally, to increase their profitability, which in turn is seen to improve economic growth in the region.[1] As such, STiNN reveals a contemporary political objective in Sami communities, namely economic growth through increased investments in tourism.

STiNN provides an important view into cultural heritage production as well as into place branding and marketing, and reflects the significance of indigenous tourism destinations related to destinations based on local or national cultures. In this chapter, I will explore what is emphasized when STiNN and the projects' entrepreneurs develop tourism experiences in Sápmi.[2] What kind of stories, experiences and sensations are produced to make Sápmi a desirable product, and what role does religion, in the form of indigenous spirituality, play in this process? My goal is to show how religion and religious values may emerge as symbolic and cultural resources (see Beckford, 1992) in contemporary indigenous tourism projects to promote a tourist site.

My argument is that focusing on the representations and staging of spirituality can deepen our understanding of the dynamics of tourism entrepreneurships, of the contemporary experience economy, as well as of how this venue of production and consumption is promoted and developed, lived and experienced (Löfgren & Willim, 2006: 2). Through the selection of symbols and stories expressed by Sami tourism entrepreneurs, boundary markers and differences are produced in a cultural production where religion is highlighted as a key cultural marker.

My analyses are based on website analysis of STiNN's homepages on the internet, on document analysis of STiNN's project reports and brochures, as well as on website analysis of the contributor's homepages.[3]

The chapter begins with a closer presentation of representations of Sami cultures in the tourism sector and of the discourse of indigenous spirituality that during the last 20 years has gradually become more visible in indigenous tourism contexts. As a point of departure for a further discussion, I present two tourist offers that are often highlighted as core Sami activities, namely reindeer experiences and Northern Lights tours. Finally, I explore the implications related to the staging of indigenous spirituality as a basis for economic development in a Sami tourism context.

Representations of Sami Pasts and Presents

In recent years, we can trace a growing awareness of how indigenous cultures are presented in tourism contexts (see Butler & Hinch, 2007; Comaroff & Comaroff, 2009). STiNN is precisely an initiative to influence and expand the pictures and notions of Sami culture presented in tourism contexts. The project is about reclaiming the right to present Sami culture and history, to attract visitors and to develop sustainable new industries in Sami communities.

The ethical awareness that is underlined in the project reports is related to the fact that indigenous tourism is often described as a double-edged sword, questioning whether tourism is primarily a means to increase financial dividends or if tourism is a new form of colonialism and a threat to traditional cultures. According to ILO Convention 169, the Sami people have the right to protect their own cultural expressions. Tourism industries that are developed to make money from Sami cultural heritage thus have to be cautious in relation to both individuals and Sami communities. They must understand and show respect for the 'raw materials', identity, cultural expressions and history that are to be marketed. One case often described as an example that violates these rights is the tourist industry in Rovaniemi, in Finnish Sápmi (Saarinen, 2001). As recently as 2008, this resulted in protests and news accounts about the tourist industry's abuse of Sami culture (see also Comaroff & Comaroff, 2009: 159).[4]

Although the prevailing narratives of Sami people and culture in the context of tourism have changed through time, they all play on closeness to nature and landscape and can be said to be deeply embedded in old traditions of Western primitivism and imaginaries about the noble savage (Mathisen, 2001). Representations and displays of Sami people and culture through varying forms and with shifting purposes have taken place over a long time span. In 1822, one of the first known exhibitions of Sami culture was arranged in the Egyptian Hall in Piccadilly, London. Together with a herd of reindeer, a Southern Sami family bemused enthusiastic visitors. In the course of the 1800s, more such exhibitions were established and all achieved great popularity and interest (Baglo, 2011). During the 1900s, Sami culture was made available for both cruise and road travelers. Sami camps were organized and equipped with sales sheds (Viken, 2002: 2).

In the marketing of Sápmi, it is primarily the nomadic herding culture, the reindeer, *lavvo*, *joik* and *gákti* (traditional Sami clothing) that have been highlighted as Sami markers and have become picturesque materializations of certain aspects of the Artic landscape and its inhabitants. According to folklorist Stein Mathisen (2015: 204), versions of Sami reindeer herding culture are 'central symbols for a way of life that represents harmony with nature and the environment and that offers a special kind of indigenous, spiritual conscience'.

As anthropologist Kjell Olsen (2006: 37) notes in a study reviewing the representation of Norwegian Sami in local and regional tourism brochures, 'these representations give an impression of the Sami that perpetuates their image as radically different from Norwegians'. Referring to the stereotypical portrayal of Sami as 'the emblematic Sami', Olsen points out that the portrayal in contemporary tourist brochures is not significantly different from representations of the Sami displayed during the world exhibition in London in 1822. Even though society has developed significantly in the approximately 200 years that have passed since the world exhibition, the representations of the Sami people have not changed appreciably (Olsen, 2003, 2006).

That these images are still centerpieces in the marketing of Sami culture we get a glimpse of when visiting Northern Norwegian Tourism's (Nord Norsk reiseliv) website. Of a total of 20 images, 14 visualize people in traditional Sami clothing (*gákti*) and over 50% of the pictures have reindeer as their theme. Further, the website markets 13 different Sami experiences and in 10 of these the reindeer is in focus branded as 'Reindeer Sledding and Northern Lights', 'Reindeer and Snowmobile' and 'Reindeer Sledding

(Raid) with Sami Lunch'.[5] The emblematic form that highlights an idea of Sami as traditional and radically different from modern Norwegian culture (Olsen, 2003) thus still prevails.

An important feature of tourism is that it often requires marketing that makes it possible for tourists to recognize the attraction. Without recognizable symbols, one fears that tourists will opt out of the destination for the benefit of another. STiNN has produced brochures that are available at Northern Norwegian Tourism's homepages; 'Sami Tourism in Northern Norway – Our Story your Experience' and 'Sápmi Activities'. The pictures in these brochures, chosen to visualize Sami culture, do not differ noticeably from the traditional presentation of Sami culture in other tourism contexts. Still, the reindeer and traditional Sami clothing take up considerable space together with bonfires, lavvus, yoik, a *noaidi* (Sami pre-Christian specialist) and traditional crafts.

Commissioned by STiNN, anthropologist Reni J. Wright (2014) has written the report 'Samisk Reiseliv og salg av kulturuttrykk' (Sami Tourism and the Selling of Cultural Expressions). Wright points out that the participants in STiNN on several occasions have expressed frustration over the unidirectional promotion of Sami cultural expressions. Particularly Sami tourism entrepreneurs in southern Sami areas point out that they do not feel that they belong in the portrayal of the 'emblematic Sami' and that the Sami culture has thus become invisible in their area. The entrepreneurs have requested a new meta-narrative that is not exclusively linked to reindeer pastoralism. The question that arises is whether STiNN can help to create a more nuanced picture of Sami culture, and if tourists are interested in such a nuancing (Wright, 2014).

Indigenous Spirituality as Meta-Narrative in Indigenous Tourism

To further develop Sami experience packages, STiNN arranged a study trip to New Zealand in November 2013 in which 25 Sami entrepreneurs took part. The objective of this trip was to learn more about branding indigenous culture in tourism contexts.

In her commissioned rapport, Wright scrutinizes the participants' impressions of Maori culture and their tourist venues. What Wright claims made a great impression during the stay were the rituals and ceremonies that Maoris include tourists in and their ability to create a link between mythical stories and present practices in tourism contexts. One of the Sami entrepreneurs underlines:

> We show the tourists *lavvus* (traditional Sami tents) and turf huts and say: 'This is how the Sami people lived in the past' while the Maoris say, 'we are so and so'. They simply have it in their hearts. Our task is

> to find a way to convey like the Maoris did. For example, to tell tourists that we are the sun's sons and the moon's daughters. We simply have to use some better narrative strategies. I think this is how we can preserve the Sami culture and disseminate it to the next generation. (Wright, 2014: 27, my translation)

Anthropologist Ronald Niezen (2003, 2009, 2012) has referred to notions of an indigenous *we* as one outcome of the processes of indigenous resistance and revival. As one of several scholars, he points to a religious dimension to indigenous people's identity-making, consisting of elements such as shamanism and animism, sacred places, environmental awareness and holism (see also Beyer, 1998; Kraft, 2009). He claims that 'the notion of "indigenous religion" has been so thoroughly conventionalized that most Euro-American lay people presented with it would likely have some notion of what is meant by it, probably by drawing upon related ideas associated with such things as shamanism and forest spirituality' (Niezen, 2012: 119). These notions have a widespread reach, but their main locations seem to be United Nations (UN) forums, academia, music, festivals, as well as tourism contexts (see Graham & Penny, 2014).

In Maori tourism, it is apparent that the religious aspect of indigenous identity-making is applied as a meta-narrative in the various tourism experience packages. During their tour, STiNN's Sami entrepreneurs took part in Maori welcome rituals, traditional Maori Karakia (prayer) to the people of the river, the Natipokiao, as well as stories connected to Maori mythology where Maoris present themselves as children of the forest and things that belong to the forest as their brothers and sisters (see Wright, 2014). According to Wright (2014), it is also this aspect that many of STiNN's entrepreneurs are inspired by and want to expand upon in their own businesses.

The relationship between tourism and religious or spiritual experiences has long been debated in anthropological, sociological as well as religious historical research on tourism (Graburn, 1989; MacCannell, 1976; Stausberg, 2011). In addition, the theories of the gurus of the modern experience economy, Pine and Gilmore (1999), are relevant to this research. Pine and Gilmore see a significant change in contemporary consumers who are no longer only concerned with buying goods and services, but who increasingly seek engaging and 'transformative' adventures. In the future, Pine and Gilmore (1999) emphasize, the ability to create personal experiences will give companies and enterprises a sustainable competitive advantage, experiences being the hottest commodities that the market has to offer.

The presentation of Maori people as children of the forest and Sami people as guardians of nature and as 'walkers with reindeer' is based on a distinct taste and a selected aesthetic. This type of aesthetic is, as shown, reflected in STiNN's brochures and photos, primarily showing humans and animals in interaction with nature. This type of portrayal, I argue, is inspired by and contributes to a discourse about indigenous spirituality that during the last 20 years has gradually become visible in the tourism industry (Fonneland, 2011, 2012a, 2012b, 2013). This discourse is highlighted as a resource for all indigenous tourism entrepreneurs, regardless of whether they are located in New Zealand or in Sápmi. The wording 'indigenous spirituality' in a tourism context, however, does not mean that the tourist taking part in the tourism offer achieves a personal spiritual or religious experience. Rather, it emphasizes the organizers' marketing strategies and their aim to create a tourist experience that breaks with everyday routines and that is both in demand and unique.

'Indigenous spirituality' is a recent term in contemporary international discourses that highlights the construction of the world's indigenous people as a people with a common colonial past, a common religious heritage and a spiritual relationship to nature and the environment (Christensen & Kraft, 2011: 18). This new global spiritual identity is partly a result of UN meetings and regulations, including laws that have helped standardize certain qualities by translating them into rights. In contrast to the 'freedom of beliefs' promoted by human rights discourse, the beliefs of indigenous people tend, for instance, to be explicitly connected to particular landscapes. To cite one example: ILO Convention 169, ratified by Norway in 1990, claims that governments must 'respect the special importance for the cultures and spiritual values of the people concerning their relationship with the lands or territories' (Article 13, 1). Further, Article 25 of the 2007 United Nations Declaration on the Rights of Indigenous Peoples (UNDRIP) asserts that indigenous people have 'the right to maintain and strengthen their distinctive spiritual relationship with their traditionally owned or otherwise occupied and used lands, territories, waters and coastal seas and other resources'. As discussed by scholars of religion, Trude Fonneland and Siv Ellen Kraft (2013: 140): 'Regular references to indigenous peoples as children of Mother Earth are similarly common in UN fora, along with references to a holistic worldview'.

The representation of indigenous people as a community with shared religious values related to nature, to the past and the past's traditions can be traced back to the 1960s countercultures, and in particular to the interplay between the environmental movement and the New Age movement. These movements emerged within the same period and have been exchanging ideas ever since. Indigenous spirituality thus is a discourse that attributes a certain kind of spirituality to indigenous people. At the same time, indigenous people are able to contribute to the shaping of this discourse and to utilize it for their own strategic purposes (Beyer, 1998, 2006; Kalland, 2003; Kraft, 2004, 2008, 2009).

We are here faced with what anthropologist Harald Prins describes as the paradox of primitivism. The primitivist discourse, according to Prins, functions in two ways. On the one hand, it reduces indigenous people to 'noble savage' – stereotypes. On the other hand, it equips indigenous people with models of self-representation that can be implemented for political purposes and in the marketing of their own cultures. He writes: 'Having become a key element in their rhetoric of self-fashioning, it shows up in their "visual performances" and thus may serve as a persuasive device in their collective quest for biological and cultural survival' (Prins, 2002: 56).

The term 'indigenous spirituality' in many ways is a cultural marker set in contrast to Western society, Western religions and Western worldviews. Values related to nature and to the landscape mark a distinction between a place-oriented, peaceful, holistic, traditional and eco-friendly indigenous culture and a modern Western capitalist society (Kraft, 2009: 188). In this context, indigenous spirituality represents a unique product that adds value to a destination by telling a different story and representing a unique experience remote from daily Western life. The performance of indigenous spirituality can be seen as a powerful instrument for creating strong and persuasive experiences. As researchers have underlined, at the heart of the tourist industry and the new economy lies a quest for what is different, and one of the primary narratives in the tourism industry concerns precisely the relationship between the modern and the traditional (Kirshenblatt-Gimblett, 1998: 153; Olsen, 2006; Pine & Gilmore, 1999; Wang, 2000).

In the following, the concept of indigenous spirituality will be contextualized through an analysis of a selection of activities, symbols and narrative productions by Sami entrepreneurs in STiNN's homepages. By focusing on the symbols and narratives chosen to market Sápmi, I will highlight the dynamics of spiritual entrepreneurships, which in turn will throw light on values and ideals in the contemporary experience economy.

Sami Experiences and Indigenous Spirituality

Reindeer migration: A tour of a lifetime

The reindeer is, as mentioned, in focus in almost every Sami activity found on Tourism in Northern Norway's homepages. This is also the case in STiNN's brochures. Here, three different tours are facilitated and connected to STiNN's entrepreneurs in the counties of Nordland, Troms and Finnmark. In all these three tours and in all three counties, a meeting with the reindeer and Sami reindeer herders is in focus and branded as 'visit a traditional reindeer herding family', 'experience reindeer sledding' and 'get an introduction to the reindeer herding Sami way of life through stories, joik and food'.[6]

In May 2014, the renowned American travel magazine *National Geographic Traveler* presented the 2014 list featuring 50 of the world's best-guided trips and tours of a lifetime. 'Reindeer Spring Migration', by an entrepreneur in STiNN, *Turgleder AS*, had the honor to be included in this list. The reindeer experience is described as a 'glimpse into one of the rare indigenous cultures left in Europe', where tourists are invited to 'live with and work alongside a Sami family in Finnmark during the annual reindeer migration from the inland winter habitat to calving grounds on the coast'.[7]

At *Turgleder As*, one can read a detailed account of the experience. Tourists are here titled as guests and helpers and welcomed to experience a 'natural miracle'. According to the tour operator, 'Mother Nature sets the itinerary' and 'time and an agenda are unknown concepts'. This is all about ignoring the clock and trusting the reindeer's instinct.[8] The portrayal signifies that entering the Sápmi landscape implies a removal from the normal pulse of everyday life and by taking part in an indigenous community, the tourist gets a taste of nature's magic and holism.

The pictures and narrative production highlighted in the description of the migration experience can be said to visualize what scholars of religion Cato Christensen and Siv Ellen Kraft (2011: 24, my translation), in their analyses of the feature film 'The Kautokeino Rebellion' describe as 'reindeer-holism'. 'It's about a lifeform where all key functional areas are interrelated, and where the reindeer is the unifying principle'. This visualization of the interaction between Sami and the reindeer underpins an understanding of Sami people as a nature people and illustrates their vulnerable position in relation to encroachment on nature and the impact from the forces of modernity (Christensen & Kraft, 2011). It further creates a picture of Sami landscapes and Sami people as holders of a spiritual knowledge, which by travelling there one will be able to take part in. The migration involves an invocation of old wisdom and mythology, a connection between place and practice and a mythical connection between Sami reindeer herders who work in the area and late modern tourists.

The commercial relationship that is presented in this tourist product is characterized by intimacy. Tourists in Sápmi, according to *Turgleder AS*, are not just tourists or anonymous customers. The tourists will also take part in family life, in the family's work, their stories and culture. The span between duties and leisure is an expression of the home's symbolic meaning. Tourists will be able to feel at home, but on holiday at the same time. Tourists and host call each other by their first names. They are also equal in being entrusted to nature's forces and the reindeer's movements and instincts. The closeness to the hosts changes the social relationship between tourism, product and seller. By repealing the traditional tourist role, the product developers make an authentic experience emerge as achievable. Authenticity is not a quality of an object, but a cultural value that is continually created and transformed in social processes (Bendix, 1997; Olsen, 2002).

The spring migration can be described as a modern pilgrimage where the meeting with other humans and environments will contribute to an inner self-development. Here, individual experiences, religious education, freedom and a break from everyday life's regular routines, but also more prosaic motives like nature experiences and socializing are at the center (Gilhus & Kraft, 2007: 15–17). The spring migration is precisely presented as a 'route only actually decided on the move'.[9] Tourists taking part in this experience are not just tourists, but travelers on a personal hike in the countryside and in history. They cross the boundaries between past and present time, between urban and rural life and between a Western culture and Sami indigenous culture.

Hunting for the Northern Lights with the Sami

Northern Lights tourism has experienced huge growth in recent years. The marketing success is among other things connected to the BBC documentary 'In the Land of the Northern Lights' (BBC One in 2008) with the famous actor Joanna Lumley. In the documentary, Lumley tells that she had dreamt of seeing the Northern Lights since she was a child. Finally, she got to live out her dream. In the BBC film, she travels across northern Norway in search of the *Aurora borealis*. Lumley, as many others who get to experience the elusive lights, describes it as an altogether emotional experience.

Northern Lights tourism is not a Sami experience, but in the context of tourism, very often Northern Lights tours are marketed with a Sami element included; a *lavvo,* Sami food or reindeer sledding. In this way, the Sami element can be said to create a wider frame for the experience, making the Aurora tour an experience package. This is also so when the mythical explanations related to the Northern Lights are marketed. The legends of the light are primarily presented as part of Sami culture and history, although the source material for this type of interpretation is sparse (see Mathisen, 2015: 75).

Northern Lights tours are portrayed as both defining and transformative experiences. These tours are all about chasing something that you cannot spot everywhere or at any time. The light is further described as personal:

> Aurora is an unpredictable lady, and you never know when she will decide to turn up. This diva keeps you waiting, so whenever you go hunting for the northern lights above the Arctic Circle, make sure you set aside the whole evening. Northern lights worshippers do everything from cross-country skiing to building snowmen in order to keep warm and entertained while outside.[10]

Chasing the female Aurora gives the experience an extra dimension. Aurora, as described in the marketing, is an unpredictable and teasing diva, playing hide and seek with the tourists. These are characteristics associated with an ancient goddess and the tourists are presented as her worshippers, at the complete mercy of nature. This implies that hunting the light reaches beyond the atmosphere of everyday life, and as in the movie with Lumley, it is emphasized that the gaze of the light is an emotional experience with the ability to affect the tourist for life. This combination of spirituality and Northern Lights experiences is clearly expressed in *Turgleder As*'s Northern Lights safaris:

> Everything and everyone has its own Háldi. Every human, every animal, every forest and every lake. Everything has its Háldi. In the old Sami religion and culture, people believed that Háldis lived under ground, often in certain mountains. If a mountain is called Halde, Háldi, Haldde or Halldo it means that this is a holy place where Háldis live.
>
> We start our northern light safari at the Engholm Husky Lodge just outside Karasjok. You will be handed out a pair of snowshoes and with these simple aids the 1,5 kilometre hike to the Halde top will be easy. Up here are no disturbing lights and you can enjoy the full view either sitting on a reindeer skin by the warming fire or sitting alone in the darkness with a nice hot drink, just listening to Aurora .... And who knows, maybe you will meet your own Háldi out there.[11]

In a New Age context, sacred geographic areas and monuments like the Halde, which is here chosen as a starting point for a gaze of the Northern Lights, are described as liminal, where the border between the world of the living and the world of the spirits is like a thin veil. These places' liminality is also imagined to make it possible for the traveler to cross the boundaries between this world and a world imbued with sacred energy (Jenkins, 2004: 201). At *Turgleder AS*, the value of the reference to the sacred power place is presented as a symbolic resource with the ability to capture both secular and spiritual tourists' interests. With its link to the Sami past, to a local landscape and to a specific place in this landscape, the Sami mythology of the Halde and the Háldi is approached as cultural capital, which will help to highlight the Sami culture as unique. The story not only establishes a connection to the past, but also creates an understanding of the present, or what might happen in the present. The figure of the Háldi belongs to a bygone age, but resurfaces in the glow of the Northern Lights, creating a meeting point between a Sami mythological figure and the tourists. This Northern Lights tour is marketed as a portal into a world of magic where the spirits are present. It creates a picture of Sami landscapes and Sami people as holders of a spiritual knowledge, which by travelling there one will be able to take part in and experience the past in the present.

The Complexities of Branding Indigenous Spirituality

The fact that Sami culture has become part of an international indigenous community in which discourses about indigenous spirituality are part of the symbolic repertoire has considerable consequences for how Sami culture is staged in the tourist industry. STiNN is a clear example of such new marketing and communication of Sami as indigenous, in which words like spirituality, holism, harmony, closeness to nature and a mythological past are central themes – in addition to the traditional markers of Sami cultures. The entrepreneurs in STiNN showcase Sápmi as a country of resources, rich in delicate ingredients, great scenery, lighting, colors and, not least, spiritual energies. The concept of indigenous spirituality is here a rhetorical strategy that focuses the tourist's attention in a certain direction, and that simultaneously helps to highlight social and ontological differences, as for example 'indigenous' to 'Western', 'us' to 'them'. It is also in this context that STiNN's study trip to New Zealand to exchange experiences with Maori tourism entrepreneurs can be understood as an important means of developing local Sami tourist businesses that involve notions of self and boundaries to (Western) others. By virtue of belonging to an indigenous community, Sami and Maori are perceived as cultures with the same kind of challenges and opportunities.

In a research context, there has been a tendency to describe romantic indigenous images as an expression of destructive exotification, largely under the auspices of New Agers. Spirituality has been on the taboo list during the entire Sami political revival and parts of the colonial legacy involved a prohibition and exclusion of Sami mythology and pre-Christian symbols (Flemmen & Kramvig, 2008: 114–117; see also Olsen, 2004). Tourism researchers Petterson and Viken (2007: 185) note: 'Developing commercial products out of important culturally based spiritual icons was (in the Sami society) seen to be morally questionable, and it certainly impacted authenticity'. A central question to be highlighted in this context is whether the presentation of an indigenous spiritual heritage underpins an emblematic stigma placing the Sami population outside modern life and development or as noble savages (Mathisen, 2015: 210).

The signposting of spirituality that has been highlighted in this chapter is a result of internal innovative processes where cultural elements from the past are retrieved and given new and positive value in order to create awareness about Sami culture in the tourism market. Still, one must be aware that what is expressed in a tourist context might be based on stereotypical images and ideas. This applies to all types of representations of cultures, not just the staging of indigenous cultures. To some extent, all forms of tourism in one way or another appropriate places, practices or artifacts that might lead to an objectifying or even folklorization of a culture and a people. Many researchers within this field have long focused on the influences and effects of the stereotypes of tourism, and the merchandising of cultural traditions (e.g. Butler & Hinch, 1996; MacCannell, 1976, 1992; Turner & Ash, 1975). However, I hold open whether the staging of indigenous spirituality produces cultural stagnation and causes distance from the power center, as long as the people who can legitimately claim ownership over cultural and religious property are not denied their rights, and as long as tangible or intangible cultural and religious property is not damaged (see also Stausberg, 2011: 222–223).

Indigenous spirituality is a discourse that some indigenous people deliberately and strategically enter into and contribute to shaping. It is also reflected in parts of their identity constructions, in the struggle for political rights and in cultural revitalization processes (Beyer, 1998, 2006; Kalland, 2003; Kraft, 2004, 2008, 2009). In the Sami experiences in focus in this chapter, indigenous spirituality, shaped by local and global influences, is used as a resource possessing an economic potential.

Conclusion

To become a destination, a place must have a story to tell, a story that fills it with sought after qualities; qualities that make it picturesque, idyllic, dramatic, sacred or magical (Selberg, 2011: 126). STiNN can be described as a story producer searching for and revealing what unique experiences travelers to Sápmi will access during their stay. The project and its entrepreneurs aim to touch potential customers' feelings and imaginations and further widen their horizon and knowledge about Sami culture. To construct a story in a tourism context is to guide customers' reading of a relevant destination and of a desired place. It is about steering the reader's perceptions of Sami culture and Sápmi as an adventure package, as an attractive and unique tourist venue.

When Sami people were incorporated into the international indigenous movement and into global discourses concerning indigenous spirituality, Sami mythology as well as pre-Christian religious symbols were included in new aesthetic frames and in processes of cultural heritage production where spirituality is highlighted as a positive cultural marker. In a tourism context, the staging of Sami indigenous spirituality is also linked to the fact that the tourism industry profits from differences and sees 'sameness' as a problem (Kirshenblatt-Gimblett, 1998: 153; Olsen, 2003; Pine & Gilmore, 1999; Wang, 2000). The tourist offerings highlighted in this chapter reflect a cultural production by which cultural and natural elements are selected and reworked for new social uses, and where a reinterpretation of the past based on contemporary issues takes place.

Indigenous spirituality has become a key tool in this production, and can be viewed as a powerful instrument for creating strong and persuasive experiences. Although there is a difference from how Maoris use myths of origin, rituals and ceremonies in tourism contexts, the understanding of indigenous spirituality plays an increasingly important role in tourist venues in Sami communities. Indigenous spirituality has become a meta-narrative that extends beyond the promotion of the reindeer and presents Sami people as part of a global indigenous community.

Notes

(1) http://www.origonord.no/autoweb/dokument/Rapport%20Samisk%20Reiseliv%20i%20Nord-Norge%20ar%201.pdf (accessed 19 February 2015).

(2) To write about Sami communities and about Sami history embodies some challenges related to the choice of words and names. Sápmi, Sábme and Saepmie are all names for the geographical area in which Sami peoples traditionally lived. Sápmi is a Northern Sami word for this area, Sábme is Lule Sami and Saepmie is Southern Sami. These three languages are all official Sami languages, but in this chapter, I choose to use the Northern Sami term 'Sápmi' as a common term for all three languages. This is in line with the practice on STiNN's homepages and brochures.

(3) The names of the reports and brochures in focus are 'Sami Tourism in Northern Norway: Our Story, Your Experiences, Sápmi Activities', 'Samisk Reiseliv og salg av kulturuttrykk' and 'Rapport Samisk Reiseliv i Nord-Norge år 1'.

(4) Finland has not yet ratified the ILO Convention. This can also be said to provide the ground for a larger scope of exotification in the Finnish tourist industry (Kramvig, 2011).

(5) http://www.nordnorge.com/en/product-event-search (accessed 20 February 2015).

(6) http://www.nordnorge.com/sites/n/nordnorge.com/files/21c3673f776838297d36c38ce18cd918.pdf (accessed 17 February 2015).

(7) http://travel.nationalgeographic.com/travel/tours/europe-tours-2014/ (accessed 17 February 2015).

(8) www.turgleder.com/eng (accessed 17 February 2015).

(9) http://www.nordnorge.com/en/sapmi/?News=429 (accessed 23 February 2015).

(10) http://www.thelocal.no/20120131/experience-the-marvel-the-northern-lights (accessed 23 February 2015).

(11) http://www.turgleder.com/?project=northern-light-safari (accessed 20 February 2015).

References

Baglo, C. (2011) På ville veger? Levende utstillinger av samer i Europa og Amerika. PhD thesis, Department of Archaeology and Anthropology, University of Tromsø.

Beckford, J. (1992) *Religion and Advanced Industrial Society*. London: Routledge.

Bendix, R. (1997) *In Search of Authenticity: The Formation of Folklore Studies*. Madison, WI: The University of Wisconsin Press.

Beyer, P. (1998) Globalization and the religion of nature. In J. Pearson and G. Samuel (eds) *Nature Religion Today: Paganism in the Modern World* (pp. 11–21). Edinburgh: Edinburgh University Press.

Beyer, P. (2006) *Religions in Global Society*. London: Routledge.

Butler, R. and Hinch, T. (1996) Indigenous tourism: A common ground for discussion. In R. Butler and T. Hinch (eds) *Tourism and Indigenous People* (pp. 3–22). London: Thomson International Business Press.

Butler, R. and Hinch, T. (2007) Introduction: Revisiting common ground. In R. Butler and T. Hinch (eds) *Tourism and Indigenous Peoples: Issues and Implications* (pp. 1–14). London: Butterworth-Heinemann.

Christensen, C. and Kraft, S.E. (2010) Religion i Kautokeino-opprøret: En analyse av Samisk urfolksspiritualitet. *Nytt norsk tidsskrift* 1, 19–27.

Comaroff, J.L. and Comaroff, J. (2009) *Ethnicity, Inc.* Chicago, IL: The University of Chicago Press.

Flemmen, A.B. and Kramvig, B. (2008) Møter: Sammenstøt av verdier i Samisk-norske hverdagsliv. In O. Leirvik and Å. Røthing (eds) *Verdier* (pp. 101–120). Oslo: Universitetsforlaget.

Fonneland, T. (2011) Sami tour: urfolksspiritualitet i ei Samisk turistnæring. *Chaos. Dansk-norsk tidsskrift for religionhistoriske studier* 55, 153–172.
Fonneland, T. (2012a) 'De syv kaffekok' – spirituelt entreprenørskap i eit samisk landskap. *Tidsskrift for kulturforskning* 11 (2), 27–44.
Fonneland, T. (2012b) Spiritual entrepreneurship in a northern landscape: Spirituality, tourism and politics. *Temenos Nordic Journal of Comparative Religion* 48 (2), 155–178.
Fonneland, T. (2013) Isogaisa: Samisk sjamanisme i festivaldrakt. *Aura. Tidsskrift for nyreligiøse studier* 5, 102–131.
Fonneland, T. and Kraft, S.E. (2013) Sami shamanism and indigenous spirituality. In S.J. Sutcliff and I.S. Gilhus (eds) *New Age Spirituality: Rethinking Religion* (pp. 132–145). Durham: Acumen Publishing.
Gilhus, I.S. and Kraft, S.E. (2007) Innledning. In I.S. Gilhus and S.E. Kraft (eds) *Religiøse reiser: Mellom gamle spor og nye mål* (pp. 11–21). Oslo: Universitetsforlaget.
Graburn, N.H.H. (1989) Tourism: The sacred journey. In V.L. Smith (ed.) *Hosts and Guests: The Anthropology of Tourism* (pp. 21–36). Philadelphia, PA: University of Pennsylvania Press.
Graham, L.R. and Penny, H.G. (2014) Performing indigeneity: Emergent identity, self-determination, and sovereignty. In L.R. Graham and H.G. Penny (eds) *Performing Indigeneity: Global Histories and Contemporary Experiences* (pp. 1–31). Lincoln, NB: University of Nebraska Press.
ILO Convention 169. http://www.regjeringen.no/nb/dep/kmd/tema/samepolitikk/midtspalte/ilokonvensjon-nr-169-om-urbefolkninger-o.html?id=451312 (accessed 8 April 2015).
Jenkins, P. (2004) *Dream Catchers. How Mainstream America Discovered Native Spirituality.* Oxford: Oxford University Press.
Kalland, A. (2003) Environmentalism and images of the Other. In H. Selin (ed.) *Nature Across Cultures: Views of Nature and the Environment in Non-Western Cultures* (pp. 1–17). Amsterdam: Kluwer Academic.
Kirshenblatt-Gimblett, B. (1998) *Destination Culture: Tourism, Museums, and Heritage.* Berkeley, CA: University of California Press.
Kraft, S.E. (2004) Et hellig fjell blir til – Om samer, OL og arktisk magi. *Nytt norsk tidsskrift* 3–4, 237–249.
Kraft, S.E. (2008) Place-making through mega-events. In J.O. Bærenholdt and B. Granås (eds) *Mobility and Place: Enacting Northern European Peripheries* (pp. 219–231). Aldershot: Ashgate.
Kraft, S.E. (2009) Sami indigenous spirituality: Religion and nation building in Norwegian Sápmi. *Temenos. Nordic Journal of Comparative Religion* 45 (2), 179–206.
Kramvig, B. (2011) Orientalisme eller kulturmøter? Tendenser i Samisk turisme og kulturnæringer. In M. Forbord, G.T. Kvam and M. Rønningen (eds) *Smak, kultur og opplevelser – nye ideer om reiseliv i Norge* (pp. 169–188). Oslo: Fagbokforlaget.
Löfgren, O. and Willim, R. (2006) Introduction. In O. Löfgren and R. Willim (eds) *The Mandrake Mode Magic: Culture and the New Economy* (pp. 1–18). Oxford: Berg Publishing.
MacCannell, D. (1976) *The Tourist: A New Theory of Leisure Class.* New York: Shocken Books.
MacCannell, D. (1992) *The Empty Meeting Grounds: The Tourist Papers.* London: Routledge.
Mathisen, S.R. (2001) Den naturlige samen: Narrative konstruksjoner av 'de andre' i norsk tradisjon. In L.A. Ytrehus (ed.) *Fortellinger om 'den andre'* (pp. 84–98). Kristiansand: Høyskoleforlaget.
Mathisen, S.R. (2015) Contextualizing exhibited versions of Sami Noaidevuohta. In T. Fonneland, S.E. Kraft and J. Lewis (eds) *Nordic Neoshamanisms* (pp. 191–213). New York: Palgrave Macmillan.
Niezen, R. (2003) *The Origins of Indigenis: Human Rights and the Politics of Identity.* Berkeley, CA: University of California Press.

Niezen, R. (2009) *The Rediscovered Self. Indigenous Identity and Cultural Justice*. Montreal and Kingston: McGill-Queen`s University Press.

Niezen, R. (2012) Indigenous religion and human rights. In J. Witte and M.C. Green (eds) *Religion and Human Rights: An Introduction* (pp. 119–134). Oxford: Oxford University Press.

Olsen, K.O.K. (2002) Authenticity as a concept in tourism research: The social organization of the experience of authenticity. *Tourist Studies* 2 (2), 159–182.

Olsen, K.O.K. (2003) The touristic construction of the 'emblematic' Sami. *Acta Borealia* 20 (1), 3–20.

Olsen, K.O.K. (2004) Heritage, religion and the deficit of meaning in institutionalized discourses. In B. Klein, S.R. Mathisen and A.L. Siikala (eds) *Creating Diversities: Folklore, Religion and the Politics of Heritage* (pp. 31–42). Helsinki: Studia Fennica.

Olsen, K.O.K. (2006) Making differences in a changing world: The Norwegian Sami in the tourist industry. *Scandinavian Journal of Hospitality and Tourism* 6, 37–53.

Origo Nord AS (undated) *Sápmi Activities*. Alta: Origo.

Origo Nord AS (undated) *Rapport Samisk Reiseliv i Nord-Norge år 1*. Alta: Origo.

Petterson, R. and Viken, A. (2007) Sami perspectives on indigenous tourism in northern Europe: Commerce or cultural development. In R. Butler and T. Hinch (eds) *Tourism and Indigenous Peoples: Issues and Implications* (pp. 176–187). Oxford: Butterworth-Heinemann.

Pine, J.B. and Gilmore, J.H. (1999) *The Experience Economy*. Boston, MA: Harvard Business Scholl Press.

Prins, H. (2002) Visual media and the primitivist perplex: Colonial fantasies, indigenous imagination and advocacy in North America. In G. Ginsburg, F.D. Abdulughod and B. Larkin (eds) *Media Worlds: Anthropology on New Terrain* (pp. 58–74). Ewing, CA: University of California Press.

Saarinen, J. (2001) *The Transformation of a Tourist Destination: Theory and Case Studies on the Production of Local Geographies in Tourism in Finnish Lapland*. Oulu: Oulu University Press.

Selberg, T. (2011) *Folkelig religiøsitet: Et kulturvitenskapelig perspektiv*. Oslo: Spartacus forlag.

Stausberg, M. (2011) *Religion and Tourism: Crossroads, Destinations and Encounters*. London: Routledge.

Turner, L. and Ash, J. (1975) *The Golden Hordes*. London: Constable.

Viken, A. (2002) Turismens Sameland: Tradisjoner i transformasjon. *Utmark* 2002 (1), n.p.

Wang, N. (2000) *Tourism and Modernity: A Sociological Analysis*. Amsterdam: Pergamon.

Wright, R.J. (2014) *Samisk Reiseliv og salg av kulturuttrykk*. Commissioned Rapport. Alta: Origo Nord AS.

Internet sources

Tourism in Northern Norway: http://www.nordnorge.com/?id=412616041 (accessed 10 April 2015).

Sami Tourism in Northern Norway: http://www.origonord.no/page.jsp?id=243 (accessed 10 April 2015).

15 Respect in the *Girdnu*: The Sami *Verdde* Institution and Tourism in Northern Norway

Gaute Svensson and Arvid Viken

Introduction

For the Sami, culture and ethnicity are closely connected to traditional industries like reindeer husbandry. Over the last decades, this industry has been extensively modernized and also challenged by a public quest for new industrial development like tourism and mining. This is, of course, also contesting traditional practices and culture. In this chapter, we will focus on a particular Sami institution and tradition, the so-called *verdde*, which has been challenged by the modern in general and by leisure and tourism activities in particular. *Verdde* translates a bit imprecisely to a *helping friend* by the Sami. Traditionally, the term denotes the mutual beneficial and social connection between the reindeer herders and the settled residents – a relationship that has been given a legal platform in the Reindeer Herding Act (LOV-2007-06-15-40). During the seasonal migration between the coast and the inland, the reindeer herders occasionally needed help, for instance from people and boats in crossing a strait or a fjord with the herd. Over time, the contact often developed into a friendship. A *verdde* relationship is a prominent form of social or cultural currency drawing on informal economic aspects. However, with modernization, new forms of assistance are needed, and new forms of payback are demanded. Hence, what is being exchanged is changing. A central empirical point is that motorized access to nature areas that otherwise would be out of reach for people outside the reindeer industry, has become an important asset for the *verdde*. Examples like this are central to the paradoxes and legal gray zones that follow the modernization of an old social institution.

Currently, the term *verdde* represents more 'beneficiary friendship' in the Sami–Norwegian relationship. It has become a popular term, and many events and institutions have adopted the name. The *verdde* institution is significant for the tourism industry: some use the institution in line with its traditional meaning, others as a platform for new tourism products, as exclusive tours to reindeer herds or exclusive hunting and angling. We will show how the *verdde* institution is dealt with in very different ways in the everyday life and leisure and tourism contexts. Although the *verdde* institution is expressed and negotiated in various situations today, tourism tends to bring out the most controversial aspects that create tension between many stakeholders. Hence, it is within tourism that this phenomenon brings up new questions about modernity and entrepreneurship that are the point of departure here.

In this chapter, we will present the *verdde* institution, its cultural base and practical significance within the reindeer herding community, and as a modernized institution. The *verdde* is a dynamic cultural institution, as are the Sami culture and society in general. Therefore, it should be contextually based not only on a theory of modernization, but also on theories of the premises and characteristics of alternative (non-)economic values. Toward this end, we will show and discuss how the modernized *verdde* seems to be a mismatch with public policies concerning land use, but in other ways has found its form and position within a modern context.

Sami Modernity

Sami society is modern. 'You are Norwegian and modern, we are Sami and modern', a Sami representative said rhetorically in a seminar presentation. The point is that Sami society has gone through the same processes as modern societies in general. This means that there is a modern formal knowledge base for the Sami society, modern technologies are part of everyday life and modern institutions govern Sami thoughts and practices. Reindeer husbandry has been modernized in many ways; today, the industry strongly relies on snowmobiles, all-terrain vehicles (ATVs) and modern communication technologies. It is still nomadic but not in the same way as before. Most Sami have a permanent residence and family life, including those living off reindeer husbandry. The herders commute to and from the herd – one period (week) with the herd within an organized unit called a 'reindeer district' or *siida*, the other period at home. Further, reindeer herding is not exclusively learned in the field and within the family, but is also taught in schools. And reindeer herding, in all its facets, is a matter of research and formal knowledge production. This, of course, has consequences for a tradition like the *verdde.* It has been shown how Sami traditions adapt to modern society, or how traditions are transformed within the reign of modernity. The *verdde*, as this chapter will show, is no exception and seems to be continually contested by modernization.

Modernity was particularly much debated in the 1980s and 1990s by scholars such as Anthony Giddens, Zygmunt Bauman, Ulrich Beck,

David Harvey, Agnes Heller and Scott Lash, among others. In many ways, modernity and enlightenment were seen as an emancipatory project in which rational thinking and science substituted for traditional belief systems. However, it can also be seen as Western dominance – a project through which the West took command over the rest. Also widely discussed is whether it is the same modernity that is valid all over, or if there is a multitude of modernities (Fourie, 2012) or alternative modernities, for instance, within the world of indigenous peoples. Latin American scholars have launched a theoretical alternative to modernity as enlightenment and a logic of rationality, seeing modern thinking as a 'Eurocentric representation of knowledge and power that suppresses and marginalized other forms of knowledge in a hegemonic project of modern development' (Stetson, 2012: 81, quoting Escobar, 2004: 217). These scholars refer to modernity as coloniality, a structure similar to that of the old colonial world. Whereas the time of colonialism has disappeared, coloniality is still a prevailing power structure; it is maintained (Stetson, 2012). In many places there have been periods of apologies for past suppression policies and reconciliation between indigenous and majority populations (Johnson, 2011). This has also happened to the relationship between the Norwegian authorities and the Sami public. However, as in many other places, land rights still have a colonial character. Until 2002, the land in the north of Norway was entirely in the hands of the state. Since then, the ownership of parts of the region – the county of Finnmark – has been transferred to a regionally based private foundation managed by a politically elected board. However, this is only a shift in ownership; the rules concerning its use are still Norwegian law. Thus, it is still the Norwegian authorities who decide how land should be used and for what purposes. It is also the Norwegian authorities who have imposed a Reindeer Herding Act where the *verdde* system formally is given its legitimacy, outrange and limitations. Thus, the regulation of the *verdde* is formal and very much a given. However, as a cultural institution, the *verdde* of course is dynamic and changes accordingly with other parts of Sami society and culture.

It is not only reindeer husbandry and the *verdde* institution that for long have been in a process of change. This has also happened to Sami and Norwegian society concerning outdoor recreation. Today, these activities are very much motorized. People use their snowmobiles, ATVs and boats for recreational purposes such as fishing, hunting, berry picking or just touring. This, of course, has raised a need for regulation. According to the Motor Travel Act (LOV 1977-06-10 nr 82), motorized vehicles are only allowed on roads and marked tracks. But there are some exceptions. One concerns reindeer herding: herders may go wherever they need to go to perform their herding activities. This is also a right, stated in the Norwegian Motor Travel Act (paragraph 4c) that can be transferred to their *verddes* – a right that has become an important part of the relation today.

Modern Economy and Sami Exchange Economy

Meat or money: Informal practices and formal regulations

As mentioned, the *verdde* institution has undergone some changes during the last decades. There are forces both within and outside the reindeer herding industry that want to redefine the *verdde* institution from something informal to something more formal. The nature management authorities call for clearer rules – for instance concerning tourism – and a more transparent system. However, there is resistance within the reindeer husbandry community to make this relationship part of a formal economy through a defined repayment. Also, the Norwegian Supreme Court supports this, stating that there is no need for a formalized contract in a *verdde* relationship. Given these premises, the theoretical platform is anchored in economic anthropology and the social aspects of economy.

The informal economy is part of social and cultural capital through which norms of interaction are negotiated and recreated. Transactions made within an informal economic system are often undefined in terms of time frame and type of repayment (Sahlins, 2004). Trust is there as an underlying premise of informal economic interaction. Barter economy (Humphrey & Hugh-Jones, 1992; Ingham, 1996), exchange (Godelier, 1999; Malinowski, 1922; Sahlins, 1972; Weiner, 1992), gift economy (Davis, 1992; Mauss, 1954) and reciprocity (Browne & Milgram, 2009; Ensminger, 2002) all revolve around the social aspects of economy. However, as pointed out by Davis (1992), this has led to a polarized understanding that could be exemplified by Sahlins' (1976: 216) famous statement: 'Money is to the West what kinship is to the rest'. What Davis (1992) is implicitly pointing out is that there is an academic reproduction creating a perception of a Western and Third World economy which is highly stereotypical as well as dichotomized. Economic anthropology has, through a focus on economic systems in pre-capitalist cultures, reproduced a perception of differences between the West and the rest. Davis' point is that in this image there is more room for differences than similarities between the West and the rest. Hence, we have failed to address questions about the numerous aspects of informal economic behavior in our own backyard. The *verdde* is just one of several examples of this.

Trust and reciprocity are prominent aspects of informal economic relationships. This is, as we will show, also the case when talking about more specific examples like *verdde*. Reciprocity has been given substantial attention in economic anthropology. Sahlins (1972) makes a much cited distinction between three different forms of reciprocity: generalized, balanced and negative. Leaning on a broad empirical base in Oseania and Melanesia, Sahlins (1972) argues that these forms are important as they point out nuances in economic life. Generalized reciprocity is, as he says, 'the solidary extreme' (Sahlins, 1972: 193). This form of reciprocity is altruistic and 'the material side of the transaction repressed by the social' and 'the counter [obligation] is not stipulated by time, quantity or quality: the expectation of reciprocity is indefinite' (Sahlins, 1972: 194). This undefined obligation to reciprocate characterizes a morality based on trust that *something* will come out of the relationship. Balanced reciprocity is 'less personal' than generalized reciprocity and 'more economic' (Sahlins, 1972: 195). The examples that Sahlins uses are direct exchange, payment, trade, buying and selling. The third form – negative reciprocity – Sahlins (1972: 195) labeled 'the unsociable extreme'. Negative reciprocity is impersonal and can be described as getting something for nothing. Haggling, gambling and theft are but a few examples of negative reciprocity. Although Sahlins (1972: 196) was arguing hard for the cultural relativistic principle, where his analytical ambitions with the distinction between different forms of reciprocity were a better understanding of what he called 'primitive societies', he also added that 'all this seems applicable to our own society'. This falls in line with Davis' (1992) argument to ask questions about the informal economic aspects of Western culture.

Despite their theoretical differences, Fredrik Barth and Marshall Sahlins share a fascination for the social, informal aspects of economic behavior. In *Process and Form in Social Life*, Barth (1981) shows how entrepreneurship that is based in social institutions like volunteer work can challenge these institutions. In the words of Sahlins, this could be an example of entrepreneurship that transforms something that is usually seen as generalized to negative reciprocity. Barth describes how the newcomer to a fur community in Darfur, Sudan, met the expectations of beer generously as he invited the community to a work session. However, instead of building a house, which would have been in line with the local custom, he asked for help to plant tomatoes.[1] No one opposed this request. However, this shift consequently challenged the volunteer institution (as well as how land was distributed) as the newcomer made a substantial profit by selling the tomatoes. The lessons learned from this are that when someone tries to make money off a social event or tradition, this might change the institution in question. The *verdde* institution represents such an arena where conversions are possible – both in legal and illegal forms. We will show how this has become an option for the tourism industry in different ways, and with different outcomes.

Multicentric market economies

Drawing on the classical work of Bohannan (1959) and Hornborg (2007), we ask if we can learn something about the Western economy by revisiting the analytical work of pre-modern cultures. Hornborg (2007: 64) states: 'The fundamental principle of any "multicentric" economy is the acknowledgment of two or more distinct and incommensurable spheres of value'. Bohannan (1959) identified three economic spheres among the Tivs of Nigeria: food, prestige goods and women. Concerning women, they are a matter of exchange within a kin context – women have value only in relation to other women. In a modern context, the multicentric system, including women, gives associations to the slave trade or trafficking. Bohannan's response would be that the Tivs had a system with its own internal logic profoundly different from the monetary, unicentric system. Bohannan argued, as pointed out by Hornborg (2007), that a multicentric distribution system marks a space for value assessment with its own internal logic, hence making it incommensurable with other spheres. In contrast, the unicentric distribution systems, which in Bohannan's (1959) analysis were represented by the British Empire, have one economic point of reference – money. Money in this sense represents a linear and singular value that separates objects from the cultural context in which meaning and value have been negotiated. The multicentric distribution system collapsed in post-war Nigeria, when the colonialist introduced money as a unifying standard. The consequences that followed from the introduction of the pound sterling were tremendous for the Tivs, according to Bohannan (1959). Bohannan's point, which is interesting when looking at the *verdde*, is that multicentrism encourages questions about other points of reference besides money or how alternative values ascribed have the power to make the implicit part of an institution like *verdde* explicit.

The distinction between unicentric and multicentric economies made by Bohannan (1959) and others is yet another example that can be read into a dichotomized understanding of the differences between the West and the rest. It is obvious that the monetary and unicentric Western dynamics impacted the diversified and multicentric system among the Nigerian Tivs. However, Davis' (1992) point about informal economies also existing in the West is again relevant. There are reasons to expect multicentric systems also in our society. The *verdde* will be just one example of a phenomenon with its own non-monetary economic logic in our society – a logic defined through alternative values. It could be argued that such alternative systems are as likely to be found within an alternative modernity as that of the indigenous world.

Methods and Data

Living in northern Norway means to a certain degree living with Sami and Sami culture, regardless of one's own ethnic background. The revitalization of Sami culture throughout the last few decades has created a space for public awareness of 'Saminess' as a cornerstone in north Norwegian modernity, although many people might argue that there is quite a distance to go to make up for past mistakes. The *verdde* institution is just one of several examples of cultural elements with symbolic value that have received public attention. The *Verdde* Industrial Park and the *Verdde* Football Tournament in Kvænangen municipality are but two examples of a recognition and revitalization of the cultural heritage of this region. However, some mythical aura still surrounds this phenomenon that represents the bridging between the Sami and Norwegian – a bridging that is constantly negotiated and recreated. This has been the empirical scope of this chapter.

The data presented in this chapter falls into two main categories. Besides being qualitative research that includes observations of a partly Sami environment, the data presented here is collected through participant observation. One of the authors has been a guest and a part-time worker at one particular camping site for more than a decade. During this period there have been many discussions about *verdde* relations and the institution as such – also with reindeer herders. In this sense, this methodological approach is reminiscent of the research principle of auto ethnography. This has contributed to identifying the interview questions that constitute the second pool of data. These were semi-structured interviews; hence, the empirical base for this chapter rests on qualitative data.

The analytical framework of this chapter is based on an interpretivistic research strategy.[2] Here, according to Blaikie (2000), one puts emphasis on the process of discovering a hypothesis (not testing it to establish knowledge as true or false). This means that the construction of research questions and locating a hypothesis have no fixed form prior to the fieldwork. Interpretivism is not to seek to test theory but is rather inviting the researcher to bring in new theoretical perspectives to highlight nuances in the phenomenon in question. This is what the Norwegian anthropologist Marianne Gullestad (1999) calls 'logic of discovery'. This strategy is considered a more 'open' approach as there is no fixed scope on any specific cultural element. In our case, we had located the phenomenon we wanted to explore, the *verdde*, prior to the fieldwork. So, the methodological implications were to bring forth the contexts that this institution appears in and analyze the context it produces. To interview different stakeholders has therefore much to do with pointing out the negotiations regarding this institution and following it as it constructs meaning in different parts of society.

The informants who are presented through interview data are different stakeholders who are involved in defining the *verdde* institution today. One couple in particular, Sigve and Inga-Maret,[3] are important informants. They work part-time as reindeer herders and owners, and both have modern jobs, one as a bureaucrat, the other as a teacher. They were contacted several times in the process. In addition to these informants, we have interviewed representatives from the various public nature management agencies who are involved in protecting, defining and enforcing the *verdde* institution today. Altogether, we interviewed six people representing five public agencies as well as the reindeer industry.[4]

The *Verdde* in Transition

The *verdde*: Historical and formal accounts

The *verdde* system emerged in tandem with the transition from hunter-gathering culture to reindeer-herding nomadism (Hansen & Olsen, 2004: 251) in the form of a symbiotic relationship; hides and reindeer meat were exchanged via boats and crew during the migration (Jåstad, 2011). An official report paraphrases this system as follows:

> The Sami term '*verdde*' is translated into Norwegian as 'guest-friend'. *Verdde*-relations (*verdde*-vuohta) must thus be translated as guest-friend relations. These were mutually beneficial relationships, which the Sami reindeer herders traditionally had with both the inland farmers (dálonat) and the farming fishermen in the coastal areas (meronat). Two people enjoying such a mutual relationship would call each other '*verdde*'. In the inland, the relationship between reindeer owners and dálon was close. Their livelihoods were similar enough (requiring sufficiently similar skills and knowledge) to make it possible to move from being a dálon to a reindeer herder and vice versa. (NOU, 2001: 34)

In 1966, the Norwegian social anthropologist, Harald Eidheim, described the system in the following way:

> Usually, the reindeer herding nomads offered small amounts of reindeer products to certain of their *verdde*s every time they had to slaughter an animal, and the locals on their end offered services to their *verdde*s when the latter needed it. Both parties appreciated the exchange of goods and services, and only rarely did people calling each other *verdde* come into conflict with each other. (Eidheim, 1966: 428)

In Eidheim's example, the nomadic Sami crossing straits with their reindeer herds received help in the form of boats and manpower. The locals also kept the equipment they did not use in storage when the winter and summer seasons ended. Eidheim points out that the *verdde* relation originally existed among Sami communities only, coastal Sami and reindeer herding Sami, but that the former group has largely been naturalized as Norwegians. He also demonstrates how ties grew weaker during this Norwegianization process. In addition, he points out that even as early as in the mid-1960s, the *verdde* system had been diluted, and the coastal dwellers gradually started charging for their services, particularly the younger generations who considered themselves Norwegian. Evjen (2009) mentions an example where reindeer herding families who had goats would have them stay at local farms during the winter season. But there was a reciprocal parallel, too: the locals would have so-called *sytingsrein*, that is, reindeers that the locals owned that were integrated within the herds and in the care of the reindeer herders. It also happened that locals cared for elderly people who had grown too frail to take part in the migration. Thus, closeness and mutuality are crucial to *verdde* relations. One of the reasons why the *verdde* relations were, and still are, important in the Sami community may be a widespread skepticism toward outsiders, maybe even toward Norwegians, because of bad experiences and past suppression by Norwegians and the Norwegian government.

The *verdde* tradition

Verdde is a Sami word that in the Norwegian translation has been *guest friend*, for instance used by Eidheim and in the state report referred to above. This may well be based on the fact that for settled people the reindeer herders were travelers with whom they became friends. However, it may also be a slightly wrong translation. According to the reindeer herder Sigve, a guest friend in Sami should be called *goussi olmmái*. A *verdde* in Sami etymology is closer (but not exactly translatable) to a *helping friend*. Sigve points out that this distinction is highly important because it used to be the need of help that was the premise for this kind of relationship. His point is simply that trust and appreciation were established through and based on qualified work abilities, skills needed by the reindeer owner in his or her everyday operations. Although the friendship was between a settler and a traveler, skilled help was a stronger premise than that one part was a traveler and often a guest. Sigve emphasizes that this help would not be asked for or offered randomly. Those involved are skilled people who the reindeer herder has known and trusted for a long period. Sigve says that things like practical skills, common friends and reciprocal respect are qualities that can open up for a *verdde* relationship. Without a practical know-how, no real *verdde* relationship will emerge. A *verdde* can inherit a status from his or her parents but how this new relationship evolves is a question of personal connection – not much different from any other friendship. But the premise for this kind of friendship is still practical skills. There are also symbols manifesting a *verdde* relationship, like becoming a godparent for a child.

Sigve points out that the *verdde* institution should not be confused with some of the other relations that are between the reindeer herders and people outside the industry. *Reangga* refers to an attendant of the reindeer herd holder, receiving a defined payment. This is a much more formalized relationship than the *verdde*. Another form of labor provided from the outside is the *vehkki* which represents a more random helper, someone at hand when needed. As opposed to the *verdde*, both the *reangga* and the *vehkki* are characterized by a more formalized and balanced reciprocal relationship.

The modernization of the reindeer industry requires immediate assistance at times. The *reangga* and the *vehkki* represent labor fulfilling such needs. Sigve adds that the *verdde* relation, despite changes in the industry, is still highly social and characterized by something more than just the work that is done. Bjørn says:

> The relation [*verdde*] incorporates a form of tradition where the relation means more than that two persons just know each other. It is more than neighborliness. To help your neighbor painting is not what *verdde* is. It is a mutual exchange of favors but it is also something that is close to a social organization. It is more than just the exchange of favors, it is a relationship that is on the brink of family. It is an extension of the content of the old siida system.

Sigve and Inga-Maret also emphasize the difference between male and female *verddes.* The male exchange can be described as also incorporating more modern favors such as helping out with practical things like the internet or the repair of cars, ATVs or snowmobiles. There is still a more traditionally oriented exchange between women, where they exchange handicraft products and meat/fish. In big events like weddings and confirmations, the female *verdde* often helps in coordinating the happening. This help is much needed as events like these can gather a thousand people or more.

The Reindeer Act allows anyone with a reindeer ear tag[5] to have *verddes.* This means that the *verdde* can acquire the same right to use motorized vehicles like ATVs and snowmobiles within the district of the reindeer herder. With the introduction of the Motor Travel Act (1977), this became rather important due to the reindeer pastoralists' free access to all nature areas including protected areas. The ranger Bjørn characterizes this as a precious asset for many locals outside the reindeer industry – an asset that they envy. And as elders see it, free access is something they 'lost' when this law was made. The exception of the reindeer herders also includes the right to hunt small game and to fish for personal use, basically closed for all others. The *verdde* status simply includes a set of exclusive privileges that no one outside the industry will have under other circumstances. All of our informants underline the importance of these privileges in the modern *verdde* relation, and that this currently constitutes a significant part of what is being exchanged.

The modernized *verdde*

With modernization, the kind of help needed in the reindeer herding industry has changed. The needs are more complex. Sigve pinpoints a change going from helpers that are good at multitasking and practical skills, to specialized knowledge. Whereas before one could have a few people who could replace the herder in most respects, there is now a need for several *verddes*, each covering a specific know-how. People who have the knowledge of how bureaucracy, applications or contract work is done represent a more recent set of knowledge that could be part of *verdde* relationships. However, it is still the practical tasks that legitimize the *verdde* status, things like transport, logistics, butchering and ear marking, tasks that would otherwise have been taken care of by the herders themselves. Those providing other services needed, such as financial, bureaucratic or consultancy assistance, are formally not qualified for a *verdde* status. However, occasionally such contacts perform (motorized) leisure activities, claiming to have a *verdde* relation, in areas where such activities are otherwise prohibited. This represents a gray zone in the eyes of the nature management representatives. They meet people insisting that they have a *verdde* relation and mission. To underline the abuse of the institution, Bjørn says: 'This is best expressed through the fact that you seldom meet the *verddes* in the mountains when the weather is bad. If we meet *verddes* when it is 37 below in February it would be the ones that are really part of it'. As representatives of various government agencies see this, the *verdde* status is built on vicarious motives. The freedom to use snowmobiles in the nearby national park (Reisa) and to hunt and fish there is seen as a substantial asset for many people. To uncover such motives for an alleged *verdde* is difficult. There have been cases where both reindeer herders and *verddes* have been prosecuted and convicted. The Supreme Court has stated that the *verdde* relation must be defined through a mutual reciprocal agreement. This is normally quite easy to obtain. This can be an oral agreement that points out what is being exchanged, and will often be based on actual need of work. The changes that reindeer husbandry has undergone has created new needs and wishes among the reindeer herders and the settled population, and broadened the herders' need of assistance in the field. The question is if the legal platform that the *verdde* rests on captures the diverse implications that follow from today's reindeer husbandry. Probably not. But the bottom line, from the authorities' perspective, is that the presence of the *verdde* should not represent additional motorized or environmental impact, according to Bjørn. He – or more correctly the Norwegian law – does not acknowledge a modernized *verdde* institution.

As the ranger Bjørn sees it, the *verdde* relation is about to move from a barter economy to exchanges more influenced by the monetary economy. He obviously sees several ways that this can be solved. Along with his colleagues, he is asking for more transparency and clearer rules – thus, a more formally defined *verdde* institution which makes it easier to deal with fakery and misuse. However, Bjørn tells a story indicating other solutions. His office had a phone call from a district saying that it had too much traffic. This simply meant that they had too many external people – *verddes* – present. The fact that this was seen as something negative was interpreted as a sign of *pro forma verdde* relations and that the *verddes* did not contribute to the work done in the area. The ranger office's response was: 'fix it yourself'. This would be to rely on self-justice which in fact would be in line with more neoliberal governance principles.

Verdde Relationships in the Tourism Industry

The *verdde* as a base for tourism operation

Reindeer pastoralism and the Sami culture have been and remain an important part of the tourism marketing of northern Norway. Tourism is not a new phenomenon within the reindeer industry. Since the late 1800s, visits to Sami camps have been a tourist activity, and reindeer herders have made money from arranging such visits and by selling self-produced traditional handicraft (*duodji*) souvenirs. Another product with a long tradition is the so-called rein raid. The rein raid is the traditional way of long-distance travel in the winter with reindeers and sledges. The touristic rein raid is currently quite short including a traditional Sami meal like a *bidos*, a stew made out of reindeer meat. However, the most common form of tourism related to reindeer herding traditions is a visit to a Sami tent – a *lavvu*, listening to storytellers, *yoik*, (traditional singing), and being served a Sami meal. This is offered both by reindeer herders and people only dealing with tourism. The most exotic product may be to pay a winter reindeer herding camp a visit or to take part in the seasonal (spring) migration. To be able to offer this, there is a need for collaboration with a reindeer herder, preferably a relative or a *verdde*. However, this is a complicated operation, where one easily ends up in legal gray zones.

To visit a reindeer winter camp, the reindeer must be situated at a roadside or on a path, which they only occasionally will be. The alternative would be to bring in the tourists by ATVs or snowmobiles, but this is forbidden because of the general prohibition on motor travel outside roads and paths. This is even forbidden for the reindeer herders because, as the law is, free access is restricted to reindeer herding. Tourism is not seen as a part of this, even if such a combination might be a smart way of making a living. What is legal is to have a tourist – as some sort of guest – on the back of the reindeer herder's snowmobile. This is not practical. Thus, tourism integrated into reindeer herding is practically impossible. One of the ways to bypass this would be to declare the tourists as *verddes*. With this status, tourists would have free access to protected areas, herds and reindeer herding camps. However, this would be a *verdde* relation built on false premises, both according to the authorities and the interviewed reindeer herders.

There have been several lawsuits and convictions of what the authorities and prosecutors claim to be *pro forma verdde* relations. What these cases have in common is that the *verdde* has no direct or indirect work tasks relating to reindeer herding; hence, the *verdde*'s presence represents an additional environmental impact. Bjørn says that he usually has a straightforward approach if he suspects someone to be claiming to be a *verdde* under false pretenses:

> If I encounter people in the national park on ATVs or snowmobiles without the normal affiliation to the reindeer herding industry, I simply ask them to point out where the reindeer are or where the herd is heading. If they do not seem to have a qualified opinion on this, I normally suspect that they sail under false colors.

This was also the case when Bjørn met some men from southern Norway on snowmobiles in Reisa national park a few years ago. The men claimed to have bought a package from the company *Sami Tours* that included transportation and ice fishing. They were ice fishing in a remote area on the border between Troms and Finnmark counties. The tourists had used snowmobiles for transportation and had no fishing license – conditions that can only be justified through a *verdde* status. This case received substantial attention from the media as the tour operator, also a reindeer herder, was a television celebrity who had taken part in several reality shows on Norwegian television. Bjørn said in a radio interview that the ranger office had been watching the activities of *Sami Tours* as they suspected illegal activities.

> A look at their [Sami Tours] web page underpins the impression that they are using the Reindeer Herding Act to create tourism products [...]. This has nothing to do with reindeer herding. The Reindeer Act is meant to be a [legal] tool for reindeer herding, not for tourism. This legal framework was never meant for tourism.

This encounter in the national park led to the conviction of the tourists as well as *Sami Tours*, even though the owner was not present. The sentence stated that the tourists had bought a trip as tourists. The *verdde* status provided a cover for an exclusive product including motorized activity and angling without licenses in the Reisa national park. The reindeer herder Sigve says that it is cases like this that weaken the legitimization of the *verdde* institution both outside and inside the reindeer herding industry. 'They do not understand where this [*verdde* institution] came from. They will gain no respect in the *girdnu*'.[6]

Krossdal camping site and the reindeer herders: A modernized *verdde* relationship

It has been said that tourists are the nomads of the present day. The point we make here is slightly different: the nomadic Sami may also be tourists. Many of them live in campers or trailer homes during the summer period, and this is particularly common during the reindeer migration. Often, the reindeer herders prefer to park their campers at a trailer park or campsite. Also, herders not doing this, but living in their own summer houses, may appreciate the existence of a nearby campsite. The following is an example of how this has grown into a modern and generalized *verdde* relation.

The example is from Krossdal, a village a few hours from Tromsø. During the summer months, several reindeer herders' families live in the village. The relations between the villagers and the reindeer herders are generally good, and there has never been any conflict between the camping site and the reindeer herding industry, but rather a great deal of mutual benefit and joy. As the reindeer herders' summer cottages are rather simple, they come to the camping site to shower and use saunas, to do laundry or other everyday tasks. They are allowed to place their boats along the pier or on the shoreline and to use the assembly rooms for meetings and such. They often take a coffee and have a chat with the camping site owner in his kitchen. In the autumn, several reindeer herding families migrating by the village move into the camping site for a week or two. This has been an ongoing practice since the camping site was established, and a tradition the current owner has inherited from his father. They do not pay up immediately; instead, the Sami come by with a slaughtered reindeer later in the fall or winter. A reindeer herder who lives in the house next door, owned by the camping site, serves as a caretaker at the camping site, fixing all sorts of issues: repairing snowmobiles and ATVs, doing carpentry, manning the reception and bringing fish and reindeer meat. Once, the camping site's computer network collapsed, and his brother, who also works in reindeer husbandry, came and fixed it. When the camping site expanded with a boathouse and a tool shed, another reindeer herder assumed responsibility for casting the floors and erecting the buildings, and brought along the family members he needed to assist him, with the camping site owner serving as an assistant. These examples reflect the fact that reindeer husbandry, to some degree, is combined with skills in a very broad variety of work and expertise. Many of the herders even have formal competence in many of the fields. In addition, the Sami connection bestows an air of Sami culture upon the camping site, representing a form of exoticization that is often lucrative in a tourism perspective. All these are examples of *verdde* relations of the original kind (Eidheim, 1966). In this case, the roles are turned around; it is the locals that need practical assistance and they get it from the migrating reindeer herders.

The fact that the mutual relations are good and that there is a great degree of trust is often emphasized. When the camping site owner turned 50, some of his Sami friends gave him a (living) reindeer (a *sytingsrein*) well placed in their herd. This is often brought up in chats around the kitchen table, years later. The reindeer is still in the herd and has calves. This is a clear demonstration of the fact that the relationship is greatly valued and is considered to be of a lasting nature. Another display of trust is the fact that the owner's daughter has repeatedly – as she grows up – been given a new Sami costume (*kofte* in Norwegian, *gákti* in Sami; the traditional Sami dress) by one of the Sami ladies familiar with the art of costume (*kofte*) making, who has lived her summers in the village her entire life. You do not give a *kofte* to just anybody! All these are signs of a strongly appreciated relationship – genuine *verdde* relation.

As we see, the camping site benefits from the relationships with both the local Sami and the reindeer herders. The camping site enables the reindeer herders to maintain a regular and modern daily life in the local community. When prompted, they often refer to the *verdde* system. Some of these ties must be characterized as being strong, and maybe the greatest benefit lies in the dyadic relations. However, the relationships, and the way the owner manages them in his major work as a journalist in a national newspaper, have probably also put him in a position of providing access to news items more easily than would have been otherwise. If the owner needs someone from the Sami community for an interview or Sami photo shoot, establishing the contacts he needs is done in a matter of minutes. The *verdde* relationships have provided him with many useful bridges into the Sami community.

Concluding Discussion

There seems to be a broad consensus between our informants that the *verdde*, as it has been up to the present, is first and foremost an informal social institution and economy. There is, nevertheless, a substantial discrepancy between what the reindeer industry ascribes to the *verdde* relationship and the interpretations made by public agencies affiliated with the industry. When the various nature management units describe how they would *like it to be*, how they would like to see it organized in order to make it more transparent, the distance is even more evident. The core of what is being negotiated here is whether or not the *verdde* exchange should include the more modern elements of the reindeer industry or if the relationship should ultimately be based on the traditional exchange of practical work and meat.

Despite the fact that economic anthropology has indirectly contributed to an economic worldview where Norway and the Western world have been seen as a market economy with little room for other elements than a balanced reciprocity and a unicentric value system, the *verdde* institution points in another direction. As the industry has been modernized, needs for assistance have changed. While the reindeer owner needed a generalist in the past, the contemporary needs are skills that are more specialized and competence in handling the bureaucratic world. However, this has not changed the reciprocal dynamic in the relationship. The objects, favors, skills, services and knowledge that lie in the relationship today are still in their making, and partly undefined and implicit. What constitutes repayment can be almost as hard to identify as the economic aspects of the friendship. It is hard to point out a single set of reasons why the reindeer owner, for instance, chooses to ask his *verdde* to become his child's godparent. It is something one simply does to a *verdde*. All this points toward something distinctly different from the balanced reciprocity that Sahlins (1972) describes. Therefore, there is a resistance in the Sami community to making the *verdde* more clearly and formally defined. Still, it should be something different from the *reangga* (part-time worker) and *vehkki* (random helper). With reference to Sahlins (1972), this could be seen as a negotiation concerning how generalized this reciprocal relationship can be. The negotiation is quite simply evolving around what legitimate exchanges within the *verdde* institution should be. This is even more evident when looking at the public response to some of the attempts to commodify the *verdde* into a part of tourism products.

Fredrik Barth's (1981) example from Darfur shows how entrepreneurship can redefine collective institutions and how individuals can create room for social change. However, the conviction of the tourism entrepreneur for misuse of the *verdde* did not represent a sphere collapse similar to Barth's (1981) example. But it does represent a contestation of the tradition. Concerning the *verdde*, the situation basically was the opposite of the case with the tomato man; it is services paid back with access to nature and leisure activities that threaten a traditional institution. But, also, when the *verdde* status was faked for a tourism entrepreneur to make money, the entrepreneur was sanctioned also by the reindeer herder community and his position as such was weakened. His loss of respect – in the *girdnu* – demonstrates the strength of the normative backbone that the *verdde* institution is based on. Also, the rangers believe that internal justice is important and a way of protecting the institution. It is extremely hard to uncover and prosecute a *pro forma verdde* relation, and internal justice and social sanctions are complementary to formal control. Thus, the *verdde* institution, or informal economic elements for that matter, should not be seen as spheres separated from other economic life. What Barth's (1981) perspective can offer is pointing out that entrepreneurs as individuals have the power to visualize and exemplify the nuances in important economic institutions in modern life in Sapmi.

Although there are some major differences between Bohannan (1959) and Barth (1981) in their understanding of economic structures and social change, both perspectives create analytical room for questioning the *verdde* institution. Bohannan's (1959) argument of the profound impact of money on the Tivs, which led to a transition from a multicentric to a unicentric economy, indicates a form of legality where money erodes all social bonds, creating an economy based on selfishness and egoism. In the case with the *verdde*, it is not entirely money, but the authorities' urge for clearer and stricter regulation that is hollowing out and, as a consequence, threatening a tradition. But also money is involved, as tourism entrepreneurs try to figure out how the institution can be used as a platform for commercial activities. Despite this, we argue that the *verdde* is just one of numerous examples of modern institutions that can be relational and personal and not merely managed by commercial values – qualities that Sahlins defines as a generalized reciprocity. What Bohannan (1959) offers through his dichotomized distinction is a theoretical tool to search for altruistic principles in economic life. This is, as mentioned, Hornborg's (2007) point as he is arguing for a multicentrism in the modern world.

In sum, there are four major outcomes of this study. First, it is demonstrated that the *verdde* institution can fit well into modern settings related to tourism. It is both challenged by tourism as well as challenging tourism. Second, in the reindeer herder community there is skepticism toward a formalized *verdde* – its values related to a generalized reciprocity should be preserved, but in an up-to-date form. Third, to declare guests – including tourists – as *verddes* is perceived, also by the court, as in conflict with the law. Fourth, and partly because of this, representatives of the authorities would prefer that the *verdde* was more clearly defined by law. We have not discussed the *verdde* with Sami politicians. With the prevailing practice, the authorities demonstrate a lack of confidence in the reindeer herder industry's ability to find a viable model for a modernized *verdde* institution. Still, we think that the survival of the *verdde* institution in a modernized form could lie in an extended system of self-justice, or in a political process or solution found within Sami governing institutions. This would at the same time be a further step away from a stage of coloniality to a more post-colonial and responsible governance in the field of Sami traditions and culture.

Notes

(1) This is why this work has been labeled and is commonly referred to as *the tomato man* in Norwegian anthropology.

(2) For Blaikie (2000), interpretivism is used synonymously with abductive research.

(3) The names of informants and places have been changed due to anonymity principles.

(4) (a) *Statskog*, the state-owned forest enterprise; (b) *Fjelltjenesten*, the ranger organization for Nordland and Troms counties; (c) *SNO*, the state nature supervision; (d) *Fylkesmannen i Troms- miljøvernavdelingen*, the county governor environmental department; and (e) *Politiets miljørepresentant*, the environmental representative in a police department.

(5) A reindeer ear tag (reinmerke) is an individual marking of each animal that a person owns. It is, furthermore, the legal term for the right to own reindeer and work with reindeer husbandry with the obligations, duties and rights that this includes.

(6) *Girdnu* is the circular fence used when the reindeers are gathered for marking. The *girdnu* represents more than just a physical fence. It is also a social room with important cultural and symbolic value where people's reputations are created and negotiated. The *girdnu* is a place for work. Discontent or bad feelings are dealt with elsewhere as there is no room for conflicts here.

References

Barth, F. (1981) *Process and Form in Social Life* (Vol. 1). London: Routledge & Kegan Paul.

Blaikie, N. (2000) *Designing Social Research: The Logic of Anticipation*. Cambridge: Polity Press.

Bohannan, P. (1959) The impact of money on an African subsistence economy. *The Journal of Economic History* 19 (4), 491–503.

Browne, K.E. and Milgram, B.L. (eds) (2009) *Economics and Morality: Anthropological Approaches*. Lanham, MD: AltaMira.

Davis, J. (1992) *Exchange*. Minneapolis, MI: University of Minnesota Press.

Eidheim, H. (1966). Lappish guest relationships under conditions of cultural change. *American Anthropologist*, 68, 426–437.

Ensminger, J. (2002) *Theory in Economic Anthropology*. Lanham, MD: Rowman & Littlefield.

Escobar, A. (2004) Beyond the Third World: Imperial globality, global coloniality and anti-globalisation social movements. *Third World Quarterly* 25 (1), 207–230.

Evjen, B. (2009) Custodial reindeer and custodial goats: Part of reindeer herding and animal husbandry. *Rangifer* 27 (2), 79–91.

Fourie, E. (2012) A future for the theory of multiple modernities: Insights from the new modernization theory. *Social Science Information* 51 (1), 52–69.

Godelier, M. (1999) *The Enigma of the Gift*. Cambridge: Polity Press.

Gullestad, M. (1999) Kunnskap for hvem? Moderne virkeligheter og utfordringer for sosial antropologien. *Norsk Antropologisk Tidsskrift* 10, 81–96.

Hansen, L.I. and Olsen, B. (2004) *Samenes historie frem til 1750*. Oslo: Cappelen.

Hornborg, A. (2007) Learning from the Tiv: Why a sustainable economy would have to be 'multicentric'. *Culture & Agriculture* 29 (2), 63–69.

Humphrey, C. and Hugh-Jones, S. (1992) *Barter, Exchange and Value: An Anthropological Approach*. Cambridge: Cambridge University Press.

Ingham, G. (1996) Money is a social relation. *Review of Social Economy* 54 (4), 507–529.

Johnson, M. (2011) Reconciliation, indigeneity, and postcolonial nationhood in settler states. *Postcolonial Studies* 14 (2), 187–201.

Jåstad, H. (2011) The effect of ethnicity and economy upon intergenerational coresidence: Northern Norway during the last part of the nineteenth century. *Journal of Family History* 36 (3), 263–285.

Malinowski, B. (1922) *Argonauts of the Western Pacific: An Account of Native Enterprise and Adventure in the Archipelagoes of Melanesian New Guinea*. New York: E.P. Dutton & Co.

Mauss, M. (1954) *The Gift: Forms and Functions of Exchange in Archaic Societies*. London: Cohen & West.

NOU (Norwegian Public Reports) (2001) *Samiske sedvaner og rettsoppfatninger*. Norges Offentlige Utredninger 2001:34. Oslo: The Ministry of Justice.

Sahlins, M.D. (1972) *Stone Age Economics*. New York: Transaction Publishers.

Sahlins, M. (1976) *Culture and Practical Reason*. Chicago, IL: University of Chicago Press.

Sahlins, M. (2004) *Stone Age Economics*. London: Routledge.

Stetson, G. (2012) Oil politics and indigenous resistance in the Peruvian Amazon: The rhetoric of modernity against the reality of coloniality. *The Journal of Environment & Development* 21 (1), 76–97.

Weiner, A.B. (1992) *Inalienable Possessions: The Paradox of Keeping-While-Giving*. Berkeley, CA: University of California Press.

Part 5

Epilogue

16 Toward a De-Essentializing of Indigenous Tourism?

Dieter K. Müller and Arvid Viken

Is it reasonable to edit a book on indigenous tourism? This was an important question for us before accepting the challenge to do this book. The question was of course contingent on a concern whether it was still timely to address indigenous tourism as something different from other forms of cultural tourism. Our Nordic experience has taught us that our indigenous neighbors, colleagues and friends most often live lives like we do and hardly deviate in any aspect. This may apply particularly in Nordic societies where Sami populations do not differ significantly from the majority population when it comes to factors like income and health, and as most other indigenous groups more or less integrated in the market economy. Hence, it is reasonable to wonder to what extent our research, but also science in general, is essentializing indigenous peoples by ascribing them unique characteristics different from the surrounding population.

In the introductory chapters of this book, we discussed concepts as 'indigenous' and 'indigeneity'. We stated that there are often reasons for using the term 'indigeneity' because it has a weaker tie to ethnic roots. Indigeneity appears and is shaped through processes of positioning, through which 'connections between group and locality' are negotiated (Merlan, 2009: 303). Thus, indigeneity relates to particular groups, indigenous or local, as outcomes of 'socio-spatial processes and practices whereby indigenous peoples and places are determined as distinct (ontologically, epistemologically, culturally, in sovereignty, etc.) to dominant universals' (Radcliff, 2015: 2). It is also an international category referring to 'peoples who have great moral claims on nation-states and on international society, often because of inhumane, unequal, and exclusionary treatment' (Radcliff, 2015: 304). But of course, those being categorized are far from being a homogenous entity. This is also the case when we are using a term as 'Arctic indigenous tourism'. There are many appearances.

Following from this, indigeneity is a relational trait defined in interaction between a series of actors: governments, industrial corporations, the civil society at large, media, non-governmental organizations (NGOs) and others (Radcliff, 2009: 3). Indigeneity is a matter of national and international politics, that has emerged from '... contexts of liberal democratic "political cultures"', Merlan (2009: 304) claims and she goes on: 'Within such cultures there are values that facilitate both recognition and regulation of those who are not only patently "different" but also marginalized and disadvantaged'. The indigenous status is an opportunity, but can also be an obstacle. Thus, indigeneity is an art of balance between on the one side indigenous or local particularities, as being historically and spatially unique, with maintained and visible traditions and a flavor of exoticism, and on the other side being an integrated part of the modern and global world (Ludlow *et al.*, 2016). This tension has been part of the background for this book. We see indigenous groups as modern, and most of them live lives similar to their majority neighbors, having equal opportunities for commercial entrepreneurship. On the other hand, there are differences in history, ethnicity, traditions, cultural expressions and mentality that should be dealt with, and for which tourism is a medium for exposure. This background can also represent obstacles – there remain leftovers from the colonial past (Viken, 2017). In the following, we will discuss the chapters of the book in light of this tension; first showing how the Arctic is balancing between being a backyard and being prosperous, thereafter discussing the indigenous tourism of the region.

The Arctic Location

The Arctic is often depicted as one region with homogenous characteristics. This is not least rooted in definitions based on scientific concepts such as latitude, tree line or temperature isotherms. However, even within social science a convenient pragmatic approach has been to mirror these scientific definitions on political maps satisfying the need to access statistical data available for predefined administrative units. Using the concept of Arctic within social sciences has not least been done for multiple geopolitical reasons (Keskitalo, 2004), but a further elaboration is beyond the scope of this chapter. Still, it is obvious that a description of the Arctic as homogenous hardly mirrors the realities in its various parts (Keskitalo *et al.*, 2013). This of course applies to indigenous peoples as well.

Particularly in a Nordic and European context, indigenous populations are neither geographically nor socially strongly disintegrated from the majority population since development did not follow the logics of a frontier (Niemi, 1997). Hence, the idea of an indigenous community as a secluded unit is alien to the Nordic situation, and in fact the Sami are a majority only in a few settlements even within their homeland Sapmi. As has been shown in this book (Smed, this volume), in Greenland the distinction between indigenous and non-indigenous populations is not applied there at all, although this has to be seen in the context of a strive for greater autonomy and even independence from Denmark, too.

The situation in North America differs. However, even here the indigenous population has gained a significant set of rights, comprising among other things even political control and self-government (Poelzer & Wilson, 2014). Hence, in relation to tourism development this means that the indigenous communities have control over tourism development, but that it is of minor importance for their development (Hull *et al.*, this volume).

It is perhaps against this background that a recent special issue on tourism and indigenous peoples in the *Journal of Sustainable Tourism* (Carr *et al.*, 2016) almost comprehensively neglected the northern situation. Instead, experiences from the southern hemisphere and a development context dominate the scientific discourse. Often, this implies a multilayered enmeshment of tourism and colonial heritage where hegemonic power structures are at the center of scientific interest.

In Sweden, for example, an important difference is that colonization has meant a discontinuation of a nomadic lifestyle and of a reindeer herding subsistence economy in favor of permanent settlement and agriculture. This inner or clone colonization did not necessarily imply the inflow of a dominant population from the outside but rather meant a transition for the local population (Bylund, 1956, 1960). Hence, as Bylund points out, 75% of the increase in the settled population was a result of this process, while only 25% in fact was due to in-migration to the area. Similarly, even in Norway the idea of a frontier development in the north has been rejected (Aas, 1998). In other words, the colonization of Sapmi has at least initially meant a clash between different livelihoods among the Sami population, rather than a clash between Sami and in-migrants. The term 'colonization' in its modern connotation is, however, justified because of previous hegemonic injustice and cruelties against the Sami population and the still prevailing power structures. Today, the governments in several lands do not recognize tourism as a way of living for reindeer herders, and tourism is not among the industries given priority in northern politics (Viken, 2016). On the other hand, many other industries and modern infrastructure (like roads, windmills and electric supply mains) are disadvantageous both for tourism development and reindeer herding in many areas with indigenous livelihood (Viken, 2016).

This history is different from other places in the Arctic and it is also not in line with indigenous experiences elsewhere. Hence, it is important to note that there is not one experience of Arctic indigeneity, but many. This is dependent on where in the Arctic the processes are located, since most places do not share a common history or presence and thus are comprised in different ways by processes of globalization and development.

Tourism Development

Scientific debate about Arctic indigenous peoples is not performed by residents of the north only (Viken, 2013), and hence a risk for stereotypical reporting is immanent. Moreover, imaginations about Arctic indigenous peoples are not only produced by science – in fact science can be expected to play only a marginal role (Müller, 2013b). Instead, media and the tourism industry are far more powerful in creating these geographical imaginations (Harvey, 2000). This justifies a fresh examination of the nexus of tourism and indigenous peoples even in an Arctic context. This is not least also because of renewed interest in Arctic resources, which once again puts pressure on indigenous industries by threatening traditional land uses.

Indigenous tourism has often been a reaction to the decline in other industries, offering employment for indigenous and non-indigenous northerners. This implies that interest in indigenous tourism is cyclical as is interest in northern resources (Müller, 2013a). This means also rapidly shifting and contrasting ideas and imaginations of the north and its peoples. While industrial development highlights modernity and is often accompanied by communities striving for de-peripheralization (Kühn, 2015), tourism tends to produce images of traditional communities, because these are expected to sell better in the marketplace (see Keskitalo, this volume; Ludlow *et al.*, 2016). This creates a situation where imaginations compete with each other; on the one hand the Arctic is the resource periphery offering multiple opportunities for exploitation and transportation, on the other hand it is the pleasure periphery and the global playground for adventure seekers (Pedersen & Viken, 1996). Moreover, the Arctic is not only within the tourism industry stereotyped as one coherent natural region and wilderness not acknowledging the variation to be found within the north (Keskitalo *et al.*, 2013). Hence, the ongoing 'Arctification'[1] of northern Europe, for example, implies not only a strategic positioning in relation to European and global political discourse in order to be seen as equal partners with other countries with Arctic claims. It also means a risk that northern areas that previously have been perceived as integrated parts of Norway, Sweden and Finland are increasingly perceived as something exotic, which can imply a re-peripheralization, which Kühn (2015) sees as a process of the production of peripheries through social relations with partly significant spatial implications. Indigenous tourism may be a contributor to this; indigenous tourism is based on otherness, supporting and reproducing the imaginaries of otherness. However, when the ethnic otherness is not articulated, as very often in Greenland (Smed, this volume) or on the Kola Peninsula (Konstantinov, this volume), it also reflects an ethnic negligence that is equally ethnopolitically problematic. Vladimirova (this volume) shows this in a discussion of a couple of indigenous festivals in the Russian North, which on the one side have the role of strengthening cultural roots, and on the other side are used to acclaim the existing political regimes. Keskitalo (this volume) calls for more studies of northern areas and peoples to become more reflective and critical, detecting the underlying social and political constructions of people and regions and the forces of globalization that comprise these regions as well.

The ethnic otherness related to indigeneity tends to be perceived as exotic. 'Exoticism' can be tied to colonialism or imperialism. The exotic is something from far away. Traditionally, exoticism has an ethnocentric base. It was the central eye that appraised the peripheral *Other*. In the 19th century, when international tourism was in its infancy, Scandinavia and particularly the Scandinavian north was regarded as a definite periphery and stood out as less modern, and was celebrated as such (Leavenworth, 2010) – as exotic – often with the indigenous Sami in focus. However, more recently the term has gone through a positive twist, and as it is, the West now tends to celebrate the exotic *Other*, and since the 1960s exotism has emerged as a resource for tourism development (Staszak, 2008: 6). This is demonstrated in several of the chapters of this book, as in that of Fonneland, where indigenous spirituality is seen as a New Age phenomenon, and in Olsen's chapter where he describes a *sieidi* that has been transformed from a sacred site into an attraction, both for the locals and for tourists. However, exoticization tends to have a colonial flavor, as the encounter with the exotic *Other* is still a way of comforting identity and superiority, as Staszak (2008) sees it. This is the dilemma of indigeneity and exoticism; they are traits that should be exposed, but at the same time not too strongly highlighted, because the displaying may produce misconception in the mind of the Other. The line between difference that attracts and difference that offends is often subtle (Conklin, 1997).

In this context, it is important to recognize that tourism first and foremost is in a play with traditional images and a colonial past. As Kramvig (this volume) points out, tourism is in itself a way of modernization, and as she claims, maybe if managed properly a way of decolonizing Sápmi. In fact, the tourism industry inherently needs modernity in order to organize its activities and to function as an industry. And, as noted by several authors (Keskitalo, this volume; Müller, this volume), Sami tourism entrepreneurs contribute to this as active partners and not least in order to make a living. However, as discussed by Svensson and Viken (this volume), modern times and technologies can also be challenging – as motorized vehicles and environmental concern – for both traditions and the self-determination of a traditional industry as reindeer herding. But also, confronted with tourism, a series of questions can be raised concerning how to deal with a traditional subsistence economy in a modern world. Should it only be traditional handicraft and modern meat production, or could it also include tourism?

Concerning the involvement of indigenous groups in tourism development, the book gives a fragmented picture. Ween and Riseth (this volume) compare tourism development in two Sami areas in Norway, one with an unregulated and chaotic development – partly due to the fact that the community in focus, Tana, is sharing its important salmon river with Finland. The other case stands out as a good model, with a development planned and well rooted in local ethno-awareness and ethnopolitics. In a third chapter, Smed (this volume) shows how nature-based tourism in Ilulisat on Greenland overshadows the cultural scene that is at play in the community. In Kilpisjärvi, Finland, Tuulentie (this volume) shows how Sami interests are more or less ignored as stakeholders in tourism development processes. Thus, development obviously strongly relies on local conditions and also on national politics. This is also emphasized by critiques of the exoticism paradigm (Kapferer, 2013); there are very many ways to perform culture. There is no essential indigeneity. Places and people are different, thus, there is a diversity or plurality to celebrate.

Therefore, it is no wonder that the situation concerning indigenous tourism varies across the Arctic. Unfortunately, there are examples of indigenous people being set aside in entrepreneurial processes, as Konstantinov (this volume) shows for Sami people living adjacent to salmon rivers on the Russian Kola Peninsula. Here, the rights to fish have been taken over by tourism entrepreneurs and international tour operators, supported by the authorities. Similarly, the Russia Nenets are struggling for real empowerment (Pashkevich, this volume), but development is also expected to be regulated by government and public stakeholders. In contrast, the situation in the Yukon is rather characterized by indigenous control of the tourism product (Hull *et al.*, this volume); however, capacity constraints imply that tourism has to compete for attention with healthcare and education. Even in the Nordic countries, degrees of freedom are great and hence tourism is not the only option for Sami. Indigenous groups are modern as are their industries. Tourism is just one of the options.

Moreover, not only are indigenous people protagonists of tourism, but they are also affected by tourism in various ways. Though the situation of indigenous peoples seems to be improved all over the Arctic, it appears that other interests often overrule indigenous interests to maintain their livelihoods. However, there is also a risk related to seeing an indigenous people as a group, and stereotype them according to history and traditions. As mentioned many times in this book, most indigenous groups are also modern, with a cultural particularity based on history and traditions. Certainly, members of the indigenous populations also act as government representatives, businessmen, miners, loggers and other land-use competitors, as well as tourist entrepreneurs. And as has been shown previously (Müller & Kuoljok Huuva, 2009), even the ambitions and agendas of those indigenous persons active within the tourism industry vary and contrast (see also Müller & Hoppstadius, this volume). In future studies, it is thus recommendable to move beyond the indigenous–non-indigenous dichotomy and to acknowledge individual agency even within indigenous tourism.

The volume also demonstrates that indigenous tourism not only faces challenges related to the indigenous dimensions of the product. Instead, geographical locations sometimes away from the beaten track and characterized by rather poor accessibility as well as core–periphery relations imply challenges for the companies. Hence, another lesson to be drawn from this book is that challenges are not always only related to indigeneity but also to the geographical characteristics of the product. Of course, these aspects are interrelated and their nexus difficult to dissect. This is not only an analytical challenge, but also something that indigenous tourism entrepreneurs have to face not least considering the obvious embeddedness of their industry in an ethnopolitical discourse. However, as Smed (this volume) shows concerning Greenland, nature experiences may overshadow culture. But on the other hand, whose nature is it they are presenting on Greenland – indigenous nature, national nature or colonial nature? Tourism scholars have counted signs of Saminess in brochures (Olsen, 2003) or at festivals (Müller & Petterson, 2006), not including 'Sami' nature. If included, the Saminess of such exposures would have been more salient.

The book further shows that tourism development in fact may be a way of empowerment of and self-reflection among indigenous populations. It provides an opportunity to highlight history, heritage and culture as it is done in cultural tourism all over the world. Hence, indigenous people seem to act like people in general and try to grasp tourism as an opportunity to achieve various individual and collective goals and objectives. From this perspective, there seems no reason to distinguish indigenous tourism from other cultural tourism. The reason to do so anyway seems to be related to exotism and a fascination for the small scale of many indigenous populations. The latter means that indigenous tourism is often rare and thus a potentially exclusive product on the international tourism market, making commodification attractive and economically promising. However, indigenous tourism in most of the countries involved – Greenland, in Norway, Sweden, Finland and Russia – seems to be a bit premature. The indigenous groups themselves are not in charge of development processes and often seem to be marginalized concerning tourism development (Tuulentie, this volume; Viken, 2016). In Sweden, a recent attempt to establish an indigenous destination marketing organization (DMO), Visit Sapmi, and a quality label, Sapmi Experiences, has been discontinued. Reasons for this are not least economic, but even the marginal role of tourism in relation to other indigenous industries has been mentioned; indigenous tourism development is simply not gaining the attention needed in order to secure funding for a systematic development work. Hence, as an area for ethnopolitics, tourism seems to be in its infancy.

It is this which still gives reasons to address indigenous tourism as a separate category of analysis, even in the context of advanced northern states and welfare systems. As long as tourists and locals, governments and the tourism industry, all continue to treat indigenous populations as different, indigenous tourism remains an important category. As several have discussed, indigenousness or indigeneity is 'embedded in the emergence of world society and its forms of communication and institutions' (Ludlow *et al.*, 2009: 3, quoting Hirtz, 2009). It is a product of modernity, defined by modern institutions dealing with politics, governance and science, to which also indigenous or local groups themselves have adapted. However, it is also because indigenous population themselves contribute to the re-production of themselves as special. Tourism is a way of negotiating indigeneity that tends to essentialize it by highlighting traditional and non-modern features. The challenge is to show how this is an integrated part of a contemporary and modern life also including the international tourism marketplace.

Conclusion

This chapter has summed up some of the findings of the book, highlighting the balance involved in performing indigeneity, and therefore also in dealing with tourism. This goes back to the fact that indigenous and other local groups tend to balance between the past and the contemporary, between cultural particularities and global similarities, between being exploited and suppressed and being in charge of their own development. This balance is also part of what makes indigenous and minority cultures dynamic, making them visible and significant in times of cultural homogenization and vanishing traditions. This balance is also part of the modernization project where contrasts identify localities and locate identities in processes of progress. Tourism is salient in such processes as a force both preserving and modernizing indigeneity. As Hirtz (2003: 2) claims, there is a series of paradoxes of indigenousness being a modern phenomenon: '... through the very process of being recognized as "indigenous," these groups enter the realm of modernity'. Tourism conveys these processes. Tourism gives and takes. Most indigenous and other groups welcome modernity, but have to cope with its deficiencies. As this book has shown, tourism may be a gentle but challenging path to modern life in the Arctic.

Note

(1) Credit for the term should be given to Professor Sverker Sörlin, KTH Royal Institute of Technology, who mentioned it at an informal work meeting at Umeå University in 2012. As it is understood here, it can be seen as a social process creating new geographical imaginations of the north of Europe as part of the Arctic and consecutively new social, economic and political relations of the area.

References

Aas, S. (1998) North Norway: The frontier of the North? *Acta Borealia* 15 (1), 27–41.

Bylund, E. (1956) *Koloniseringen av Pite lappmark till och med år 1867*. Uppsala: Geografiska institutionen.

Bylund, E. (1960) Theoretical considerations regarding the distribution of settlement in inner north Sweden. *Geografiska Annaler* 42 (4), 225–231.

Carr, A., Ruhanen, L. and Whitford, M. (eds) (2016) Sustainable tourism and indigenous peoples. Special issue. *Journal of Sustainable Tourism* 24 (8–9), 1067–1079.

Conklin, B.A. (1997) Body, paint, feathers and VCRs: Aesthetics and authenticity in Amazonian activism. *American Ethnologist* 24, 711–737.

Harvey, D. (2001) *Spaces of Capital: Towards a Critical Geography*. London: Routledge.

Hirtz, F. (2003) It takes modern means to be traditional: On recognizing indigenous cultural communities in the Philippines. *Development and Change* 34, 887–914.

Kapferer, B. (2013) How anthropologists think: Configurations of the exotic. *Journal of Royal Anthropological Institute* 19, 813–836.
Keskitalo, E.C.H. (2004) *Negotiating the Arctic: The Construction of an International Region.* London: Routledge.
Keskitalo, E.C.H., Malmberg, G., Westin, K., Wiberg, U., Müller, D.K. and Pettersson, Ö. (2013) Contrasting Arctic and mainstream Swedish descriptions of northern Sweden: The view from established domestic research. *Arctic* 66 (3), 351–365.
Kühn, M. (2015) Peripheralization: Theoretical concepts explaining socio-spatial inequalities. *European Planning Studies* 23 (2), 367–378.
Leavenworth, M.L. (2010) Exotic norths? Representations of Northern Scandinavia in S.H. Kent's 'Within the Arctic Circle' and Baeyard Taylor's 'Northern Travel'. *Nordlit* 26, 1–14.
Ludlow, F., Baker, L., Brock, S., Hebdon, C. and Dove, M. (2016) The double binds of indigeneity and indigenous resistance. *Humanities* 5 (53), 1–19.
Merlan, F. (2009) Indigeneity, global and local. *Current Anthropology*, 50, 303–333.
Müller, D.K. (2013a) Hibernating economic decline? Tourism and labor market change in Europe's northern periphery. In G. Visser and S. Ferreira (eds) *Tourism and Crisis* (pp. 113–128). London: Routledge.
Müller, D.K. (2013b) Tourism and the definition of the Arctic. In R.H. Lemelin, P. Maher and D. Liggett (eds) *From Talk to Action: How Tourism is Changing the Polar Regions* (pp. 9–20). Thunder Bay: Lakehead University Centre for Northern Studies.
Müller, D.K. and Pettersson, R. (2006) Sami heritage at the winter festival in Jokkmokk, Sweden. *Scandinavian Journal of Hospitality and Tourism* 6 (1), 54–69.
Müller, D.K. and Kuoljok Huuva, S. (2009) Limits to Sami tourism development: The case of Jokkmokk, Sweden. *Journal of Ecotourism* 8 (2), 115–127.
Niemi, E. (1997) Sami history and the frontier myth: A perspective on northern Sami spatial and rights history. In H. Gaski (ed.) *Sami Culture in a New Era: The Norwegian Sami Experience* (pp. 62–85). Kárásjohka/Karasjok: Davvi Girji.
Olsen, K. (2003) The touristic construction of the 'emblematic' Sami. *Acta Borealia* 20, 3–20.
Pedersen, K. and Viken, A. (1996) From Sami nomadism to global tourism. In M.F. Price (ed.) *People and Tourism in Fragile Environment* (pp. 69–88). Chichester: Wiley.
Poelzer, G. and Wilson, G.N. (2014) Governance in the Arctic: Political systems and geopolitics. In J. Nymand Larsen and G. Fondahl (eds) *Arctic Human Development Report: Regional Processes and Global Linkages* (pp. 185–222). Copenhagen: Nordic Council of Ministers.
Radcliffe, S.A. (2015) Geography and indigeneity I: Indigeneity, coloniality and knowledge. *Progress in Human Geography* 1–10.
Staszak, J.-F. (2008) Other/otherness. In R. Kitchin and N. Thrift (eds) *International Encyclopedia of Human Geography Vol. 8* (pp. 43–47). New York: Elsevier.
Viken, A. (2013) What is Arctic tourism, and who should define it? In D.K. Müller, L. Lundmark and R.H. Lemelin (eds) *New Issues in Polar Tourism* (pp. 37–50). Amsterdam: Springer.
Viken, A. (2016) Reiseliv i samiske områder; neoliberalisme og marginalisering? In E. Angell, S. Eikeland and P. Selle (eds) *Brems eller drivkraft? Samepolitikken i den nye utviklinga i nord* (in print). Oslo: Gyldendal Akademisk.

Index

For Product Safety Concerns and Information please contact our EU Authorised Representative:

Easy Access System Europe

Mustamäe tee 50

10621 Tallinn

Estonia

gpsr.requests@easproject.com

www.ingramcontent.com/pod-product-compliance
Ingram Content Group UK Ltd.
Pitfield, Milton Keynes, MK11 3LW, UK
UKHW021828100926
8556IPUK00003B/222

9781845416096